Applied Probability and Statistics (Continued)

CHAKRAVARTI, LAHA and ROY · Handbook of Methods of Applied Statistics, Vol. II
CHERNOFF and MOSES · Elementary Decision Theory
CHEW · Experimental Designs in Industry
CHIANG · Introduction to Stochastic Processes in Biostatistics
CLELLAND, deCANI, BROWN, BURSK, and MURRAY · Basic Statistics with Business Applications
COCHRAN · Sampling Techniques, *Second Edition*
COCHRAN and COX · Experimental Designs, *Second Edition*
COX · Planning of Experiments
COX and MILLER · The Theory of Stochastic Processes
DEMING · Sample Design in Business Research
DODGE and ROMIG · Sampling Inspection Tables, *Second Edition*
DRAPER and SMITH · Applied Regression Analysis
GOLDBERGER · Econometric Theory
GUTTMAN and WILKS · Introductory Engineering Statistics
HALD · Statistical Tables and Formulas
HALD · Statistical Theory with Engineering Applications
HANSEN, HURWITZ, and MADOW · Sample Survey Methods and Theory, Volume I
HOEL · Elementary Statistics, *Second Edition*
JOHNSON and LEONE · Statistics and Experimental Design: In Engineering and the Physical Sciences, Volumes I and II
KEMPTHORNE · An Introduction to Genetic Statistics
MEYER · Symposium on Monte Carlo Methods
PRABHU · Queues and Inventories: A Study of Their Basic Stochastic Processes
SARHAN and GREENBERG · Contributions to Order Statistics
TIPPETT · Technological Applications of Statistics
WILLIAMS · Regression Analysis
WOLD and JURÉEN · Demand Analysis
YOUDEN · Statistical Methods for Chemists

Tracts on Probability and Statistics

BILLINGSLEY · Ergodic Theory and Information
CRAMÉR and LEADBETTER · Stationary and Related Stochastic Processes
RIORDAN · Combinatorial Identities
TAKÁCS · Combinatorial Methods in the Theory of Stochastic Processes

Introduction to Stochastic Processes in Biostatistics

A WILEY PUBLICATION IN APPLIED STATISTICS

Introduction to Stochastic Processes in Biostatistics

CHIN LONG CHIANG

Professor of Biostatistics
University of California, Berkeley

John Wiley & Sons, Inc.
New York · London · Sydney

LIBRARY OF CONGRESS CATALOG CARD NUMBER: 68-21178
GB 471 15500X
PRINTED IN THE UNITED STATES OF AMERICA

In Memory of My Parents

Preface

Time, life, and risks are three basic elements of stochastic processes in biostatistics. Risks of death, risks of illness, risks of birth, and other risks act continuously on man with varying degrees of intensity. Long before the development of modern probability and statistics, men were concerned with the chance of dying and the length of life, and they constructed tables to measure longevity. But it was not until the advances in the theory of stochastic processes made in recent years that empirical processes in the human population have been systematically studied from a probabilistic point of view.

The purpose of this book is to present stochastic models describing these empirical processes. Emphasis is placed on specific results and explicit solutions rather than on the general theory of stochastic processes. Those readers who have a greater curiosity about the theoretical arguments are advised to consult the rich literature on the subject.

A basic knowledge of probability and statistics is required for a profitable reading of the text. Calculus is the only mathematics presupposed, although some familiarity with differential equations and matrix algebra is needed for a thorough understanding of the material.

The text is divided into two parts. Part 1 begins with one chapter on random variables and one on probability generating functions for use in succeeding chapters. Chapter 3 is devoted to basic models of population growth, ranging from the Poisson process to the time-dependent birth-death process. Some other models of practical interest that are not included elsewhere are given in the problems at the end of the chapter.

Birth and death are undoubtedly the most important events in the human population, but the illness process is statistically more complex. Illnesses are potentially concurrent, repetitive, and reversible, and consequently analysis is more challenging. In this book illnesses are treated as discrete entities, and a population is visualized as consisting of discrete states of illness. An individual is said to be in a particular state of illness if he is affected with the corresponding diseases. Since he may leave one illness state for another or enter a death state, consideration of illness opens up a new domain of interest in multiple transition probability and multiple transition time. A basic and important case is that in which there are two illness states. Two chapters (Chapters 4 and 5) are devoted to this simple illness-death process.

In dealing with a general illness-death process that considers any finite number of illness states, I found myself confronted with a finite Markov process. To avoid repetition and to maintain a reasonable graduation of mathematical involvement, I have interrupted the development of illness processes to discuss the Kolmogorov differential equations for a general situation in Chapter 6. This chapter is concerned almost entirely with the derivation of explicit solutions of these equations. For easy reference a section (Section 3) on matrix algebra is included.

Once the Kolmogorov differential equations are solved in Chapter 6, the discussion on the general illness-death process in Chapter 7 becomes straightforward; however, the model contains sufficient points of interest to require a separate chapter. The general illness-death process has been extended in Chapter 8 in order to account for the population increase through immigration and birth. These two possibilities lead to the emigration-immigration process and the birth-illness-death process, respectively. But my effort failed to provide an explicit solution for the probability distribution function in the latter case.

Part 2 is devoted to special problems in survival and mortality. The life table and competing risks are classical and central topics in biostatistics, while the follow-up study dealing with truncated information is of considerable practical importance. I have endeavored to integrate these topics as thoroughly as possible with probabilistic and statistical principles. I hope that I have done justice to these topics and to modern probability and statistics.

It should be emphasized that although the concept of illness processes has arisen from studies in biostatistics, the models have applications to other fields. Intensity of risk of death (force of mortality) is synonymous with "failure rate" in reliability theory; illness states may be alternatively interpreted as geographic locations (in demography), compartments (in compartment analysis), occupations, or other defined conditions. Instead of the illness of a person, we may consider whether a person is unemployed, or whether a gene is a mutant gene, a telephone line is busy, an elevator is in use, a mechanical object is out of order, and so on.

This book was written originally for students of biostatistics, but it may be used for courses in other fields as well. The following are some suggestions for teaching plans:

1. For a one-semester course in stochastic processes: Chapters 2 through 8.

2. For a year course in biostatistics: Chapters 1 and 2 followed by Chapters 10 through 12, and then by Chapters 3 through 8. In this arrangement, a formal introduction of the pure death process is necessary at the beginning of Chapter 10.

3. For a year course in demography: Plan 2 above may be followed, except that the term "illness process" might be more appropriately interpreted as "internal migration process."

4. As a supplementary text for courses in biostatistics or demography: Chapters 9 through 12.

If it is used as a general reference book, Chapter 9 may be omitted.

The book is an outgrowth partly of my own research, some of which appears here for the first time (e.g., Chapter 5 and parts of Chapter 6), and partly of lecture notes for courses in stochastic processes, for which I am grateful to the many contributors to the subject. I have used this material in my teaching at the Universities of California (Berkeley), Michigan, Minnesota, and North Carolina; at Yale and Emory Universities; and at the London School of Hygiene, University of London.

This work could not have been completed without the aid of a number of friends, to whom I am greatly indebted. It is my pleasure to acknowledge the generous assistance of Mrs. Myra Jordan Samuels and Miss Helen E. Supplee, who have read early versions and made numerous constructive criticisms and valuable suggestions. Their help has tremendously improved the quality of the book. I am indebted to the School of Public Health, University of California, Berkeley, and the National Institutes of Health, Public Health Service, for financial aid under Grant No. 5-SO1-FR-05441-06 to facilitate my work. An invitation from Peter Armitage to lecture in a seminar course at the London School of Hygiene gave me an opportunity to work almost exclusively on research projects associated with this book. I also wish to express my appreciation to Richard J. Brand and Geoffrey S. Watson who read some of the chapters and provided useful suggestions. My thanks are also due to Mrs. Shirley A. Hinegardner for her expert typing of the difficult material; to Mrs. Dorothy Wyckoff for her patience with the numerical computations; and to Mrs. Lois Karp for secretarial assistance.

CHIN LONG CHIANG

University of California, Berkeley
May, 1968

Contents

PART 1

* This chapter may be omitted without loss of continuity.

Introduction to Stochastic Processes in Biostatistics

Part 1

Part I

CHAPTER 1

Random Variables

1. INTRODUCTION

A large body of probability theory and statistics has been developed for the study of phenomena arising from random experiments. Some studies take the form of mathematical models constructed to describe observable events, while others are concerned with statistical inference regarding random experiments. A familiar random experiment is the tossing of dice. When a die is tossed, there are six possible outcomes and it is not certain which one will occur. In a laboratory determination of antibody titer the result will also vary from one trial to another, even if the same blood specimen is used and the laboratory conditions are kept constant. Examples of random experiments can be found almost everywhere; in fact, the concept of random experiment may be extended so that any phenomenon may be thought of as the result of some random experiment, be it real or hypothetical.

As a framework for discussing random phenomena, it is convenient to represent each conceivable outcome of a random experiment by a point, called a *sample point*, denoted by s. The totality of all sample points for a particular experiment is called the *sample space*, denoted by S. Events may be represented by subsets of S; thus an event A consists of a certain collection of possible outcomes s. If two subsets contain no points s in common, they are said to be disjoint, and the corresponding events are said to be *mutually exclusive*: they cannot both occur as a result of a single experiment.

The probabilities of the various events in S are the starting point for analysis of the experiment represented by S. Denoting the probability of an event A by $\Pr\{A\}$, we may state the three fundamental assumptions about these probabilities as follows.

(i) Probabilities of events are nonnegative:

$$\Pr\{A\} \geq 0 \qquad \text{for any event } A \tag{1.1}$$

(ii) The probability of the whole sample space is unity:

$$\Pr\{S\} = 1 \tag{1.2}$$

(iii) The probability that one of a sequence of mutually exclusive events $\{A_i\}$ occurs is

$$\Pr\{A_1 \text{ or } A_2 \text{ or } \cdots\} = \sum_{i=1}^{\infty} \Pr(A_i). \tag{1.3}$$

This is called the countably additive assumption. In the case of two mutually exclusive events, A_1 and A_2, we have

$$\Pr\{A_1 \text{ or } A_2\} = \Pr\{A_1\} + \Pr\{A_2\}. \tag{1.4}$$

2. RANDOM VARIABLES

Any single-valued numerical function $X(s)$ defined on a sample space S will be called a *random variable*; thus, a random variable associates with each point s in the sample space a unique real number, called its value at s. The probability of an event in S can then be translated into the probability that the value of the random variable will lie in a certain interval or other set of real numbers. The most common types of random variables are discrete random variables and continuous random variables. A *discrete random variable* $X(s)$ assumes a finite or denumerable number of values; for each possible value x_i there is a unique probability

$$\Pr\{X(s) = x_i\} = p_i, \qquad i = 0, 1, \ldots, \tag{2.1}$$

that the random variable assumes the value x_i. The sequence $\{p_i\}$ is called the *probability distribution* of $X(s)$, and the cumulative probability

$$\Pr\{X(s) \leq x\} = \sum_{x_i \leq x} p_i = F(x), \qquad -\infty < x < \infty \tag{2.2}$$

is called the *distribution function* of $X(s)$.

If $X(s)$ assumes only one value, say x_k, with $p_k = 1$, then $X(s)$ is called a *degenerate random variable*, and is, in effect, a constant.

$X(s)$ is a *proper random variable* if its distribution $\{p_i\}$ satisfies the condition

$$\sum_i p_i = 1. \tag{2.3}$$

If there is a subset of S where $X(s)$ is not defined or is infinite so that

$$\sum_i p_i < 1, \tag{2.4}$$

then $X(s)$ is an *improper random variable;* the difference $1 - \sum_i p_i$ is the probability that $X(s)$ assumes no (finite) value at all. We shall encounter some improper random variables in the later chapters.

Most of statistical theory can be developed in terms of random variables without explicit reference to the sample space. Therefore, from now on, we shall write simply X instead of $X(s)$.

Example 1. In throwing a fair die, the sample space S consists of six points, one for each face of the die. Let a random variable X be the number of dots on the face shown; then X takes on values 1, 2, 3, 4, 5, 6, with the corresponding probabilities $p_i = \frac{1}{6}$, $i = 1, \ldots, 6$. Now we define a random variable Y so that $Y = 0$ when the number of dots showing is even and $Y = 1$ when the number of dots showing is odd. The corresponding probabilities are $\Pr\{Y = 0\} = \frac{1}{2}$ and $\Pr\{Y = 1\} = \frac{1}{2}$. If we let another random variable Z assume the value 0 or 1 according to whether the number of dots showing is ≤ 3 or >3, then $\Pr\{Z = 0\} = \frac{1}{2}$ and $\Pr\{Z = 1\} = \frac{1}{2}$. Thus X, Y, and Z are different random variables defined on the same sample space; furthermore, Y and Z have the same probability distribution, although they are different random variables. ▶

Example 2. In tossing a fair coin twice, the sample space S consists of four points: (H, H), (H, T), (T, H), (T, T). On this sample space let us define the random variable X as the number of heads obtained. The values that X can take are 0, 1, 2, with the corresponding probabilities $p_0 = \frac{1}{4}$, $p_1 = \frac{1}{2}$, and $p_2 = \frac{1}{4}$.

Tossing a coin is a simple example of a Bernoulli trial and the binomial distribution. *Bernoulli trials* are repeated independent trials, each trial having two possible outcomes with the corresponding probabilities remaining the same for all trials. A more general example follows. ▶

Example 3. Consider a sequence of n independent trials, each trial resulting either in a "success" with probability p or in a "failure" with probability $1 - p$. If X is the number of successes in the n trials, then X has a *binomial distribution* with the probability

$$\Pr\{X = k\} = \binom{n}{k} p^k (1 - p)^{n-k}, \qquad k = 0, 1, \ldots, n. \tag{2.5}$$

In formula (2.5) the quantity $p^k(1 - p)^{n-k}$ is the probability that k specified trials result in success and the remaining $n - k$ trials result in failure; for example, the *first* k trials may be successes and the last $n - k$ trials will be failures as represented by the sequence (SS $\cdots$ SFF $\cdots$ F).

The combinatorial factor (or *binomial coefficient*)

$$\binom{n}{k} = \frac{n!}{k!\,(n-k)!} \tag{2.6}$$

is the number of possible ways in which k successes can be obtained in n trials. ▶

A random variable X is called a *continuous random variable* if there exists a nonnegative function f such that for any $a \le b$,

$$\Pr\{a < X \le b\} = \int_a^b f(x)\,dx. \tag{2.7}$$

The function $f(x)$ is called the *probability density function* (or *density function*) of X. The *distribution function* of X is

$$F(x) = \Pr\{X \le x\} = \int_{-\infty}^{x} f(t)\,dt \tag{2.8}$$

so that

$$dF(x) = f(x)\,dx$$

and

$$\Pr\{a < X \le b\} = F(b) - F(a). \tag{2.9}$$

We note that for a continuous random variable X, $\Pr\{X = x\} = 0$ for any x, and that the values of the density function $f(x)$ are not probabilities; they are nonnegative but need not be less than unity. The density function is merely a tool which by (2.7) will yield the probability that X lies in any interval.

As in the discrete case, a continuous random variable is proper if its density function satisfies

$$\int_{-\infty}^{\infty} f(x)\,dx = 1$$

and improper if the integral is less than one.

Example 4. The *exponential distribution* has the density function

$$f(x) = \begin{cases} \mu e^{-\mu x} & x \ge 0 \\ 0 & x < 0 \end{cases} \tag{2.10}$$

and the distribution function

$$F(x) = \begin{cases} 1 - e^{-\mu x} & x \ge 0 \\ 0 & x < 0 \end{cases} \tag{2.11}$$

where the parameter μ is a fixed positive number. The exponential distribution occurs in life table studies, where μ is interpreted as the force of mortality, and in reliability theory, where μ is the failure rate. ▶

Example 5. The standard *normal distribution* has the density function

$$f(x) = \frac{1}{\sqrt{2\pi}} e^{-x^2/2}, \qquad -\infty < x < \infty. \tag{2.12}$$

The distribution function

$$F(x) = \int_{-\infty}^{x} \frac{1}{\sqrt{2\pi}} e^{-y^2/2} dy \tag{2.13}$$

has no closed form. This function is of great importance in the theory of probability and statistics as well as in applications to practical problems. The density function $f(x)$ has the maximum of $f(0) = (2\pi)^{-1/2}$ at $x = 0$. It is symmetrical with respect to the origin, having two points of inflection at $x = \pm 1$, and approaching zero as $x \to \pm\infty$. ▶

3. MULTIVARIATE PROBABILITY DISTRIBUTIONS

Let X and Y be two discrete proper random variables defined on the same sample space; then their *joint* (or *bivariate*) *probability distribution* $\{p_{ij}\}$ is defined by

$$p_{ij} = \Pr\{X = x_i \text{ and } Y = y_j\}, \qquad i, j = 0, 1, \ldots, \tag{3.1}$$

with

$$\sum_i \sum_j p_{ij} = 1. \tag{3.2}$$

Let

$$p_i = \sum_j p_{ij} \tag{3.3}$$

and

$$q_j = \sum_i p_{ij}; \tag{3.4}$$

then it follows by (3.1) that

$$p_i = \sum_j p_{ij} = \sum_j \Pr\{X = x_i \text{ and } Y = y_j\} = \Pr\{X = x_i\}. \tag{3.5}$$

and, similarly, that

$$q_j = \sum_i p_{ij} = \Pr\{Y = y_j\}. \tag{3.6}$$

The sequences $\{p_i\}$ and $\{q_j\}$ are called the *marginal distributions* of X and Y, respectively. They are, of course, simply the probability distributions of X and Y; the adjective "marginal" has meaning only in relation to the joint probability distribution $\{p_{ij}\}$. It is clear that the marginal distributions $\{p_i\}$ and $\{q_j\}$ may be determined from the joint distribution $\{p_{ij}\}$; however, the joint distribution cannot, in general, be determined from the marginal distributions.

For improper random variables, a joint distribution can be defined exactly as in (3.1), but the equations (3.2), (3.5), and (3.6) are no longer valid. In the remainder of this chapter we consider only proper random variables, since the concepts to be introduced are not useful for improper random variables.

The *conditional probability distribution* of Y given $X = x_i$ is defined by

$$\Pr\{Y = y_j \mid X = x_i\} = \frac{\Pr\{X = x_i \text{ and } Y = y_j\}}{\Pr\{X = x_i\}} = \frac{p_{ij}}{p_i}, \qquad p_i > 0. \tag{3.7}$$

The conditional distribution of X given $Y = y_j$ is defined analogously. It follows from (3.6) and (3.7) that

$$\Pr\{Y = y_j\} = q_j = \sum_i \Pr\{Y = y_j \mid X = x_i\}\Pr\{X = x_i\}.$$

Two random variables X and Y are said to be independently distributed, or *stochastically independent*, if

$$p_{ij} = p_i q_j \tag{3.8}$$

for all i and j. Formula (3.8) is equivalent to the more intuitive condition that

$$\Pr\{Y = y_j \mid X = x_i\} = \Pr\{Y = y_j\}. \tag{3.9}$$

Similar definitions and relations hold when X and Y are continuous proper random variables. The *joint density function* $h(x, y)$ is defined by the property that

$$\Pr\{a < X \le b \text{ and } c < Y \le d\} = \int_c^d \int_a^b h(x, y)\, dx\, dy \tag{3.10}$$

with $h(x, y)$ satisfying

$$\int_{-\infty}^{\infty} \int_{-\infty}^{\infty} h(x, y)\, dx\, dy = 1. \tag{3.11}$$

Thus, as in the discrete case, the *marginal density functions* defined by

$$f(x) = \int_{-\infty}^{\infty} h(x, y)\, dy \qquad \text{and} \qquad g(y) = \int_{-\infty}^{\infty} h(x, y)\, dx \tag{3.12}$$

satisfy

$$\Pr\{a < X \le b\} = \int_a^b f(x)\, dx \qquad \text{and} \qquad \Pr\{c < Y \le d\} = \int_c^d g(y)\, dy. \tag{3.13}$$

The *conditional density functions* are defined by

$$f(x \mid y) = \frac{h(x, y)}{g(y)} \qquad \text{and} \qquad g(y \mid x) = \frac{h(x, y)}{f(x)} \tag{3.14}$$

for $g(y) > 0$, $f(x) > 0$, and satisfy

$$f(x) = \int_{-\infty}^{\infty} f(x \mid y) g(y)\, dy \qquad \text{and} \qquad g(y) = \int_{-\infty}^{\infty} g(y \mid x) f(x)\, dx. \tag{3.15}$$

The random variables X and Y are said to be *stochastically independent* or independently distributed if for all x and y

$$h(x, y) = f(x)\, g(y). \tag{3.16}$$

If X and Y are independent random variables, then $\phi(X)$ and $\psi(Y)$, functions of X and Y, respectively, are also independent random variables.

The above definitions and relations for the bivariate case extend readily to the *multivariate* case of several random variables. For example, for three discrete random variables X, Y, and Z with the joint probability distribution

$$p_{ijk} = \Pr\{X = x_i, Y = y_j, Z = z_k\} \tag{3.17}$$

there are two "levels" of marginal distributions, one typified by

$$p_{ij} = \Pr\{X = x_i, Y = y_j\} = \sum_k p_{ijk} \tag{3.18}$$

and the other by

$$p_i = \Pr\{X = x_i\} = \sum_j p_{ij} = \sum_j \sum_k p_{ijk}. \tag{3.19}$$

Two levels of conditional probability can also be defined in the same way. Independence of X, Y, and Z is defined by the relation

$$p_{ijk} = \Pr\{X = x_i\} \Pr\{Y = y_j\} \Pr\{Z = z_k\}. \tag{3.20}$$

The extensions to continuous random variables and to more than three random variables are exactly analogous.

Notice that a set of random variables $(X_1, \ldots, X_n)$, defined on the same sample space, can equivalently be regarded as a *random vector*

$$\mathbf{X} = \begin{pmatrix} X_1 \\ \cdot \\ \cdot \\ \cdot \\ X_n \end{pmatrix}$$

defined on that same sample space; in the later chapters we shall find the vector notation convenient.

4. MATHEMATICAL EXPECTATION

The *mathematical expectation* of a (proper) random variable X is defined as

$$E(X) = \sum_i x_i p_i \tag{4.1}$$

if X is discrete with probability distribution $\{p_i\}$, and as

$$E(X) = \int_{-\infty}^{\infty} x f(x)\,dx \tag{4.2}$$

if X is continuous with density function $f(x)$. Thus, the mathematical expectation of a random variable is the weighted mean of the values that the random variable assumes weighted by the corresponding probabilities. The mathematical expectation $E(X)$ exists and is finite if the sum in (4.1), or the integral in (4.2), converges absolutely. The terms *mean value*, *expected value*, and *expectation* have the same meaning as *mathematical expectation.*

Example 1. Let X have the *Poisson* distribution

$$\Pr\{X = k\} = \frac{e^{-\lambda}\lambda^k}{k!} \qquad k = 0, 1, \ldots. \tag{4.3}$$

The expectation of X is

$$E(X) = \sum_{k=0}^{\infty} k\,\frac{e^{-\lambda}\lambda^k}{k!} = \lambda. \tag{4.4}$$

It is apparent that, since the parameter λ is not necessarily an integer, the expectation of a discrete random variable need not coincide with any possible value of the random variable. ▶

Example 2. Let X have the exponential distribution

$$f(x)\,dx = \mu e^{-\mu x}\,dx. \tag{4.5}$$

The expectation of X is

$$E(X) = \int_0^{\infty} x\mu e^{-\mu x}\,dx = \frac{1}{\mu}. \qquad ▶ \tag{4.6}$$

It is clear that any function $\phi(X)$ of a random variable X is also a random variable. The expectation of the function $\phi(X)$, if it exists, is given by

$$E[\phi(X)] = \sum_i \phi(x_i) p_i \tag{4.7}$$

in the discrete case and by

$$E[\phi(X)] = \int_{-\infty}^{\infty} \phi(x) f(x)\, dx \tag{4.8}$$

in the continuous case. Expressions of the type $E[\phi(X)]$ will sometimes be written $E\phi(X)$ when there is no ambiguity.

These definitions extend immediately to multivariate distributions. For example, if X and Y are discrete random variables with joint distribution $\{p_{ij}\}$ and if $\psi(X, Y)$ is any function of X and Y, then the mathematical expectation of ψ is

$$E[\psi(X, Y)] = \sum_i \sum_j \psi(x_i, y_j) p_{ij}. \tag{4.9}$$

The most important property of mathematical expectation is that it is a *linear* operator. Let X and Y be discrete random variables with joint distribution $\{p_{ij}\}$ and marginal distributions $\{p_i\}$ and $\{q_j\}$, respectively. Then, for any constants a, b, and c, we may use the definition of expectation to write

$$\begin{aligned} E(a + bX + cY) &= \sum_i \sum_j (a + bx_i + cy_j) p_{ij} \\ &= a \sum_i \sum_j p_{ij} + b \sum_i x_i \sum_j p_{ij} + c \sum_j y_j \sum_i p_{ij} \\ &= a + b \sum_i x_i p_i + c \sum_j y_j q_j \\ &= a + bE(X) + cE(Y). \end{aligned}$$

In general we have

Theorem 1. *The expectation of a linear function of random variables is equal to the same linear function of the expectations, so that if $X_1, \ldots, X_n$ are random variables and $c_0, c_1, \ldots, c_n$ are constants, then*

$$E(c_0 + c_1 X_1 + \cdots + c_n X_n) = c_0 + c_1 E(X_1) + \cdots + c_n E(X_n). \tag{4.10}$$

Notice that independence of $X_1, \ldots, X_n$ is *not* assumed. The proofs for the continuous case and for several random variables are left to the reader.

It is clear from the theorem that the expectation of a constant is equal to the constant itself. ▶

4.1. A Useful Inequality

Theorem 1 states that the equality

$$E[\phi(X_1, \ldots, X_n)] = \phi[E(X_1), \ldots, E(X_n)]$$

holds true when ϕ is a linear function. This equality does not hold true for arbitrary functions ϕ; in fact, we have a useful inequality for the case $\phi(X) = 1/X$:

Theorem 2. *If a nondegenerate random variable X assumes only positive values, then the expectation of the reciprocal of X is greater than the reciprocal of the expectation; that is,*

$$E\left(\frac{1}{X}\right) > \frac{1}{E(X)}. \tag{4.11}$$

The reverse inequality holds true if X assumes only negative values.

Proof. Let

$$X = E(X)\left(1 + \frac{X - E(X)}{E(X)}\right) = E(X)(1 + \Delta), \tag{4.12}$$

so that

$$E(\Delta) = E\left(\frac{X - E(X)}{E(X)}\right) = 0. \tag{4.13}$$

From (4.12),

$$E\left(\frac{1}{X}\right) = E\left(\frac{1}{E(X)(1 + \Delta)}\right) = \frac{1}{E(X)}\, E\left(\frac{1}{1 + \Delta}\right). \tag{4.14}$$

Substituting the identity

$$\frac{1}{1 + \Delta} = 1 - \Delta + \frac{\Delta^2}{1 + \Delta} \tag{4.15}$$

in (4.14) yields

$$\begin{aligned} E\left(\frac{1}{X}\right) &= \frac{1}{E(X)}\, E\left(1 - \Delta + \frac{\Delta^2}{1 + \Delta}\right) \\ &= \frac{1}{E(X)}\left(1 + E\left(\frac{\Delta^2}{1 + \Delta}\right)\right) > \frac{1}{E(X)} \end{aligned} \tag{4.16}$$

since the random variables Δ^2 and $1 + \Delta$, and hence the expectation $E[\Delta^2/(1 + \Delta)]$, are positive. ▶

4.2. Conditional Expectation

The conditional expectation of Y given x is the expectation of Y with respect to the conditional distribution of Y given x.

$$E(Y \mid x_i) = \sum_j y_j \frac{p_{ij}}{p_i}, \qquad p_i > 0 \tag{4.17}$$

for the discrete case and

$$E(Y \mid x) = \int_{-\infty}^{+\infty} y \frac{h(x, y)}{f(x)}\, dy, \qquad f(x) > 0 \tag{4.18}$$

for the continuous case. In order that these conditional expectations be meaningful it is necessary that $p_i > 0$ in (4.17) and $f(x) > 0$ in (4.18). The conditional expectations in (4.17) and (4.18) are constant for a given value x (or x_i); if x (or x_i) is regarded as a random variable X, then both expectations are written as $E(Y \mid X)$ and are themselves random variables. $E(Y \mid X)$ takes on the value $E(Y \mid x_i)$ when X assumes the value x_i.

Conditional expectations of functions, such as $E[\phi(X, Y) \mid X]$, are defined analogously, with the random variable upon which the expectation is conditioned treated as a constant when calculating the expectation; thus, for example,

$$E[(X + Y) \mid X] = X + E(Y \mid X) \tag{4.19}$$

$$E[XY \mid X] = XE[Y \mid X]. \tag{4.20}$$

Since expressions such as $E(Y \mid X)$ are random variables, they themselves have expectations; in fact, we have

Theorem 3. *For any two random variables X and Y,*[1]

$$E[E(Y \mid X)] = E(Y). \tag{4.21}$$

Corollary.

$$E(XY) = E[XE(Y \mid X)]. \tag{4.22}$$

Proof. For the discrete case we have

$$\begin{aligned} E[E(Y \mid X)] &= \sum_i p_i \left(\sum_j y_j \frac{p_{ij}}{p_i} \right) \\ &= \sum_i \sum_j y_j p_{ij} = \sum_j y_j \sum_i p_{ij} = \sum_j y_j p_j = E(Y). \quad (4.23 \end{aligned}$$

The proof for the continuous case is left to the reader. The corollary follows from (4.20) and from the theorem. ▶

An important consequence of the independence of random variables is that their expectations are multiplicative:

Theorem 4. *If $X, Y, \ldots, Z$ are independent random variables, then*

$$E(XY \cdots Z) = E(X) \cdot E(Y) \cdots E(Z). \tag{4.24}$$

Proof. We prove the theorem for the case of two random variables X and Y; extension to the general case is immediate by induction. It follows from the definition of independence and from (4.17) and (4.18) that if X and Y are independent, then $E(Y \mid X) = E(Y)$; consequently, (4.22) implies that $E(XY) = E(X)E(Y)$. ▶

---

[1] As an aid in interpreting repeated expectations, remember that $E(Y \mid X)$ is a function of X, so the outer expectation is taken with respect to the distribution of X.

5. MOMENTS, VARIANCE, AND COVARIANCE

Expectations in the distribution of a random variable X that are of particular interest are:

(i) The *moments*

$$E[X^r] \qquad r = 1, 2, \ldots \tag{5.1}$$

(ii) The *factorial moments*

$$E[X(X-1)\cdots(X-r+1)] \qquad r = 1, 2, \ldots \tag{5.2}$$

(iii) The *central moments*

$$E[X - E(X)]^r \qquad r = 1, 2, \ldots. \tag{5.3}$$

The first moment, which is nothing but the expectation $E(X)$, is called the *mean* of the distribution and is a commonly used index of its location. The second central moment

$$\sigma_X^2 = E[X - E(X)]^2 \tag{5.4}$$

$$= E(X^2) - [E(X)]^2 \tag{5.4a}$$

is known as the *variance* of the distribution [the notation Var(X) will also be used] and its square root σ_X is called the *standard deviation.* The variance and standard deviation are often used as measures of the spread, or dispersion, of the distribution. It is therefore reasonable that the variance vanishes for a degenerate random variable or a constant:

$$\sigma_c^2 = 0 \tag{5.5}$$

and that it depends on the scale of measurement but not on the origin of measurement:

$$\sigma_{a+bX}^2 = b^2\sigma_X^2. \tag{5.6}$$

Proof of these properties is left to the reader.

Example 1. Let X be a Poisson random variable with the distribution given by (4.3). Then we have

$$E(X^2) = \sum_{k=0}^{\infty} k^2 \frac{e^{-\lambda}\lambda^k}{k!} = \lambda(\lambda + 1) \tag{5.7}$$

and using (4.4) and (5.4a)

$$\sigma_X^2 = \lambda. \qquad \blacktriangleright \tag{5.8}$$

Example 2. Let X have the exponential distribution given in (2.10). Then the variance of X is

$$\sigma_X^2 = \int_0^{\infty} \left(x - \frac{1}{\mu}\right)^2 \mu e^{-\mu x}\, dx = \frac{1}{\mu^2}. \qquad \blacktriangleright \tag{5.9}$$

Consider two random variables X and Y having a joint distribution. The moments of the marginal distributions describe the properties of X and Y separately; however, in order to relate X and Y, we define the *product-moment* or *covariance* of X and Y as

$$\sigma_{X,Y} = E\{[X - E(X)][Y - E(Y)]\} \tag{5.10}$$

$$= E(XY) - E(X)E(Y). \tag{5.10a}$$

Notice that $\sigma_{X,X} = \sigma_X{}^2$. The notation $\text{Cov}(X, Y)$ will also be used for $\sigma_{X,Y}$. If X and Y are stochastically independent, then $\sigma_{X,Y}$ vanishes because of Theorem 4; but a zero covariance between random variables does not imply that they are independently distributed. An example to the contrary follows.

Example 3. Suppose that we have two urns. The ith urn contains a proportion p_i of white balls and q_i of black balls with $p_i + q_i = 1, i = 1, 2$. From the ith urn n_i balls are drawn with replacement, of which X_i are white. The number n_1 is fixed in advance, but the number n_2 is a random variable and is set equal to the number of white balls drawn from the first urn; in other words, $n_2 = X_1$. The proportions of white balls drawn

$$\hat{p}_1 = \frac{X_1}{n_1} \quad \text{and} \quad \hat{p}_2 = \frac{X_2}{X_1} \tag{5.11}$$

are estimators of the corresponding proportions of white balls in the urns. *It can be shown that $\hat{p}_1$ and $\hat{p}_2$ have zero covariance and they are not independently distributed.* See problem 16 for proof.

This urn scheme is quite useful for illustrative purposes in the life table analysis, where n_1 is taken as the number of individuals starting the first time interval, and X_1 and X_2 as the survivors of the first and the second time intervals. A more general model is given in Section 2.1 of Chapter 10. ▶

5.1. Variance of a Linear Function of Random Variables

Using the properties of expectation we have, for any random variables X, Y and any constants a, b, c,

$$\begin{aligned}
\text{Var}(a + bX + cY) &= E\{a + bX + cY - E(a + bX + cY)\}^2 \\
&= E\{b[X - E(X)] + c[Y - E(Y)]\}^2 \\
&= b^2E[X - E(X)]^2 + c^2E[Y - E(Y)]^2 \\
&\qquad + 2bcE[X - E(X)][Y - E(Y)] \\
&= b^2\sigma_X{}^2 + c^2\sigma_Y{}^2 + 2bc\sigma_{X,Y}.
\end{aligned}$$

In general, we have

Theorem 5. *The variance of the linear function* $c_0 + c_1X_1 + \cdots + c_nX_n$ *is*

$$\operatorname{Var}(c_0 + c_1X_1 + \cdots + c_nX_n) = \sum_{i=1}^{n} c_i^2\sigma_{X_i}^2 + \sum_{\substack{i=1\\ i\neq j}}^{n}\sum_{j=1}^{n} c_ic_j\sigma_{X_i,X_j}. \tag{5.12}$$

In particular, if $X_1, \ldots, X_n$ are independent random variables, then

$$\operatorname{Var}(c_1X_1 + \cdots + c_nX_n) = c_1^2\sigma_{X_1}^2 + \cdots + c_n^2\sigma_{X_n}^2 \tag{5.13}$$

and if $X_1, \ldots, X_n$ are independent and have the same variance σ_X^2, then the variance of $\bar{X} = (1/n)(X_1 + \cdots + X_n)$ is

$$\operatorname{Var}(\bar{X}) = \frac{1}{n}\sigma_X^2. \quad ▶ \tag{5.14}$$

Example 4. Consider a sequence of n independent trials, each with probability p of success and $1 - p$ of failure; the random variable X_i, defined as the number of successes in the ith trial, has the *Bernoulli distribution*

$$\Pr\{X_i = 0\} = 1 - p, \qquad \Pr\{X_i = 1\} = p$$

so that

$$E(X_i) = p \tag{5.15}$$

$$\sigma_{X_i}^2 = p(1 - p). \tag{5.16}$$

Set

$$Z = X_1 + \cdots + X_n.$$

Then Z represents the number of successes in n trials and hence has the binomial distribution (2.5). Theorem 1 and (5.15) imply that

$$E(Z) = np \tag{5.17}$$

and, since $X_1, \ldots, X_n$ are independent, Theorem 5 and (5.16) imply that

$$\sigma_Z^2 = np(1 - p). \tag{5.18}$$

Note that this approach to $E(Z)$ and σ_Z^2 is much simpler than direct computation from (2.5). ▶

Example 5. The *multinomial distribution* is a direct generalization of the binomial distribution. Consider an experiment that results in the mutually exclusive outcomes $E_0, E_1, \ldots, E_m$ with probabilities $p_0, p_1, \ldots, p_m$, respectively, where $\sum_{i=0}^{m} p_i = 1$. Then the numbers $X_1, \ldots, X_m$

of occurrences of $E_1, \ldots, E_m$ in n independent repetitions of the experiment have jointly the multinomial distribution given by[2]

$$\Pr\{X_1 = k_1, \ldots, X_m = k_m\} = \frac{n!}{k_1! \cdots k_m!\,(n-k)!}\, p_1^{k_1} \cdots p_m^{k_m} p_0^{(n-k)}$$

$$k_i = 0, 1, \ldots, n; \qquad i = 1, \ldots, m, \tag{5.19}$$

where $k = k_1 + \cdots + k_m$. We are interested in the covariance between X_i and X_j. It is clear that each X_i has the binomial distribution with the variance

$$\sigma_{X_i}^{\ 2} = np_i(1 - p_i) \tag{5.20}$$

and that the sum $X_i + X_j$ also has a binomial distribution with the variance

$$\sigma^2_{X_i+X_j} = n(p_i + p_j)[1 - (p_i + p_j)]. \tag{5.21}$$

But from Theorem 5 we can write

$$\sigma^2_{X_i+X_j} = \sigma_{X_i}^{\ 2} + \sigma_{X_j}^{\ 2} + 2\sigma_{X_i,X_j} \tag{5.22}$$

Using (5.20) and (5.21) in (5.22) gives the covariance

$$\sigma_{X_i,X_j} = -np_ip_j. \qquad \blacktriangleright \tag{5.23}$$

5.2. Covariance Between Two Linear Functions of Random Variables

Let $U = a_1X_1 + \cdots + a_mX_m$ and $V = b_1Y_1 + \cdots + b_nY_n$ be two linear functions of the random variables X_i and Y_j, respectively, where a_i and b_j are constants. The covariance between U and V is given by

$$\sigma_{U,V} = \sum_{i=1}^{m}\sum_{j=1}^{n} a_ib_j\sigma_{X_i,Y_j}. \tag{5.24}$$

The proof of (5.24) is left to the reader.

5.3. Variance of a Product of Random Variables

We have an exact formula for the variance of the product of random variables that are independently distributed. In the case of two random variables,

$$\mathrm{Var}(XY) = \sigma_X^{\ 2}\sigma_Y^{\ 2} + [E(Y)]^2\sigma_X^{\ 2} + [E(X)]^2\sigma_Y^{\ 2} \tag{5.25}$$

---

[2] A random variable X_0 corresponding to the occurrences of E_0 is not explicitly included in (5.19) because the value of X_0 is completely determined by the relation $X_0 = n - X_1 - \cdots - X_m$.

which may be proved by direct computation: using the definition in (5.4*a*) we have

$$\begin{aligned}\operatorname{Var}(XY) &= E[(XY)^2] - [E(XY)]^2 \\ &= E(X^2)E(Y^2) - [E(X)]^2[E(Y)]^2. \qquad (5.26)\end{aligned}$$

Using (5.4*a*) again for $E(X^2)$ and $E(Y^2)$ in (5.26), we obtain (5.25). In general, the variance of the product of n independent random variables is

$$\operatorname{Var}(X_1 \cdots X_n) = \prod_{i=1}^{n} \{\sigma_{X_i}{}^2 + [E(X_i)]^2\} - \prod_{i=1}^{n} [E(X_i)]^2. \qquad (5.27)$$

5.4. Approximate Variance of a Function of Random Variables

When random variables are not independently distributed, there is no general exact formula expressing the variance of their product in terms of the individual variances and covariances. However, Taylor's formula can be used to obtain an approximate formula for the variance of the product, or any function, of random variables. Let $\Phi(X)$ be a function of random variables $X = (X_1, \ldots, X_n)$. Taylor's expansion about the expected value $\mu = (\mu_1, \ldots, \mu_n)$ gives the first two terms

$$\Phi(X) \doteq \Phi(\mu) + \sum_{i=1}^{n} (X_i - \mu_i)\Phi_i(\mu) \qquad (5.28)$$

where

$$\Phi_i(\mu) = \frac{\partial}{\partial X_i}\Phi(X)\big|_{X=\mu}. \qquad (5.29)$$

Applying Theorem 5 to (5.28), we have the approximate variance of $\Phi(X)$

$$\operatorname{Var}[\Phi(X)] = \sum_{i=1}^{n} [\Phi_i(\mu)]^2\sigma_{X_i}{}^2 + \sum_{\substack{i=1 \\ i \neq j}}^{n} \sum_{j=1}^{n} [\Phi_i(\mu)][\Phi_j(\mu)]\sigma_{X_i,X_j}. \qquad (5.30)$$

The necessity for caution in applying this approximation is illustrated by the following example.

Example 6. Consider the product $\Phi = XY$ of two independent random variables X and Y so that the covariance $\sigma_{X,Y} = 0$. Formula (5.30) gives the approximation

$$\operatorname{Var}(XY) = [E(Y)]^2\sigma_X{}^2 + [E(X)]^2\sigma_Y{}^2. \qquad (5.31)$$

Comparing (5.31) with the exact formula in (5.25) shows that the error is in the missing term $\sigma_X{}^2\sigma_Y{}^2$. ▶

5.5. Conditional Variance and Covariance

Variances and covariances can be conditioned on other random variables just as expectations were in Section 4.2. The conditional variance of X given Z is defined as

$$\sigma^2_{X|Z} = E\{[X - E(X \mid Z)]^2 \mid Z\} \tag{5.32}$$

and the conditional covariance of X and Y given Z defined as

$$\sigma_{X,Y|Z} = E\{[X - E(X \mid Z)][Y - E(Y \mid Z)] \mid Z\}. \tag{5.33}$$

The conditional variance and covariance are, of course, random variables; their moments satisfy

Theorem 6. *For any random variables X, Y, and Z*

$$\sigma_X{}^2 = E[\sigma^2_{X|Z}] + \sigma^2_{E(X|Z)} \tag{5.34}$$

$$\sigma_{X,Y} = E[\sigma_{X,Y|Z}] + \sigma_{E(X|Z),E(Y|Z)}. \tag{5.35}$$

Proof. To prove (5.34), we use the definition of variance to write

$$\begin{aligned}\sigma_X{}^2 &= E[X - E(X)]^2 = E[X - E(X \mid Z) + E(X \mid Z) - E(X)]^2 \\ &= E[X - E(X \mid Z)]^2 + E[E(X \mid Z) - E(X)]^2 \\ &\qquad + 2E\{[X - E(X \mid Z)][E(X \mid Z) - E(X)]\}. \end{aligned} \tag{5.36}$$

Applying Theorem 3, the first term after the last equality sign becomes

$$E(E\{[X - E(X \mid Z)]^2 \mid Z\}) = E[\sigma^2_{X|Z}],$$

the second term becomes

$$E\{E(X \mid Z) - E[E(X \mid Z)]\}^2 = \sigma^2_{E(X|Z)},$$

and the third term is

$$\begin{aligned}2E(E\{[X - E(X \mid Z)][E(X \mid Z) - E(X)] \mid Z\}) \\ = 2E([E(X \mid Z) - E(X)]E\{[X - E(X \mid Z)] \mid Z\}) \\ = 0\end{aligned}$$

since $E\{[X - E(X \mid Z)] \mid Z\} = 0$, and (5.34) is proved. The proof of (5.35) is similar. ▶

5.6. Correlation Coefficient

Consider two random variables X and Y with standard deviations σ_X and σ_Y. Let

$$U = \frac{X - E(X)}{\sigma_X} \quad \text{and} \quad V = \frac{Y - E(Y)}{\sigma_Y}. \tag{5.37}$$

It is easy to see that

$$E(U) = 0, \qquad E(V) = 0 \tag{5.38}$$

and that

$$\sigma_U^2 = E(U^2) = 1, \qquad \sigma_V^2 = E(V^2) = 1. \tag{5.39}$$

For these reasons, U and V are called *standardized random variables*. The correlation coefficient between X and Y, denoted by ρ_{XY}, is defined as the covariance between U and V,

$$\rho_{XY} = \sigma_{U,V} = E(UV) = E\left(\frac{X - E(X)}{\sigma_X}\right)\left(\frac{Y - E(Y)}{\sigma_Y}\right) \tag{5.40}$$

which can also be written as

$$\rho_{XY} = \frac{\sigma_{X,Y}}{\sigma_X \sigma_Y}. \tag{5.41}$$

Since the correlation coefficient is calculated from standardized random variables, its value is not affected by changes of scale and of origin, that is,

$$\rho_{a+bX,c+dY} = \rho_{XY}. \tag{5.42}$$

The correlation coefficient is often used as a measure of the extent to which X and Y obey a linear relation. If the linear relation $Y = a + bX$ holds true exactly, Y and X are said to be *perfectly correlated*; in this case, the correlation coefficient between X and Y is either -1 or $+1$. It is interesting to notice, however, that X and Y may obey a perfect non-linear relation and, nevertheless, have a very small correlation coefficient. In any case, the inequality

$$-1 \le \rho_{XY} \le 1 \tag{5.43}$$

is always true, since

$$E(U \pm V)^2 \ge 0 \tag{5.44}$$

so that

$$E(U^2) \pm 2E(UV) + E(V^2) = 1 \pm 2\rho_{XY} + 1 \ge 0. \tag{5.45}$$

The last inequality rearranges to yield (5.43).

PROBLEMS

1. Let X be a binomial random variable with the probability distribution given in (2.5) and let $Y = n - X$. Find the covariance $\sigma_{X,Y}$.

2. Show that a Poisson distribution with the parameter λ has both an expectation and variance equal to λ.

3. Prove that

$$E(c_0 + c_1X_1 + \cdots + c_nX_n) = c_0 + c_1E(X_1) + \cdots + c_nE(X_n) \tag{4.10}$$

for the continuous case.

4. If a, b, and c are constant and the random variable X has finite variance, show that

$$\sigma_c^2 = 0 \tag{5.5}$$

and

$$\sigma_{a+bX}^2 = b^2\sigma_X^2. \tag{5.6}$$

5. Let X have an exponential distribution with the probability density function

$$f(x) = \mu e^{-\mu x}. \tag{2.10}$$

Show that $F_X(\infty) = 1$, and compute the expectation and variance of X.

6. A lot contains m defective units and $N - m$ acceptable units. The units are tested one at a time without replacement. Let X be the number of defectives found before the first acceptable one. Derive the probability distribution and the expectation of X.

7. A gamma function is defined by

$$\Gamma(n) = \int_0^\infty x^{n-1}e^{-x}\,dx.$$

(a) Show that $\Gamma(n + 1) = n\Gamma(n)$, $\Gamma(1) = 1$, $\Gamma(0) = \infty$, $\Gamma(-1) = -\infty$.

(b) Find $\Gamma(-n)$ for $n = 2, 3, 4$.

(c) Show that $\Gamma(\frac{1}{2}) = \sqrt{\pi}$. (Hint. Use polar coordinates to prove that $[\Gamma(\frac{1}{2})]^2 = \pi$.)

8. *Normal distribution.* For the standard normal distribution

$$f(x) = \frac{1}{\sqrt{2\pi}}e^{-x^2/2}$$

(a) Determine the points of inflection.

(b) Show that the integral

$$\int_{-\infty}^{+\infty} f(x)\,dx = 1$$

so that the random variable is proper.

9. *Continuation.* Suppose that X has a normal distribution with the expectation μ, variance σ^2, and the probability density function

$$f(x) = \frac{1}{\sqrt{2\pi}\sigma}\exp\left\{-\frac{(x-\mu)^2}{2\sigma^2}\right\}.$$

Show that the standardized form

$$Z = \frac{X - \mu}{\sigma}$$

has the standard normal distribution in Problem 8.

10. *Continuation.* Suppose X and Y have the bivariate normal distribution with the joint probability density function given by

$$f(x, y) = \frac{1}{2\pi\sigma_X\sigma_Y\sqrt{1-\rho^2}} \exp\left\{-\frac{1}{2(1-\rho^2)}\left[\left(\frac{x-\mu}{\sigma_X}\right)^2 - 2\rho\left(\frac{x-\mu}{\sigma_X}\right)\left(\frac{y-\nu}{\sigma_Y}\right) + \left(\frac{y-\nu}{\sigma_Y}\right)^2\right]\right\}.$$

(a) Show that the density function can be rewritten as

$$f(x, y) = \frac{1}{\sqrt{2\pi}\sigma_X} \exp\left\{-\frac{1}{2}\left(\frac{x-\mu}{\sigma_X}\right)^2\right\} \times \frac{1}{\sqrt{2\pi}\,\sigma_Y\sqrt{1-\rho^2}} \exp\left\{\frac{-1}{2(1-\rho^2)\sigma_Y^2}\left[(y-\nu) - \rho\frac{\sigma_Y}{\sigma_X}(x-\mu)\right]^2\right\}.$$

(b) Using (a), verify that the marginal density functions of X and Y are normal with means zero and variances σ_X^2 and σ_Y^2, respectively, and that the covariance between X and Y is $\rho\sigma_X\sigma_Y$, so that ρ is the correlation coefficient.

(c) Using (a), show that the conditional density function $g(y \mid x)$ is a normal distribution with the expectation

$$E(Y \mid x) = \nu + \rho\frac{\sigma_Y}{\sigma_X}(x - \mu)$$

and the variance $(1 - \rho^2)\sigma_Y^2$. The line represented by this equation is called the regression line of Y on X. How does the value of ρ affect the closeness of the points of the random variables (X, Y) to this line? Discuss the meaning of correlation coefficients in this light. (*Note.* In the case of a bivariate normal distribution, correlation coefficient of zero implies independence of X and Y.)

(d) Verify for the bivariate normal case that $E[E(Y \mid X)] = E(Y)$ and $E(XY) = E[XE(Y \mid X)]$.

(e) Use (b) and (c) to verify that

$$\sigma_Y^2 = E[\sigma^2_{Y|X}] + \sigma^2_{E(Y|X)}.$$

11. Prove the following inequality for a nondegenerate positive random variable X

$$E\left(\frac{1}{X}\right) > \frac{1}{E(X)}$$

(a) using the Schwarz inequality, and
(b) using the concept of convexity.

12. Let $X_1, \ldots, X_n$ be a sample of independent random variables having the common expectation $E(X_i) = \mu$ and the common variance $\text{Var}(X_i) = \sigma^2$. Let

$$\bar{X} = \frac{X_1 + \cdots + X_n}{n} \quad \text{and} \quad S_X^2 = \frac{\sum_{i=1}^{n}(X_i - \bar{X})^2}{n-1}$$

be the sample mean and sample variance. Show that

$$E(\bar{X}) = \mu \quad \text{and} \quad E(S_X{}^2) = \sigma^2.$$

In this sense $\bar{X}$ and $S_X{}^2$, as defined above, are unbiased estimators of μ and σ^2, respectively.

13. Prove the following identities for the continous case

$$\sigma_X{}^2 = E[\sigma^2_{X|Z}] + \sigma^2_{E(X|Z)} \tag{5.34}$$

and

$$\sigma_{X,Y} = E[\sigma_{X,Y|Z}] + \sigma_{E(X|Z),E(Y|Z)}. \tag{5.35}$$

14. Let $U = a_1X_1 + \cdots + a_mX_m$ and $V = b_1Y_1 + \cdots + b_nY_n$ as given in Section 5.2. Show that the covariance between U and V is given by

$$\sigma_{U,V} = \sum_{i=1}^{m}\sum_{j=1}^{n} a_ib_j\sigma_{X_i,Y_j}. \tag{5.24}$$

15. Show that, if $X_1, \ldots, X_n$ are independent random variables, then

$$\text{Var}(X_1 \cdots X_n) = \prod_{i=1}^{n} \{\sigma_{X_i}{}^2 + [E(X_i)]^2\} - \prod_{i=1}^{n} [E(X_i)]^2. \tag{5.27}$$

16. Referring to (5.11), show that the two proportions $\hat{p}_1$ and $\hat{p}_2$ of white balls drawn have zero covariance but that they are not independently distributed. (*Hint:* Compare $\Pr\{\hat{p}_2 = 1\}$ with $\Pr\{\hat{p}_2 = 1 \mid \hat{p}_1 = 1\}$.)

17. Show that if $Y = a + bX$, then $\rho_{X,Y} = \pm 1$.

18. Let Y be a nonnegative random variable and c be any positive constant. Show that

$$\Pr\{Y \geq c\} \leq \frac{E(Y)}{c}.$$

19. *Chebyshev's inequality*. Let X be a random variable with expectation $E(X) = \mu$ and variance σ^2. Then for any positive number t,

$$\Pr\{|X - \mu| \geq t\} \leq \frac{\sigma^2}{t^2}.$$

Use the result of Problem 18 to prove the inequality.

20. Compute the expectation $E(X_iX_j)$ for the multinomial distribution (5.19) and use the result to show that $\sigma_{X_i,X_j} = np_ip_j$.

CHAPTER 2

Probability Generating Functions

1. INTRODUCTION

The method of *generating functions* is the most important tool in the study of stochastic processes with a discrete sample space. It has been used in differential and integral calculus and in combinatorial analysis. The generating function of an integral-valued random variable completely determines its probability distribution and provides convenient ways to obtain the moments of the distribution; furthermore, certain important relations among random variables may be very simply expressed in terms of generating functions. Detailed treatments of the subject are given in Feller [1957] and Riordan [1958].

2. GENERAL PROPERTIES

Definition. Let $a_0, a_1, \ldots,$ be a sequence of real numbers. If the power series

$$A(s) = a_0 + a_1 s + a_2 s^2 + \cdots \tag{2.1}$$

converges in some interval $-s_0 < s < s_0$, then the function $A(s)$ is called the generating function of the sequence $\{a_k\}$. Formula (2.1), in fact, defines a transformation that carries the sequence $\{a_k\}$ into the function $A(s)$.

If X is an integral-valued random variable with the probability distribution

$$\Pr\{X = k\} = p_k, \qquad k = 0, 1, \ldots, \tag{2.2}$$

then the power series

$$g_X(s) = p_0 + p_1 s + p_2 s^2 + \cdots \tag{2.3}$$

is the probability generating function (p.g.f.) of X (or of the sequence $\{p_k\}$). Since all the p_k are less than unity, comparison with the geometric series shows that (2.3) converges at least for $|s| < 1$; furthermore, since (2.3) also converges for $|s| = 1$, it converges uniformly for $|s| \leq 1$, and its sum is therefore continuous in that interval. Thus, for every probability distribution $\{p_k\}$, there exists a unique and continuous function $g_X(s)$. The usefulness of (2.3) lies in the converse fact that every p.g.f. $g_X(s)$ determines a unique probability distribution $\{p_k\}$; in fact, the individual probabilities p_k can be obtained from the function $g_X(s)$ by the relation

$$p_k = \frac{1}{k!} \frac{d^k}{ds^k} g_X(s) \bigg|_{s=0}, \qquad k = 0, 1, \ldots, \tag{2.4}$$

which follows when (2.3) is differentiated term by term.[1]

When X is a proper random variable so that

$$g_X(1) = \sum_{k=0}^{\infty} p_k = 1, \tag{2.5}$$

the power series in (2.3) defines an expectation

$$g_X(s) = E(s^X). \tag{2.6}$$

In this case, we may use the p.g.f. also to obtain the expectation and high-order factorial moments of X by taking appropriate derivatives:[2]

$$E(X) = p_1 + 2p_2 + 3p_3 + \cdots$$

$$= \frac{d}{ds} g_X(s) \bigg|_{s=1} \tag{2.7}$$

$$E[X(X-1)] = \frac{d^2}{ds^2} g_X(s) \bigg|_{s=1} \tag{2.8}$$

$$E[X(X-1)\cdots(X-r+1)] = \frac{d^r}{ds^r} g_X(s) \bigg|_{s=1}. \tag{2.9}$$

The variance of X may be obtained from

$$\sigma_X{}^2 = E[X(X-1)] + E(X) - [E(X)]^2. \tag{2.10}$$

---

[1] If a power series $\sum_k^{\infty} c_k s^k$ converges to a function $f(s)$ for $|s| < s_0$, then the differentiated series $\sum_k^{\infty} k c_k s^{k-1}$ converges at least for $|s| < s_0$, and is equal to $f'(s)$ in that interval. Repeated application of this argument justifies (2.4). Furthermore, if $\sum_k^{\infty} k c_k s_0^{k-1}$ converges, then it is equal to $f'(s_0)$. See Buck [1965] or Rudin [1953].

[2] If any one of the derivatives in (2.7) to (2.9) is infinite, then the corresponding series diverges and the moment is said to be infinite.

3. CONVOLUTIONS

Let $\{a_k\}$ and $\{b_k\}$ be two sequences of real numbers. The sequence $\{c_k\}$ with elements

$$c_k = a_0 b_k + a_1 b_{k-1} + \cdots + a_k b_0 \tag{3.1}$$

is called the *convolution* of $\{a_k\}$ and $\{b_k\}$; this relation will be denoted by

$$\{c_k\} = \{a_k\}*\{b_k\}. \tag{3.2}$$

Let

$$g_a(s) = \sum_{k=0}^{\infty} a_k s^k$$

and

$$g_b(s) = \sum_{k=0}^{\infty} b_k s^k$$

be the generating functions of $\{a_k\}$ and $\{b_k\}$. Multiplying the infinite sums for $g_a(s)$ and $g_b(s)$ together and grouping terms with equal powers of s, we find that

$$g_a(s)g_b(s) = \sum_{k=0}^{\infty} c_k s^k = g_c(s). \tag{3.3}$$

Thus *the generating function of the convolution of two sequences is the product of the generating functions of the sequences.* This relation extends immediately by induction to an arbitrary number of sequences; furthermore, the order in which the convolutions are performed is immaterial, since the associativity and commutativity of the multiplication are carried over to the convolution.

These results gain intuitive content when the sequences are probability distributions. Suppose two random variables X and Y have probability distributions

$$\Pr\{X = k\} = p_k \qquad \text{and} \qquad \Pr\{Y = k\} = q_k \tag{3.4}$$

and that X and Y are independent, so that their joint distribution is

$$\Pr\{X = i \text{ and } Y = j\} = p_i q_j. \tag{3.5}$$

Then the sum $Z = X + Y$ is a new random variable with

$$\Pr\{Z = k\} = r_k = p_0 q_k + p_1 q_{k-1} + \cdots + p_k q_0, \tag{3.6}$$

so that the sequence $\{r_k\}$ is the convolution of $\{p_k\}$ and $\{q_k\}$. Thus the above result (3.3) implies that the p.g.f. of the sum of two independent random variables is the product of the individual p.g.f.'s. Generalizing by induction to an arbitrary number of random variables, we have proved

Theorem 1. *Let $X_1, \ldots, X_n$ be independent random variables with p.g.f.'s $g_1(s), \ldots, g_n(s)$, respectively, and let $Z = X_1 + \cdots + X_n$. Then*

the p.g.f. of Z is

$$G_Z(s) = g_1(s)g_2(s) \cdots g_n(s). \tag{3.7}$$

A more direct proof of the theorem is available when $X_1, \ldots, X_n$ are proper random variables, since in this case

$$\begin{aligned} G_Z(s) = E(s^Z) &= E(s^{X_1+X_2+\cdots+X_n}) \\ &= E(s^{X_1}s^{X_2}\cdots s^{X_n}) \\ &= E(s^{X_1})E(s^{X_2})\cdots E(s^{X_n}) \\ &= g_1(s)g_2(s)\cdots g_n(s). \end{aligned}$$ ►

Of particular interest in statistical analysis is the case in which $X_1, \ldots, X_n$ have the same probability distribution $\{p_k\}$ and hence the same p.g.f. $g(s)$. In this case the distribution of Z is the n-fold convolution of $\{p_k\}$ with itself, denoted by $\{p_k\}^{n*}$, and the p.g.f. of Z is

$$G_Z(s) = [g(s)]^n. \tag{3.8}$$

4. EXAMPLES

4.1. Binomial Distribution

A Bernoulli random variable X has the probability distribution (see Example 4, Section 5, Chapter 1)

$$\Pr\{X = 0\} = 1 - p \qquad \Pr\{X = 1\} = p \tag{4.1}$$

so that its p.g.f. is

$$g_X(s) = (1 - p) + ps. \tag{4.2}$$

The number Z of successes (occurrences of $X = 1$) in n Bernoulli trials is the sum of n independent Bernoulli random variables; hence, by (3.8), Z has the p.g.f.

$$G_Z(s) = [(1 - p) + ps]^n. \tag{4.3}$$

It is readily verified by direct computation that (4.3) is the p.g.f. of the binomial distribution

$$\Pr\{Z = k\} = \binom{n}{k} p^k(1 - p)^{n-k}, \qquad k = 0, 1, \ldots, n. \tag{4.4}$$

Consider now a more general sequence of random variables $X_1, \ldots, X_n$ having the distributions

$$\Pr\{X_i = k\} = \binom{m_i}{k} p^k(1 - p)^{m_i-k}, \quad k = 0, 1, \ldots, m_i; \quad i = 1, \ldots, n; \tag{4.5}$$

and the p.g.f.'s

$$g_i(s) = [(1-p) + ps]^{m_i}, \qquad i = 1, \ldots, n. \tag{4.6}$$

The p.g.f. of the sum $Z = X_1 + \cdots + X_n$ is

$$G_Z(s) = [(1-p) + ps]^{m_1 + \cdots + m_n} = [(1-p) + ps]^N, \tag{4.7}$$

which is the p.g.f. of the binomial distribution with $N = m_1 + \cdots + m_n$ trials. Hence the sum of independent binomial random variables with the same parameter p is also a binomial random variable. ▶

4.2. Poisson Distribution

The p.g.f. of the Poisson distribution

$$\Pr\{X = k\} = \frac{e^{-\lambda}\lambda^k}{k!} \tag{4.8}$$

is

$$g_X(s) = \sum_{k=0}^{\infty} s^k \frac{e^{-\lambda}\lambda^k}{k!} = e^{-\lambda(1-s)}. \tag{4.9}$$

Consider now another Poisson random variable Y, independent of X, with the distribution

$$\Pr\{Y = k\} = \frac{e^{-\mu}\mu^k}{k!} \tag{4.10}$$

and p.g.f.

$$g_Y(s) = \sum_{k=0}^{\infty} s^k \frac{e^{-\mu}\mu^k}{k!} = e^{-\mu(1-s)}. \tag{4.11}$$

The p.g.f. of the sum $Z = X + Y$ is

$$g_Z(s) = g_X(s)g_Y(s) = e^{-(\lambda+\mu)(1-s)}. \tag{4.12}$$

Thus, Z also has a Poisson distribution regardless of the values of the parameters λ and μ. ▶

4.3. Geometric and Negative Binomial Distributions

Let X be the number of failures preceding the first success in an infinite sequence of Bernoulli trials. Obviously, X has the *geometric distribution*

$$\Pr\{X = k\} = q^k p, \qquad k = 0, 1, \ldots, \tag{4.13}$$

with the p.g.f.

$$g_X(s) = \sum_{k=0}^{\infty} s^k q^k p = \frac{p}{1 - qs}, \tag{4.14}$$

and expectation

$$E(X) = \frac{d}{ds} g_X(s)\bigg|_{s=1} = \frac{q}{p}, \tag{4.15}$$

and variance

$$\sigma_X{}^2 = \frac{q}{p^2}. \tag{4.16}$$

Let Z_r be the number of failures preceding the rth success in an infinite sequence of Bernoulli trials. It is clear that Z_r is the sum of r independent geometric random variables, so that its mean and variance are

$$E(Z_r) = r\frac{q}{p} \qquad \sigma_{Z_r}{}^2 = r\frac{q}{p^2} \tag{4.17}$$

and its p.g.f. is

$$G_Z(s) = \left[\frac{p}{1 - qs}\right]^r. \tag{4.18}$$

Equation (4.18) may be verified by direct computation as follows: when $Z_r = k$, there are altogether $k + r$ trials, with the last trial resulting in a success (the rth success). Since there are $\binom{k+r-1}{k}$ ways of having k failures and $r - 1$ successes in the first $(k + r - 1)$ trials, the distribution of Z_r is given by

$$\Pr\{Z_r = k\} = \binom{k+r-1}{k} q^k p^r, \qquad k = 0, 1, \ldots. \tag{4.19}$$

We now find it useful to recall *Newton's binomial expansion*

$$(1 + t)^x = \sum_{k=0}^{\infty} \binom{x}{k} t^k, \qquad |t| < 1 \tag{4.20}$$

which holds true for any real number x if the binomial coefficients are defined by

$$\binom{x}{k} = \frac{x(x-1)\cdots(x-k+1)}{k!}. \tag{4.21}$$

Notice that (4.21) reduces to (2.6) in Chapter 1 if x is a positive integer. From (4.21) it follows that

$$\binom{-r}{k} = (-1)^k \binom{r+k-1}{k}. \tag{4.22}$$

Using (4.22) and (4.20), we obtain the p.g.f. of (4.19) as

$$p^r \sum_{k=0}^{\infty} \binom{k+r-1}{k} (qs)^k = \left(\frac{p}{1 - qs}\right)^r \tag{4.23}$$

and (4.18) is verified.

In deriving (4.19) we assumed that the parameter r is a positive integer. The student may readily verify that, with the extended definition (4.21) of the binomial coefficients, (4.19) is a proper probability distribution with p.g.f. (4.18) and moments (4.17) for *any* positive real number r. With an arbitrary value of $r > 0$, (4.19) is known as the *negative binomial distribution*; it arises frequently in biological problems.

It is clear from (4.18) that the sum of two independent negative binomial random variables with the same parameter p is also a negative binomial random variable. ▶

5. PARTIAL FRACTION EXPANSIONS

Individual probabilities may be obtained from the p.g.f. by taking repeated derivatives as in formula (2.4), but in many cases the necessary calculations are so overwhelming as to render this approach undesirable, and we must resort to some other means to expand the p.g.f. in a power series. The most useful method is that of the partial fraction expansion, which we present for the common case where $G(s)$ is a rational function.

Suppose that the p.g.f. of a random variable can be expressed as a ratio

$$G(s) = \frac{U(s)}{V(s)} \tag{5.1}$$

where $U(s)$ and $V(s)$ are polynomials with no common roots. If $U(s)$ and $V(s)$ have a common factor, it should be canceled before the method is applied. For simplicity, we shall assume that the degree of $V(s)$ is m and the degree of $U(s)$ is less than m. Suppose that the equation $V(s) = 0$ has m distinct roots, $s_1, \ldots, s_m$, so that

$$V(s) = (s - s_1)(s - s_2) \cdots (s - s_m). \tag{5.2}$$

Then the ratio in (5.1) can be decomposed into partial fractions

$$G(s) = \frac{a_1}{s_1 - s} + \frac{a_2}{s_2 - s} + \cdots + \frac{a_m}{s_m - s}, \tag{5.3}$$

where the constants $a_1, \ldots, a_m$ are determined from

$$a_r = \frac{-U(s_r)}{V'(s_r)}, \qquad r = 1, \ldots, m. \tag{5.4}$$

Equation (5.3) is well known in algebra. To prove (5.4) for $r = 1$, we substitute (5.2) in (5.1) and multiply by $(s_1 - s)$ to obtain

$$(s_1 - s)G(s) = \frac{-U(s)}{(s - s_2)(s - s_3) \cdots (s - s_m)}. \tag{5.5}$$

It is evident from (5.3) that as $s \to s_1$, the left side of (5.5) tends to a_1, while the numerator on the right side tends to $-U(s_1)$ and the denominator to $(s_1 - s_2)(s_1 - s_3) \cdots (s_1 - s_m)$, which is the same as $V'(s_1)$; hence, (5.4) is proved for $r = 1$. Since the same argument applies to all roots, (5.4) is true for all r.

Now formula (5.3) can be used to derive the coefficient of s^k in $G(s)$. We first write

$$\frac{1}{s_r - s} = \frac{1}{s_r}\left(\frac{1}{1 - s/s_r}\right) = \frac{1}{s_r}\left[1 + \frac{s}{s_r} + \left(\frac{s}{s}\right)^2 + \cdots\right]. \tag{5.6}$$

Substituting (5.6) for $r = 1, \ldots, m$ in (5.3) and selecting s such that $|s/s_r| < 1$, we find the exact expression for the coefficient p_k of s^k,

$$p_k = \frac{a_1}{s_1^{k+1}} + \frac{a_2}{s_2^{k+1}} + \cdots + \frac{a_m}{s_m^{k+1}}, \qquad k = 0, 1, \ldots. \tag{5.7}$$

Strict application of this method requires calculation of all m roots, and this is usually prohibitive. However, from (5.7), it appears that the value of p_k is dominated by the term corresponding to the root with the smallest absolute value. Suppose that s_1 is smaller in absolute value than any other root. Then

$$p_k \doteq \frac{a_1}{s_1^{k+1}} \tag{5.8}$$

represents an approximation to p_k and the approximation becomes better as k increases. Furthermore, even if the roots $\{s_r\}$ are not all distinct, (5.8) remains valid as long as s_1 is itself a simple root.

6. MULTIVARIATE PROBABILITY GENERATING FUNCTIONS

Let us consider a vector of m integral-valued random variables

$$\mathbf{X} = \begin{pmatrix} X_1 \\ \cdot \\ \cdot \\ \cdot \\ X_m \end{pmatrix} \tag{6.1}$$

with a joint probability distribution

$$\Pr\{X_1 = k_1, \ldots, X_m = k_m\} = p_{k_1 \cdots k_m}, \qquad k_i = 0, 1, \ldots; \quad i = 1, \ldots, m. \tag{6.2}$$

$\mathbf{X}$ is called a *proper random vector* if

$$\sum_{k_1} \cdots \sum_{k_m} p_{k_1 \cdots k_m} = 1 \tag{6.3}$$

and an *improper random vector* if the sum is less than one. The p.g.f. of $\mathbf{X}$ is defined, for $|s_1| < 1, \ldots, |s_m| < 1$, by

$$G_{\mathbf{X}}(\mathbf{s}) = G_{\mathbf{X}}(s_1, \ldots, s_m) = \sum_{k_1} \cdots \sum_{k_m} s_1^{k_1} \cdots s_m^{k_m} p_{k_1 \cdots k_m}, \tag{6.4}$$

which becomes

$$G_{\mathbf{X}}(\mathbf{s}) = E(s_1^{X_1} \cdots s_m^{X_m}) \tag{6.5}$$

when $\mathbf{X}$ is a proper random vector. As in the case of a single random variable, when the p.g.f. $G_{\mathbf{X}}(s_1, \ldots, s_m)$ is expanded in a power series of $s_1, \ldots, s_m$, the coefficients are the corresponding joint probabilities in (6.2). It is clear from (6.5) that for a proper random vector $\mathbf{X}$ the p.g.f. of the marginal distribution $\Pr\{X_i = k\}$ is

$$E(s_i^{X_i}) = G_{\mathbf{X}}(1, \ldots, s_i, \ldots, 1) \tag{6.6}$$

so that

$$E(X_i) = \left. \frac{\partial G_{\mathbf{X}}(s_1, \ldots, s_m)}{\partial s_i} \right|_{s=1} \tag{6.7}$$

and

$$E[X_i(X_i - 1)] = \left. \frac{\partial^2 G_{\mathbf{X}}(s_1, \ldots, s_m)}{\partial s_i^2} \right|_{s=1}.$$

Analogously, it is again evident from (6.5) that the p.g.f. of the joint probability distribution of any two components X_i, X_j of a proper random vector can be obtained from $G_{\mathbf{X}}(s_1, \ldots, s_m)$ by setting all the s's equal to unity except s_i and s_j, and also that

$$E(X_i X_j) = \left. \frac{\partial^2}{\partial s_i \partial s_j} G_{\mathbf{X}}(s_1, \ldots, s_m) \right|_{s=1}. \tag{6.8}$$

The covariance between X_i and X_j can be found from

$$\sigma_{X_i, X_j} = E(X_i X_j) - E(X_i)E(X_j). \tag{6.9}$$

Equation (6.5) also shows that the p.g.f. of the sum $Z = X_1 + \cdots + X_m$ is

$$E(s^Z) = E(s^{X_1} \cdots s^{X_m}) = G_{\mathbf{X}}(s, \ldots, s). \tag{6.10}$$

Furthermore, $X_1, \ldots, X_m$ are mutually independent random variables if, and only if,

$$E(s_1^{X_1} \cdots s_m^{X_m}) = E(s_1^{X_1}) \cdots E(s_m^{X_m}). \tag{6.11}$$

The multivariate analog of Theorem 1 is simply stated: the p.g.f. of the sum of independent random vectors is the product of the p.g.f.'s. Explicitly, we state

Theorem 2. *If n random vectors*

$$\mathbf{Y}_j = \begin{pmatrix} Y_{1j} \\ \cdot \\ \cdot \\ \cdot \\ Y_{mj} \end{pmatrix}, \qquad j = 1, \ldots, n,$$

have p.g.f.'s $g_j(s_1, \ldots, s_m)$ *and are independently distributed, then the p.g.f. of the vector sum*

$$\mathbf{X} = \sum_{j=1}^{n} \mathbf{Y}_j = \begin{pmatrix} \sum_{j=1}^{n} Y_{1j} \\ \cdot \\ \cdot \\ \cdot \\ \sum_{j=1}^{n} Y_{mj} \end{pmatrix}$$

is

$$G_{\mathbf{X}}(s_1, \ldots, s_m) = \prod_{j=1}^{n} g_j(s_1, \ldots, s_m). \tag{6.12}$$

In particular, if the $\mathbf{Y}_j$ *are identically distributed with p.g.f.* $g(s_1, \ldots, s_m)$, *then*

$$G_{\mathbf{X}}(s_1, \ldots, s_m) = [g(s_1, \ldots, s_m)]^n. \qquad \blacktriangleright \tag{6.13}$$

Both methods of proof used for Theorem 1 extend immediately to this case; details are left to the reader.

Example 1. *Multinomial distribution.* The multinomial distribution given by Equation (5.19) in Chapter 1 has the p.g.f.

$$G_{\mathbf{X}}(s_1, \ldots, s_m) = (p_0 + p_1 s_1 + \cdots + p_m s_m)^n. \tag{6.14}$$

This is most easily seen by the device of writing each multinomial variable X_i as the sum of n independent and identically distributed random variables and applying (6.13).

Applying (6.8) to (6.14), we have

$$E(X_i X_j) = n(n-1)p_i p_j. \tag{6.15}$$

The p.g.f. of the marginal distribution of X_i is

$$E[s_i^{X_i}] = G_{\mathbf{X}}(1, \ldots, s_i, \ldots, 1) = [1 - p_i + p_i s_i]^n, \tag{6.16}$$

which is consistent with the fact that each X_i is binomially distributed. Using $E(X_i) = np_i$ and (6.15) gives

$$\mathrm{Cov}(X_i, X_j) = -np_ip_j, \tag{6.17}$$

which agrees with Equation (5.23) in Chapter 1. ▶

Example 2. *Negative multinomial distribution* (*multivariate Pascal distribution*). Consider a sequence of independent trials, each of which results in a success, S, or in one of the failures, $F_1, F_2, \ldots$, or F_m, with corresponding probabilities $p, q_1, q_2, \ldots, q_m$, so that

$$p + q_1 + q_2 + \cdots + q_m = 1. \tag{6.18}$$

Let X_i be the number of failures of type F_i, for $i = 1, \ldots, m$, preceding the rth success in an infinite sequence of trials. The joint probability distribution of $X_1, \ldots, X_m$ is

$$\Pr\{X_1 = k_1, \ldots, X_m = k_m\} = \frac{(k + r - 1)!}{k_1! \cdots k_m!\,(r-1)!}\, q_1^{k_1} \cdots q_m^{k_m} p^r \tag{6.19}$$

where $k = k_1 + \cdots + k_m$, and the corresponding p.g.f. is

$$\begin{aligned} G_{\mathbf{X}}(s_1, \ldots, s_m) &= \sum_{k_1=0}^{\infty} \cdots \sum_{k_m=0}^{\infty} s_1^{k_1} \cdots s_m^{k_m} \frac{(k + r - 1)!}{k_1! \cdots k_m!\,(r-1)!} \\ &\quad \times q_1^{k_1} \cdots q_m^{k_m} p^r \\ &= \left(\frac{p}{1 - q_1 s_1 - \cdots - q_m s_m}\right)^r. \end{aligned} \tag{6.20}$$

As with (6.13), the form of the p.g.f. shows that each X_i can be regarded as the sum of r independent and identically distributed random variables. From (6.20) we use (6.7), (6.8), and (6.9) to compute the expectation

$$E(X_i) = r\frac{q_i}{p}, \tag{6.21}$$

the variance

$$\sigma_{X_i}^2 = r\frac{q_i}{p}\left(1 + \frac{q_i}{p}\right) \tag{6.22}$$

and the covariance

$$\sigma_{X_i, X_j} = r\frac{q_i q_j}{p^2}. \tag{6.23}$$

The marginal distribution of any of the random variables X_i and the distribution of the sum $X_1 + \cdots + X_m$ are negative binomial distributions.

This is evident from the definitions of the random variables; we may see it explicitly by writing the p.g.f. of the marginal distribution of X_i as

$$E(s_i^{X_i}) = G_{\mathbf{X}}(1, \ldots, s_i, \ldots, 1) = \left(\frac{p}{1 - q_1 - \cdots - q_i s_i - \cdots - q_m}\right)^r$$

$$= \left(\frac{p}{p + q_i - q_i s_i}\right)^r = \left(\frac{P_i}{1 - Q_i s_i}\right)^r, \tag{6.24}$$

where

$$P_i = \frac{p}{p + q_i}, \qquad Q_i = \frac{q_i}{p + q_i}, \qquad P_i + Q_i = 1 \tag{6.25}$$

and the p.g.f. of $Z = X_1 + \cdots + X_m$ as

$$E(s^Z) = G_{\mathbf{X}}(s, \ldots, s) = \left(\frac{p}{1 - q_1 s - \cdots - q_m s}\right)^r = \left(\frac{p}{1 - qs}\right)^r \tag{6.26}$$

where $q = q_1 + \cdots + q_m$. Notice that the expectation and the variance of X_i in (6.21) and (6.22) can also be obtained from (6.24). Also, since $Z = X_1 + \cdots + X_m$, we find the following relation

$$\sigma_Z^{\,2} = \sum_{i=1}^{m} \sigma_{X_i}^{\,2} + \sum_{\substack{i=1 \\ i \neq j}}^{m} \sum_{j=1}^{m} \sigma_{X_i, X_j} \tag{6.27}$$

or, by substituting (4.17), (6.22), and (6.23),

$$r\frac{q}{p^2} = \sum_{i=1}^{m} r\frac{q_i}{p}\left(1 + \frac{q_i}{p}\right) + \sum_{\substack{i=1 \\ i \neq j}}^{m} \sum_{j=1}^{m} r\frac{q_i q_j}{p^2} \tag{6.28}$$

which may be verified by direct computation. ▶

7. SUM OF A RANDOM NUMBER OF RANDOM VARIABLES

If a probability distribution is altered by allowing one of its parameters to behave as a random variable, the resulting distribution is said to be *compound.* An important compound distribution is that of the sum of a random number of random variables.

Theorem 3. *Let $\{X_k\}$ be a sequence of independent and identically distributed proper random variables with the common p.g.f.*

$$g(s) = E(s^{X_i}), \qquad i = 1, 2, \ldots, \tag{7.1}$$

and let

$$Z_N = X_1 + \cdots + X_N \tag{7.2}$$

where N is also a random variable, with p.g.f.

$$h(s) = E(s^N). \tag{7.3}$$

Denote the p.g.f. of the compound distribution of Z_N by

$$G(s) = E(s^{Z_N}). \tag{7.4}$$

Then

$$G(s) = h[g(s)].$$

Proof. Since

$$E(s^{Z_N}) = E[E(s^{Z_N} \mid N)] \tag{7.5}$$

and

$$E(s^{Z_N} \mid N) = E(s^{X_1 + \cdots + X_N} \mid N) = \{g(s)\}^N, \tag{7.6}$$

we have

$$G(s) = E[\{g(s)\}^N] = h[g(s)]. \qquad \blacktriangleright \tag{7.7}$$

This simple functional relation between the p.g.f. $G(s)$ of the compound distribution of Z_N and the p.g.f.'s $h(s)$ and $g(s)$ is quite useful, as we shall illustrate with an example in branching processes in Section 8.

Example. Let X_i have a logarithmic distribution with

$$\Pr\{X_i = k\} = -\frac{p^k}{k \log(1-p)}, \qquad k = 1, 2, \ldots, \tag{7.8}$$

where $0 < p < 1$, and p.g.f.

$$g(s) = \frac{\log(1-ps)}{\log(1-p)}. \tag{7.9}$$

Let N have a Poisson distribution with p.g.f.

$$h(s) = e^{-(1-s)\lambda}. \tag{7.10}$$

According to the theorem, the p.g.f. of $Z_N = X_1 + \cdots + X_N$ is

$$G(s) = e^{-[1-\log(1-ps)/\log(1-p)]\lambda}$$

$$= e^{-\lambda}(1-ps)^{\lambda/\log(1-p)}. \tag{7.11}$$

Now let

$$r = -\frac{\lambda}{\log(1-p)} \tag{7.12}$$

so that

$$-\lambda = r \log(1-p) \qquad \text{and} \qquad e^{-\lambda} = e^{r\log(1-p)} = (1-p)^r. \tag{7.13}$$

Substituting (7.12) and (7.13) in (7.11) gives

$$G(s) = \left(\frac{1-p}{1-ps}\right)^r \tag{7.14}$$

which is the p.g.f. of the negative binomial distribution in (4.18) with p replaced by $(1-p)$. ▶

8. A SIMPLE BRANCHING PROCESS

The idea of branching processes was introduced first by Francis Galton and the Reverend H. W. Watson in 1874, when they published their solution to the problem of the extinction of family names. Their mathematical model and its generalizations have also been used to study frequencies of mutant genes (Fisher [1922a] and [1930]), other problems in genetics (Haldane [1927]), epidemics (Neyman and Scott [1964]), nuclear chain reactions, and similar problems. For an extensive theoretical treatment of the subject, see Harris [1963].

The basic mechanism of a branching process is as follows: an individual (the 0th generation) is capable of producing 0, 1, 2, . . . offspring to form the first generation; each of his offspring in turn produces offspring, which together constitute the second generation; and so on. Let the number of individuals in the nth generation be Z_n. If we now impose a probability structure on the process of reproduction, then $Z_1, Z_2, \ldots$ are random variables whose probability distributions can be calculated.

We shall assume the simplest reproductive structure, namely, (i) that the number X of offspring produced by an individual has the probability distribution

$$\Pr\{X = k\} = p_k \qquad k = 0, 1, \ldots \tag{8.1}$$

which is the same for each individual in a given generation; (ii) that this probability distribution remains fixed from generation to generation; and (iii) that individuals produce offspring independently of each other. Thus we are dealing with independent and identically distributed random variables.

Let

$$g(s) = \sum_{k=0}^{\infty} p_k s^k \tag{8.2}$$

be the p.g.f. of X, and let $g_n(s)$ be the p.g.f. of Z_n, $n = 1, 2, \ldots$. Since $Z_0 = 1$, the size of the first generation Z_1 has the same probability distribution as X,

$$\Pr\{Z_1 = k\} = p_k, \tag{8.3}$$

and the same p.g.f. $g(s)$. The second generation consists of the direct descendants of the Z_1 members of the first generation, so that Z_2 is the sum of Z_1 independent random variables, each of which has the probability distribution $\{p_k\}$ and the p.g.f. $g(s)$. Therefore Z_2 has a compound distribution with the p.g.f. obtained from formula (7.7):

$$g_2(s) = g[g(s)]. \tag{8.4}$$

Similarly, the $(n + 1)$th generation consists of the direct descendants of the Z_n members of the nth generation, so that Z_{n+1} is the sum of the Z_n independent random variables, each with p.g.f. $g(s)$. Hence by (7.7) the p.g.f. of Z_{n+1} is

$$g_{n+1}(s) = g_n[g(s)]. \tag{8.5}$$

But the $(n + 1)$th generation consists of the nth generation descendants of the Z_1 members of the first generation. Therefore the p.g.f. of Z_{n+1} can also be written as

$$g_{n+1}(s) = g[g_n(s)], \tag{8.6}$$

an alternative form of (8.5).

The explicit form for the p.g.f. of Z_n depends upon the probability distribution $\{p_k\}$. Consider, for example, organisms which can either die or split into two so that

$$\begin{aligned} p_0 &= (1 - p) \\ p_2 &= p \\ p_k &= 0, \qquad \text{for} \qquad k = 1, 3, 4, \ldots . \end{aligned}$$

The p.g.f. of X is $g(s) = (1 - p) + ps^2$ and the p.g.f.'s of the $\{Z_n\}$ are

$$\begin{aligned} g_1(s) &= (1 - p) + ps^2 \\ g_2(s) &= (1 - p) + p[(1 - p) + ps^2]^2 \\ g_3(s) &= (1 - p) + p\{(1 - p) + p[(1 - p) + ps^2]^2\}^2, \text{ etc.} \end{aligned}$$

This simple example demonstrates that, although the probability $\Pr\{Z_n = k\}$ can be obtained from $g_n(s)$, actual computations are quite involved when n is large.

Probability of Extinction. An interesting problem with a particularly nice solution is the question of the ultimate extinction of the population. The probability that a population starting with a single ancestor will become extinct at or before the nth generation is

$$q_n = \Pr\{Z_n = 0\} = g_n(0). \tag{8.7}$$

We wish to investigate the limit of q_n as n tends to infinity. If $p_0 = 1$ the population will never start, and if $p_0 = 0$ it will never become extinct; therefore we shall assume that $0 < p_0 < 1$. Then the generating function $g(s)$ in (8.2) is a strictly monotone increasing function of s with

$$0 < g(0) = p_0 < 1. \tag{8.8}$$

Using (8.5) and (8.8) and the monotonicity property of $g(s)$ we have

$$q_{n+1} = g_{n+1}(0) = g_n[g(0)] > g_n(0) = q_n. \tag{8.9}$$

The sequence $\{q_n\}$ is bounded above by unity, and the inequality (8.9) shows that it is monotonically increasing; therefore the sequence will tend to a limit ζ as n tends to infinity. On the other hand, (8.6) implies the relation

$$q_{n+1} = g(q_n). \tag{8.10}$$

Taking limits of both sides of (8.10) we see that ζ satisfies the equation

$$\zeta = g(\zeta). \tag{8.11}$$

In fact, the limit ζ is the smallest root of the equation. To prove this we let x be an arbitrary positive root of the equation $x = g(x)$; then

$$q_1 = g(0) < g(x) = x \tag{8.12}$$

and hence

$$q_2 = g(q_1) < g(x) = x. \tag{8.13}$$

By induction

$$q_{n+1} = g(q_n) < g(x) = x, \tag{8.14}$$

demonstrating that $\zeta < x$.

We may now study the behavior of the function $y = g(s)$ for $0 \leq s \leq 1$. If $p_0 + p_1 = 1$, then $g(s)$ is linear and, since $p_0 > 0$, the line $y = g(s)$ intersects the line $y = s$ only at the point (1, 1). Hence, in this case, $\zeta = 1$ and ultimate extinction is certain. We now consider the case $p_0 + p_1 < 1$; it is apparent from (8.2) that in this case the derivative $g'(s)$ is a strictly increasing function of s. Thus the curve $y = g(s)$ is convex and can intercept the line $y = s$ in at most two points. One of these is (1, 1) and therefore the equation (8.11) has at most one root between 0 and 1. Whether such a root exists depends entirely upon the derivative $g'(1)$. If $g'(1) > 1$, the root $\zeta < 1$ exists; for, tracing the curve $y = g(s)$ backward from the point (1, 1), we find that it must fall below the line $y = s$, and eventually cross the line to reach the point $(0, p_0)$. If, on the other hand, $g'(1) \leq 1$, then the curve must be entirely above the line and there is no root less than unity, so that ζ must be unity (Figure 1).

Now the derivative $g'(1)$ is the expected number of offspring of any individual in the population. According to the preceding argument, if this expected number is greater than one, the probability of extinction tends to a quantity ζ that is less than one; if the expected number of offspring is less than or equal to one, the probability that the population will eventually become extinct tends to unity.

Further investigation of the case $g'(1) > 1$ yields even more dramatic results. It is easily seen from the graph that

$$\begin{aligned} g(s) &\geq s \quad \text{for} \quad 0 \leq s \leq \zeta \\ g(s) &\leq s \quad \text{for} \quad \zeta \leq s < 1. \end{aligned} \tag{8.15}$$

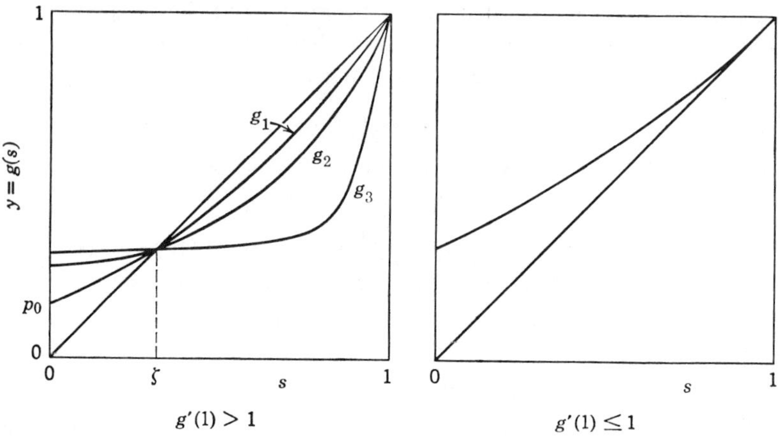

Figure 1

Hence, by arguments similar to those used above to show the convergence of q_n, we have

$$\begin{aligned} s \leq g_1(s) \leq g_2(s) \leq \cdots \leq \zeta \qquad & 0 \leq s \leq \zeta \\ s \geq g_1(s) \geq g_2(s) \geq \cdots \geq \zeta \qquad & \zeta \leq s < 1. \end{aligned} \tag{8.16}$$

Equations (8.16) show that the limit

$$\lim_{n \to \infty} g_n(s) = G(s) \tag{8.17}$$

must exist for all s, and that $G(s) = 1$ only for $s = 1$. Passing to the limit in (8.6) shows that

$$G(s) = \begin{cases} \zeta & 0 \leq s < 1 \\ 1 & s = 1. \end{cases} \tag{8.18}$$

This limiting function $G(s)$ is clearly not a p.g.f., since it is discontinuous at $s = 1$. If, however, we redefine $G(1)$ to be equal to ζ, then we may interpret $G(s)$ as the p.g.f. of an improper random variable Z, the "ultimate size" of the population, which has the probability distribution

$$\begin{aligned} \Pr\{Z = 0\} &= \zeta \\ \Pr\{Z = k\} &= 0 \qquad k = 1, 2, \ldots. \\ \Pr\{Z = \infty\} &= 1 - \zeta. \end{aligned} \tag{8.19}$$

Thus the population either becomes extinct with probability ζ, or grows without bound with probability $1 - \zeta$; no intermediate course is possible. Of course, in practical applications to the growth of real populations, the assumptions of the branching process model are violated long before an infinite population is approached; on the other hand, the investigation of extinction phenomena through the use of branching processes is more useful, since extinction often occurs early in the process. ▶

PROBLEMS

1. Derive the p.g.f. for a binomial distribution in n trials with probability p.

2. In an infinite sequence of Bernoulli trials with probability p of success in any trial, if the "success" occurs at least r times in succession, we say that there is a run of r successes. Let p_n be the probability that a run of r successes will occur in n trials. Find the p.g.f. of p_n and from which the probability p_n for $n \geq r$.

3. *Newton's binomial expansion.* Verify the equation

$$(1+t)^x = \sum_{k=0}^{\infty} \binom{x}{k} t^k, \qquad |t| < 1$$

for any real number x by the method of undetermined coefficients.

4. *Continuation.* Newton's binomial formula in Problem 3 can be written as

$$(a_0 + a_1)^x = \sum_{k_1=0}^{\infty} \binom{x}{k_1} a_1^{k_1} a_0^{x-k_1}, \qquad |a_1| < |a_0|.$$

Generalize this formula by induction to the multinomial expansion:

$$(a_0 + a_1 + \cdots + a_m)^x = \sum_{k_1=0}^{m} \cdots \sum_{k_m=0}^{\infty} \binom{x}{k_1, \ldots, k_m} a_1^{k_1} \cdots a_m^{k_m} a_0^{n-k}$$

where $\left| \sum_{i=1}^{m} a_i \right| < |a_0|$, $k = k_1 + \cdots + k_m$, and the multinomial coefficient is defined by

$$\binom{x}{k_1, \ldots, k_m} = \frac{x(x-1)\cdots(x-k+1)}{k_1! \cdots k_m!}$$

for all real x. Use this multinomial expansion to verify the p.g.f. (6.14).

5. Prove the identity

$$\binom{-r}{k} = (-1)^k \binom{r+k-1}{k} \tag{4.22}$$

and use it to verify the identity

$$\sum_{k=0}^{\infty} \binom{r+k-1}{k} (qs)^k = (1-qs)^{-r}$$

6. Let Z_r have the negative binomial distribution in (4.19). Show that Z_r is a proper random variable, or

$$\sum_{k=0}^{\infty}\binom{k+r-1}{k}q^k p^r = 1,$$

and find the expectation $E(Z_r)$ from

$$E(Z_r) = \sum_{k=0}^{\infty} k\binom{k+r-1}{k}q^k p^r.$$

7. Prove Theorem 2.

8. Use Theorem 2 to derive the p.g.f. of the multinomial distribution in (6.14).

9. *Negative multinomial distribution.* In formula (6.19) let $m = 3$ so that $p + q_1 + q_2 + q_3 = 1$. Find the marginal probability distributions: $\Pr\{X_1 = k_1\}$ and $\Pr\{X_1 = k_1, X_2 = k_2\}$.

10. *Continuation.* Find the conditional probability distributions: $\Pr\{X_3 = k_3 \mid k_1, k_2\}$, $\Pr\{X_2 = k_2, X_3 = k_3 \mid k_1\}$, and $\Pr\{X_2 = k_2 \mid k_1\}$.

11. *Continuation.* Derive the p.g.f. for each of the probability distributions in Problems 9 and 10.

12. *Continuation.* Use the results in Problem 11 and Equation (4.22) in Chapter 1 to derive the p.g.f. for the joint probability distribution of X_1, X_2, and X_3.

13. *Multinomial distribution.* Solve Problems 9 through 12 for a multinomial distribution with $p_0 + p_1 + p_2 + p_3 = 1$.

14. Let a random variable X have the p.g.f. $G_X(s)$ and let a and b be constants. Find the p.g.f. for the linear function $a + bX$.

15. *Continuity theorem.* For each n let $\{p_{k,n}\}$ be a probability distribution so that

$$p_{k,n} \geq 0 \quad \text{and} \quad \sum_{k=0}^{\infty} p_{k,n} = 1,$$

and let

$$g_n(s) = \sum_{k=0}^{\infty} p_{k,n} s^k$$

be the corresponding p.g.f. Suppose that there exists a probability distribution $\{p_k\}$ with p.g.f.

$$G(s) = \sum_{k=0}^{\infty} p_k s^k.$$

For each k

$$\lim_{n\to\infty} p_{k,n} = p_k$$

if, and only if,

$$\lim_{n\to\infty} g_n(s) = G(s)$$

whatever may be $0 \leq s < 1$. Prove the theorem.

16. *Continuation.* In a binomial distribution with probability p, if $n \to \infty$ in such a way that $np \to \lambda$, then the probability function

$$\lim_{n\to\infty} \binom{n}{k} p^k(1-p)^{n-k} = \frac{e^{-\lambda}\lambda^k}{k!}$$

and the p.g.f.

$$\lim_{n\to\infty} [1-p+ps]^n = e^{-\lambda(1-s)}.$$

Prove the equalities.

17. *Continuation.* In the negative binomial distribution (4.19), if $r \to \infty$ in such a way that $rq \to \lambda$, then the probability function

$$\lim_{r\to\infty} \binom{k+r-1}{k} q^k p^r = \frac{e^{-\lambda}\lambda^k}{k!}$$

and the p.g.f.

$$\lim_{r\to\infty} \left(\frac{p}{1-qs}\right)^r = e^{-\lambda(1-s)}.$$

Prove the equalities.

18. *Gambler's ruin.* In the classical gambler's-ruin problem, gambler A plays a series of games with an adversary B, with the probability p of winning a single game and probability $q = 1 - p$ of losing a single game. If they play for a constant stake of one dollar per game, and if the initial capitals of A and B are a dollars and b dollars, respectively, what is the probability that gambler A will be ruined if no limit is set for the number of games? Answer the question for $p \neq q$ and for $p = q = \frac{1}{2}$.

Find the probability that gambler A will be ruined if his adversary B has an infinite amount of capital.

19. *Continuation.* Suppose, in Problem 18, that players A and B have initial capitals of a and b, respectively. What is the probability that A (or B) will be ruined at the nth game?

20. *Continuation.* Suppose that, when gambler A begins with x dollars $0 < x < a + b$, the series of games has a finite expectation D_x. Justify that

$$D_x = pD_{x+1} + qD_{x-1} + 1.$$

From the above equation show that, for $p \neq q$,

$$D_a = \frac{a}{q-p} - \frac{a+b}{q-p}\left[\frac{1-(q/p)^a}{1-(q/p)^{a+b}}\right].$$

21. *Branching process.* In the simple branching process, let the expectation and variance of X be μ and σ^2, respectively. Use

$$E[Z_{n+1}] = E[E(Z_{n+1} \mid Z_n)],$$

and

$$\operatorname{Var}(Z_{n+1}) = E[\sigma^2_{Z_{n+1}|Z_n}] + \sigma^2_{E(Z_{n+1}|Z_n)},$$

to show that $E(Z_n) = \mu^n$ and

$$\begin{aligned} \operatorname{Var}(Z_n) &= \frac{\mu^{n-1}(\mu^n - 1)}{\mu - 1}\sigma^2 & \mu \neq 1 \\ &= n\sigma^2 & \mu = 1. \end{aligned}$$

22. *Continuation.* Use the relation $g_{n+1}(s) = g_n[g(s)]$ to derive the formulas for $E(Z_n)$ and $\operatorname{Var}(Z_n)$ in Problem 21.

23. Verify the following identity for the negative multinomial distribution:

$$r\frac{q}{p^2} = \sum_{i=1}^{m} r\frac{q_i}{p}\left[1 + \frac{q_i}{p}\right] + \sum_{\substack{i=1 \\ i \neq j}}^{m} \sum_{j=1}^{m} r\frac{q_i q_j}{p^2} \tag{6.28}$$

where $p + q = 1$ and $q = q_1 + \cdots + q_m$.

24. Let X be a random variable with p.g.f. $g_X(s) = \Sigma p_k s^k$. Suppose the series converges for $|s| < s_0$ for some $s_0 > 1$, show that the moments $E(X^r)$ exist for all $r = 1, 2, \ldots$.

25. The *moment generating function* (m.g.f.) of a random variable X is defined by

$$M_X(s) = E(e^{sX}).$$

Derive from the m.g.f. the moment $E(X^r)$.

26. *Continuation.* The m.g.f. and the p.g.f. of a random variable X have the relation

$$M_X(s) = g_X(e^s).$$

Use this relation to write down the m.g.f. of:

(i) The binomial distribution (4.4).
(ii) The Poisson distribution (4.8).
(iii) The negative binomial distribution (4.19).

Derive from each of the above the corresponding expectations $E(X)$ and $E(X^2)$.

27. *Continuation.* Find the m.g.f.'s for:

(i) The exponential distribution with probability density function

$$\begin{aligned} f(x) &= \mu e^{-\mu x} & x \geq 0 \\ &= 0 & x < 0. \end{aligned}$$

(ii) The uniform distribution with probability density function

$$\begin{aligned} f(x) &= 1/a & 0 \leq x \leq a \\ &= 0 & \text{elsewhere.} \end{aligned}$$

(iii) The normal distribution with the probability density function

$$f(x) = \frac{1}{\sigma\sqrt{2\pi}} e^{-(x-\mu)^2/2\sigma^2}, \qquad -\infty < x < \infty.$$

From the m.g.f.'s, find the corresponding expectations and variances.

CHAPTER 3

Some Stochastic Models of Population Growth

1. INTRODUCTION

Since the early work of Kolmogorov [1931] and Feller [1936], the theory of stochastic processes has developed rapidly. It has been used to describe empirical phenomena and to solve many practical problems. Systematic treatments of the subject are given by Chung [1960], Doob [1953], Feller [1957, 1966], Harris [1963]; reference should also be made to Bartlett [1956], Bharucha-Reid [1960], Bailey [1964], Cox and Miller [1965], Karlin [1966], Kendall [1948], Parzen [1962], Takács [1959].

A stochastic process $\{X(t)\}$ is a family of random variables describing an empirical process whose development is governed by probabilistic laws. The parameter t, which is often interpreted as time, is real, but may be either discrete or continuous. The random variable $X(t)$ may be real-valued or complex-valued, or it may take the form of a vector. In diffusion processes, for example, both $X(t)$ and t are continuous variables, whereas in Markov chains, $X(t)$ and t take on discrete values. In the study of population growth we shall have a continuous time parameter t, but the random variable $X(t)$ will have a discrete set of possible values, namely, the nonnegative integers. Our main interest is the probability distribution

$$p_k(t) = \Pr\{X(t) = k\} \qquad k = 0, 1, \ldots. \tag{1.1}$$

As with any attempt to describe empirical phenomena mathematically, the formulation of a practical problem in terms of stochastic processes necessarily involves some simplifying assumptions. On the other hand, such formulation often produces enlightening and useful answers to the problem. In this chapter we introduce a few basic, simple, but widely applicable stochastic processes that are commonly used as models of population growth.

2. THE POISSON PROCESS

In many practical problems the occurrence of a random event at a particular moment is independent of time and of the number of events that have already taken place. Examples of this kind are telephone calls coming into a switchboard, radioactive disintegrations, and accidents. Let $X(t)$ be the total number of events that occur within the time interval $(0, t)$. Basic assumptions underlying the Poisson process are (i) for any $t \geq 0$, the probability that an event occurs during the time interval $(t, t + \Delta)$ is $\lambda\Delta + o(\Delta)$[1], where the constant λ does not depend on t or on the number of events occurring in $(0, t)$; and (ii) the probability that more than one event occurs in $(t, t + \Delta)$ is $o(\Delta)$. Therefore the probability of no change in $(t, t + \Delta)$ is $1 - \lambda\Delta - o(\Delta)$.

In order to derive differential equations for the probabilities (1.1), we extend the interval $(0, t)$ to a point $t + \Delta$ to analyze the probability $p_k(t + \Delta)$ by enumerating all possibilities leading to the occurrence of k events during the interval $(0, t + \Delta)$. This is a standard approach in stochastic processes and will be repeatedly, but not necessarily explicitly, used in this book. Consider, then, the two adjacent intervals $(0, t)$ and $(t, t + \Delta)$. The occurrence of exactly k events during the interval $(0, t + \Delta)$ can be realized in three mutually exclusive ways: (i) all k events will occur in $(0, t)$ and none in $(t, t + \Delta)$ with probability $p_k(t)[1 - \lambda\Delta - o(\Delta)]$; (ii) exactly $k - 1$ events will occur in $(0, t)$ and one event in $(t, t + \Delta)$ with probability $p_{k-1}(t)[\lambda\Delta + o(\Delta)]$; and (iii) exactly $k - j$ events will occur in $(0, t)$ and j events in $(t, t + \Delta)$, where $2 \leq j \leq k$, with probability $o(\Delta)$. Hence, considering all these possibilities and combining all quantities of order $o(\Delta)$, we have for $k \geq 1$,

$$p_k(t + \Delta) = p_k(t)[1 - \lambda\Delta] + p_{k-1}(t)\lambda\Delta + o(\Delta) \tag{2.1}$$

and, for $k = 0$,

$$p_0(t + \Delta) = p_0(t)[1 - \lambda\Delta] + o(\Delta). \tag{2.2}$$

Transposing $p_k(t)$ to the left side of (2.1), dividing the resulting equation through by Δ, and passing to the limit as $\Delta \to 0$, we find that the probabilities satisfy the system of differential equations

$$\frac{d}{dt} p_k(t) = -\lambda p_k(t) + \lambda p_{k-1}(t), \qquad k \geq 1, \tag{2.3a}$$

[1] The standard notation $o(\Delta)$ represents any function of Δ which tends to 0 faster than Δ, that is any function such that $[o(\Delta)/\Delta] \to 0$ as $\Delta \to 0$.

and similarly

$$\frac{d}{dt}\,p_0(t) = -\lambda p_0(t); \tag{2.3b}$$

clearly the initial conditions are

$$p_0(0) = 1, \qquad p_k(0) = 0, \qquad k \geq 1. \tag{2.4}$$

We shall solve (2.3*a*) and (2.3*b*) successively, starting with $k = 0$. Integrating (2.3*b*) and using the condition $p_0(0) = 1$ yields

$$p_0(t) = e^{-\lambda t}. \tag{2.5}$$

Setting $k = 1$ in (2.3*a*) and multiplying through by $e^{\lambda t}$ gives

$$e^{\lambda t}\frac{d}{dt}\,p_1(t) + \lambda e^{\lambda t}p_1(t) = \frac{d}{dt}\,[e^{\lambda t}p_1(t)] = \lambda; \tag{2.6}$$

integrating and using $p_1(0) = 0$, we have

$$p_1(t) = \lambda t e^{-\lambda t}. \tag{2.7}$$

Repeated applications of the same procedure yield the general formula

$$p_k(t) = \frac{e^{-\lambda t}(\lambda t)^k}{k!}, \qquad k = 0, 1, \ldots. \tag{2.8}$$

Equation (2.8) is, of course, a Poisson distribution with the parameter λt. Thus we see that the Poisson distribution, which is often derived as a limiting case of the binomial distribution, also arises as a consequence of reasonable assumptions about a simple stochastic process. ▶

2.1. The Method of Probability Generating Functions

The recursive system of differential equations (2.3) was solved successively without difficulty. However, in many stochastic processes the differential equations are quite complicated and such a direct approach is fruitless. We can then resort to the method of p.g.f.'s wherein a differential equation is first derived and solved to obtain the p.g.f. of $X(t)$, and then the probability distribution is derived from the p.g.f. To illustrate the technique, we reconsider the Poisson process.

Let the p.g.f. of the probability distribution $\{p_k(t)\}$ be defined as in Chapter 2 by

$$G_X(s; t) = \sum_{k=0}^{\infty} p_k(t)s^k \qquad \begin{array}{c} |s| \leq 1 \\ 0 \leq t < \infty. \end{array} \tag{2.9}$$

Differentiating (2.9) under the summation sign[2] and using (2.3) we obtain

$$\frac{\partial}{\partial t} G_X(s;t) = \sum_{k=0}^{\infty}\left[\frac{d}{dt} p_k(t)\right] s^k \tag{2.10}$$

$$= -\lambda \sum_{k=0}^{\infty} p_k(t)s^k + \lambda s \sum_{k=1}^{\infty} p_{k-1}(t)s^{k-1}. \tag{2.11}$$

The lower limit of the last summation is unity because of (2.3*b*). Since each of the last two summations is equal to the p.g.f. (2.9), we have the differential equation for the p.g.f.

$$\frac{\partial}{\partial t} G_X(s;t) = -\lambda(1-s)G_X(s;t) \tag{2.12}$$

with the initial condition

$$G_X(s;0) = s^0 = 1. \tag{2.13}$$

If (2.12) is rewritten as

$$\partial \log G_X(s;t) = -\lambda(1-s)\,\partial t \tag{2.14}$$

direct integration gives

$$G_X(s;t) = c(s)e^{-\lambda t(1-s)}. \tag{2.15}$$

Setting $t = 0$ in (2.15) and using the initial condition (2.13), we determine that the constant of integration is $c(s) = 1$. Therefore

$$G_X(s;t) = e^{-\lambda t(1-s)} \tag{2.16}$$

which is the p.g.f. of the Poisson distribution (2.8). ▶

2.2. Some Generalizations of the Poisson Process

Generalizations of the Poisson process can be made in many ways, of which we shall mention two.

(i) Time-Dependent Poisson Process. We may replace the constant λ by a function of time $\lambda(t)$ in (2.1) and (2.2). Then (2.14) becomes

$$\partial \log G_X(s;t) = -\lambda(t)(1-s)\,\partial t$$

---

[2] Since $|p_k(t)s^k| < |s|^k$, each of the infinite series in (2.11) converges uniformly in t for $|s| < 1$; hence $\sum (d/dt)p_k(t)s^k$ converges uniformly in t and the term-by-term differentiation is justified for $|s| < 1$. Letting $s \to 1$, we have $\lim_{s\to 1} G_X(s;t) = G_X(1;t)$. Thus our argument applies also for $s = 1$. Justification for term-by-term differentiation in the remainder of this book is similar and will not be given explicitly.

which gives the p.g.f.

$$G_X(s; t) = \exp\left\{-(1 - s)\int_0^t \lambda(\tau)\, d\tau\right\}, \tag{2.17}$$

the probabilities

$$p_k(t) = \frac{\exp\left\{-\int_0^t \lambda(\tau)\, d\tau\right\}\left[\int_0^t \lambda(\tau)\, d\tau\right]^k}{k!} \tag{2.18}$$

and the expectation

$$E[X(t)] = \int_0^t \lambda(\tau)\, d\tau. \tag{2.19}$$

We see that the generalization does not affect the form of the distribution, but allows variation to occur as determined by the function $\lambda(\tau)$. ▶

(ii) Weighted Poisson Process. The Poisson process describes the frequency of occurrence of an event to an individual with "risk" parameter λ; if we are sampling from a population of individuals, then variability of individuals with respect to this risk should be taken into account. Proneness to accidents, for example, varies throughout the population according to a density function $f(\lambda)$. Then the probability distribution (2.8) must be reinterpreted as the conditional distribution $p_{k|\lambda}(t)$ of $X(t)$ given λ, and the probability that an individual chosen at random from the population will experience k events in a time interval of length t is

$$p_k(t) = \int_0^\infty p_{k|\lambda}(t) f(\lambda)\, d\lambda = \int_0^\infty \frac{e^{-\lambda t}(\lambda t)^k}{k!} f(\lambda)\, d\lambda. \tag{2.20}$$

Suppose, for example, that λ has a *gamma distribution*

$$\begin{aligned} f(\lambda) &= \frac{\beta^\alpha}{\Gamma(\alpha)} \lambda^{\alpha-1} e^{-\beta\lambda} & \lambda > 0, \alpha > 0, \beta > 0 \\ &= 0 & \lambda \le 0 \end{aligned} \tag{2.21}$$

where the gamma function $\Gamma(\alpha)$ is defined by

$$\Gamma(\alpha) = \int_0^\infty y^{\alpha-1} e^{-y}\, dy. \tag{2.22}$$

Substituting (2.21) in (2.20), we have

$$\begin{aligned} p_k(t) &= \frac{\beta^\alpha t^k}{\Gamma(\alpha) k!} \int_0^\infty \lambda^{k+\alpha-1} e^{-\lambda(\beta+t)}\, d\lambda \\ &= \frac{\Gamma(k+\alpha)}{k!\,\Gamma(\alpha)} \beta^\alpha t^k (\beta + t)^{-(k+\alpha)}, \qquad k = 0, 1, \ldots. \end{aligned} \tag{2.23}$$

Now the gamma function satisfies the recursion equation

$$\Gamma(\alpha + 1) = \alpha\Gamma(\alpha) \tag{2.24}$$

which follows readily upon integrating (2.22) by parts. Using (2.24) above and (4.21) in Chapter 2, equation (2.23) may be rewritten as

$$p_k(t) = \binom{k + \alpha - 1}{k}\left(\frac{t}{\beta + t}\right)^k\left(\frac{\beta}{\beta + t}\right)^\alpha \tag{2.25}$$

which is for each t a negative binomial distribution with parameters

$$r = \alpha,\ p = \frac{\beta}{\beta + t} \tag{2.26}$$

[cf. Equation (4.19) in Chapter 2]. Thus we have the expected number of events occurring to a randomly chosen individual during the interval $(0, t)$

$$E[X(t)] = \alpha\frac{t}{\beta} \tag{2.27}$$

and the variance

$$\sigma^2_{X(t)} = \alpha\frac{t}{\beta}\left(1 + \frac{t}{\beta}\right). \qquad \blacktriangleright \tag{2.28}$$

In the original Poisson model with λ constant, the expectation and variance of $X(t)$ were both equal to λt. In the present generalization the expectation is still a linear function of λt, but the variance is now a quadratic function of λt. The present model incorporating variability of risk has been found useful in the study of accident proneness (Bates and Neyman [1952]) and in the incidence of illness (Lundberg [1940], Chiang [1966]).

3. PURE BIRTH PROCESSES

In the study of growth in a broad sense, "birth" may be liberally interpreted as an event whose probability is dependent upon the number of "parent" events already in existence. It may refer to a literal birth, to a new case in an epidemic, to the appearance of a new tumor cell in a tissue, etc. More precisely, we assume that given $X(t) = k$, (i) the conditional probability that a new event will occur during $(t, t + \Delta)$ is $\lambda_k\Delta + o(\Delta)$, where λ_k is some function of k; and (ii) the conditional probability that more than one event will occur is $o(\Delta)$. Following the same procedure as in the preceding section, we find that the probabilities $p_k(t)$ satisfy the system of differential equations

$$\frac{d}{dt} p_{k_0}(t) = -\lambda_{k_0} p_{k_0}(t), \tag{3.1a}$$

$$\frac{d}{dt} p_k(t) = -\lambda_k p_k(t) + \lambda_{k-1} p_{k-1}(t), \qquad k > k_0. \tag{3.1b}$$

Here $k_0 = X(0)$, the number of events existing at $t = 0$, so that the initial conditions are

$$p_{k_0}(0) = 1, \qquad p_k(0) = 0, \qquad \text{for} \qquad k \neq k_0. \tag{3.2}$$

The differential equations (3.1) have a recursive relation and can be solved successively. If the values of λ_i are all different, the solution is

$$p_k(t) = (-1)^{k-k_0} \lambda_{k_0} \cdots \lambda_{k-1} \left[\sum_{i=k_0}^{k} \frac{e^{-\lambda i t}}{\prod_{\substack{j=k_0 \\ j\neq i}}^{k} (\lambda_i - \lambda_j)} \right]. \tag{3.3}$$

Inductive proof of (3.3) requires the following identity. Whatever may be distinct $\lambda_{k_0}, \ldots, \lambda_k$,

$$\sum_{i=k_0}^{k} \frac{1}{\prod_{\substack{j=k_0 \\ j\neq i}}^{k} (\lambda_i - \lambda_j)} = 0. \tag{3.4}$$

This identity is a special case of the lemma in Section 3.3 of Chapter 6; the reader may refer to equation (3.33) in that chapter for verification.

We now prove (3.3) by induction. Equation (3.1*a*) has the solution

$$p_{k_0}(t) = e^{-\lambda_{k_0} t},$$

hence (3.3) is true for $k = k_0$. Suppose (3.3) is true for $k - 1$, that is,

$$p_{k-1}(t) = (-1)^{k-1-k_0} \lambda_{k_0} \cdots \lambda_{k-2} \left[\sum_{i=k_0}^{k-1} \frac{e^{-\lambda_i t}}{\prod_{\substack{j=k_0 \\ j\neq i}}^{k-1} (\lambda_i - \lambda_j)} \right]. \tag{3.5}$$

We shall derive (3.3) from (3.1*b*). Following the usual procedure for linear first-order differential equations, we multiply (3.1*b*) by $e^{\lambda_k t}$ to obtain

$$e^{\lambda_k t} \frac{d}{dt} p_k(t) + e^{\lambda_k t} \lambda_k p_k(t) = e^{\lambda_k t} \lambda_{k-1} p_{k-1}(t)$$

or

$$\frac{d}{dt}[e^{\lambda_k t}p_k(t)] = \lambda_{k-1}e^{\lambda_k t}p_{k-1}(t) \tag{3.6}$$

which, upon substitution of (3.5), becomes

$$\frac{d}{dt}[e^{\lambda_k t}p_k(t)] = (-1)^{k-1-k_0}\lambda_{k_0}\cdots\lambda_{k-1}\left[\sum_{i=k_0}^{k-1}\frac{1}{\prod\limits_{\substack{j=k_0\\ j\neq i}}^{k-1}(\lambda_i-\lambda_j)}e^{(\lambda_k-\lambda_i)t}\right]$$

$$= (-1)^{k-1-k_0}\lambda_{k_0}\cdots\lambda_{k-1}\left[\sum_{i=k_0}^{k-1}\frac{1}{\prod\limits_{\substack{j=k_0\\ j\neq i}}^{k-1}(\lambda_i-\lambda_j)}\frac{(d/dt)e^{(\lambda_k-\lambda_i)t}}{(\lambda_k-\lambda_i)}\right].$$

Therefore

$$e^{\lambda_k t}p_k(t) = (-1)^{k-k_0}\lambda_{k_0}\cdots\lambda_{k-1}\left[\sum_{i=k_0}^{k-1}\frac{1}{\prod\limits_{\substack{j=k_0\\ j\neq i}}^{k}(\lambda_i-\lambda_j)}e^{(\lambda_k-\lambda_i)t}+c\right]. \tag{3.7}$$

At $t = 0$, $p_k(0) = 0$ for $k > k_0$. Using identity (3.4), we see that the constant c is given by

$$c = \frac{1}{\prod\limits_{j=k_0}^{k-1}(\lambda_k-\lambda_j)}. \tag{3.8}$$

Substituting (3.8) in (3.7) and dividing the resulting equation through by $e^{\lambda_k t}$ yield (3.3), completing the inductive proof. ▶

3.1. The Yule Process

Suppose that each of k individuals alive at time t has the same probability $\lambda\Delta + o(\Delta)$ of giving birth to another individual in the interval $(t, t + \Delta)$ and that individuals give birth independently of each other. Then the total number of births during $(t, t + \Delta)$ has a binomial distribution

$$\Pr\{j \text{ births during } (t, t+\Delta)\} = \binom{k}{j}[\lambda\Delta + o(\Delta)]^j[1 - \lambda\Delta - o(\Delta)]^{k-j}.$$

It follows that the probability of exactly one birth among the k individuals in $(t, t + \Delta)$ is $k\lambda\Delta + o(\Delta)$ and the probability of more than one birth is $o(\Delta)$. Thus we have a simple birth process with $\lambda_k = k\lambda$ and system (3.1) becomes

$$\frac{d}{dt}\,p_{k_0}(t) = -k_0\lambda p_{k_0}(t) \tag{3.9a}$$

$$\frac{d}{dt}\,p_k(t) = -k\lambda p_k(t) + (k-1)\lambda p_{k-1}(t), \qquad k > k_0 \tag{3.9b}$$

with the initial conditions (3.2). System (3.9) is easily solved; for $k \geq k_0$

$$p_k(t) = \binom{k-1}{k-k_0} e^{-k_0\lambda t}(1 - e^{-\lambda t})^{k-k_0}. \tag{3.10}$$

Formula (3.10) can be proved by induction, but we shall derive it from (3.3).

The system of differential equations (3.9) is a special case of (3.1) where $\lambda_k = k\lambda$. Therefore, (3.10) can be obtained from (3.3) by making the same substitution. This is demonstrated as follows. When $\lambda_k = k\lambda$, (3.3) becomes

$$p_k(t) = (-1)^{k-k_0}[k_0\lambda]\cdots[(k-1)\lambda]\left[\sum_{i=k_0}^{k} \frac{e^{-i\lambda t}}{\prod\limits_{\substack{j=k_0\\ j\neq i}}^{k}(i\lambda - j\lambda)}\right], \tag{3.11}$$

where

$$[k_0\lambda]\cdots[(k-1)\lambda] = \lambda^{k-k_0}\binom{k-1}{k-k_0}(k-k_0)! \tag{3.12}$$

and

$$\prod_{\substack{j=k_0\\ j\neq i}}^{k}(i\lambda - j\lambda) = \lambda^{k-k_0}(-1)^{k-i}\binom{k-k_0}{i-k_0}^{-1}(k-k_0)!\,. \tag{3.13}$$

Therefore, (3.11) can be rewritten as

$$\begin{aligned} p_k(t) &= \binom{k-1}{k-k_0} e^{-k_0\lambda t}\left[\sum_{i=k_0}^{k}\binom{k-k_0}{i-k_0}(-e^{-\lambda t})^{i-k_0}\right] \\ &= \binom{k-1}{k-k_0} e^{-k_0\lambda t}[1 - e^{-\lambda t}]^{k-k_0} \end{aligned} \tag{3.14}$$

since

$$\sum_{i=k_0}^{k}\binom{k-k_0}{i-k_0}(-e^{-\lambda t})^{i-k_0} = [1 - e^{-\lambda t}]^{k-k_0},$$

and we have (3.10). ▶

This process was first considered by Yule in a mathematical study of evolution, with $X(t)$ representing the total number of species in some genus at time t. It is instructive to consider the number $Y(t)$ of new species generated in the interval $(0, t)$; that is, $Y(t) = X(t) - k_0$. From (3.10), we have the distribution of $Y(t)$

$$\Pr\{Y(t) = k\} = p_{k+k_0}(t) = \binom{k + k_0 - 1}{k} e^{-\lambda t k_0}(1 - e^{-\lambda t})^k \tag{3.15}$$

which is easily recognized as a negative binomial distribution, and hence

$$E[X(t)] = k_0 + E[Y(t)] = k_0 e^{\lambda t}$$
$$\sigma^2_{X(t)} = \sigma^2_{Y(t)} = k_0 e^{\lambda t}[e^{\lambda t} - 1].$$

From the form of the negative binomial p.g.f. [see equation (4.18) in Chapter 2], we see that $Y(t)$ is the sum of k_0 independent and identically distributed random variables, which are, of course, the species generated by each of the k_0 initial species.

3.2. Time-Dependent Yule Process

As with the Poisson process, we may generalize the Yule process by allowing λ to be a function of time, $\lambda(t)$. Then the system (3.9) is replaced by

$$\frac{d}{dt} p_{k_0}(t) = -k_0\lambda(t)p_{k_0}(t) \tag{3.16a}$$

$$\frac{d}{dt} p_k(t) = -k\lambda(t)p_k(t) + (k-1)\lambda(t)p_{k-1}(t), \qquad k > k_0. \tag{3.16b}$$

The system (3.16) could be solved successively; instead, we shall use the method of p.g.f.'s. Let

$$G_X(s; t) = \sum_{k=k_0}^{\infty} p_k(t)s^k \tag{3.17}$$

with

$$G_X(s; 0) = s^{k_0}. \tag{3.18}$$

Differentiating (3.17) term by term with respect to t and using (3.16b), we have

$$\frac{\partial}{\partial t} G_X(s; t) = \sum_{k=k_0}^{\infty} \frac{d}{dt} p_k(t)s^k$$
$$= -\lambda(t)s \sum_{k=k_0}^{\infty} kp_k(t)s^{k-1} + \lambda(t)s^2 \sum_{k=k_0+1}^{\infty} (k-1)p_{k-1}(t)s^{k-2}. \tag{3.19}$$

Each of the summations on the right is equal to the derivative of $G_X(s; t)$ with respect to s; hence

$$\frac{\partial}{\partial t} G_X(s; t) + \lambda(t)s(1 - s)\frac{\partial}{\partial s} G_X(s; t) = 0. \tag{3.20}$$

To solve the linear first-order partial differential equation (3.20), we use the standard method (see Ford [1933], Miller [1941]) and write the auxiliary equations

$$\frac{dt}{1} = \frac{ds}{\lambda(t)s(1-s)}, \qquad dG_X(s;t) = 0. \tag{3.21}$$

We have from the first auxiliary equation

$$\frac{s}{1-s}\exp\left\{-\int_0^t \lambda(\tau)\,d\tau\right\} = \text{const.} \tag{3.22}$$

and from the second

$$G_X(s;t) = \text{const.} \tag{3.23}$$

Therefore the general solution of (3.20) is

$$G_X(s;t) = \Phi\left[\frac{s}{1-s}\exp\left\{-\int_0^t \lambda(\tau)\,d\tau\right\}\right] \tag{3.24}$$

where Φ is an arbitrary differentiable function. To obtain the particular solution corresponding to the initial condition (3.18), we set $t = 0$ in (3.24) to obtain

$$\Phi\left[\frac{s}{1-s}\right] = G_X(s;0) = s^{k_0}. \tag{3.25}$$

Equation (3.25) holds true at least for all s with $|s| < 1$; hence, for any θ such that $|\theta/(1+\theta)| < 1$,

$$\Phi\{\theta\} = \left[\frac{\theta}{1+\theta}\right]^{k_0}. \tag{3.26}$$

Letting

$$\theta = \frac{s}{1-s}\exp\left\{-\int_0^t \lambda(\tau)\,d\tau\right\}$$

we have, from (3.24) and (3.26), the solution

$$G_X(s;t) = s^{k_0}\left\{\frac{\exp\left\{-\int_0^t \lambda(\tau)\,d\tau\right\}}{1 - s\left(1 - \exp\left\{-\int_0^t \lambda(\tau)\,d\tau\right\}\right)}\right\}^{k_0}. \tag{3.27}$$

The second factor on the right side is recognized as the p.g.f. of a negative binomial distribution with parameters k_0 and $\exp\left\{-\int_0^t \lambda(\tau)\,d\tau\right\}$; the first factor informs us that $X(t)$ is obtained from the negative binomial random variable by adding the constant k_0. Hence the distribution and moments of X are the same as for the simple Yule process, except that λt is replaced by $\int_0^t \lambda(\tau)\, d\tau$. ▶

3.3. Joint Distribution in the Time-Dependent Yule Process

Consider two points t_1 and t_2 on the time axis, with $t_1 < t_2$. We wish to investigate the joint probability distribution of the corresponding random variables $X(t_1)$ and $X(t_2)$ in the case of the time-dependent simple birth process of Section 3.2.

In deriving the equations for the birth process, we have treated $k_0 = X(0)$ as a constant; however, the probabilities and p.g.f.'s can equivalently be regarded as conditional on $X(0)$. Taking this approach, we define the conditional bivariate p.g.f.

$$G(s_1, s_2; t_1, t_2) = E[s_1^{X(t_1)} s_2^{X(t_2)} \mid X(0)]. \tag{3.28}$$

The expectation on the right side may be rewritten as

$$E[s_1^{X(t_1)} s_2^{X(t_2)} \mid X(0)] = E[s_1^{X(t_1)} E\{s_2^{X(t_2)} \mid X(t_1)\} \mid X(0)]. \tag{3.29}$$

The inner conditional expectation in the last expression is the p.g.f. of $X(t_2)$ given $X(t_1)$ and, in view of (3.27), is given by

$$E\{s_2^{X(t_2)} \mid X(t_1)\} = \left\{\frac{s_2 \exp\left\{-\int_{t_1}^{t_2} \lambda(\tau)\, d\tau\right\}}{1 - s_2\left(1 - \exp\left\{-\int_{t_1}^{t_2} \lambda(\tau)\, d\tau\right\}\right)}\right\}^{X(t_1)}. \tag{3.30}$$

Substituting (3.30) in (3.29) and introducing

$$z_1 = \frac{s_1 s_2 \exp\left\{-\int_{t_1}^{t_2} \lambda(\tau)\, d\tau\right\}}{1 - s_2\left(1 - \exp\left\{-\int_{t_1}^{t_2} \lambda(\tau)\, d\tau\right\}\right)} \tag{3.31}$$

(3.28) becomes

$$G(s_1, s_2; t_1, t_2) = E[z_1^{X(t_1)} \mid X(0)]. \tag{3.32}$$

Since, clearly, $|z_1| < 1$, we can use (3.27) once again to obtain

$$G(s_1, s_2; t_1, t_2) = \left\{\frac{z_1 \exp\left\{-\int_0^{t_1} \lambda(\tau)\, d\tau\right\}}{1 - z_1\left(1 - \exp\left\{-\int_0^{t_1} \lambda(\tau)\, d\tau\right\}\right)}\right\}^{X(0)}. \tag{3.33}$$

Substituting (3.31) in (3.33) and letting

$$\pi_1 = \exp\left\{-\int_0^{t_1} \lambda(\tau)\, d\tau\right\}, \qquad \pi_2 = \exp\left\{-\int_{t_1}^{t_2} \lambda(\tau)\, d\tau\right\}, \tag{3.34}$$

we have the desired form for the generating function

$$G(s_1, s_2; t_1, t_2) = \left\{\frac{s_1\pi_1 s_2\pi_2}{1 - s_1(1-\pi_1)s_2\pi_2 - s_2(1-\pi_2)}\right\}^{X(0)}. \tag{3.35}$$

From (3.35), direct computations yield the covariance

$$\operatorname{Cov}[X(t_1), X(t_2) \mid X(0)] = X(0) \exp\left\{\int_0^{t_2} \lambda(\tau)\, d\tau\right\}\left(\exp\left\{\int_0^{t_1} \lambda(\tau)\, d\tau\right\} - 1\right). \tag{3.36}$$

The joint probability distribution of $X(t_1)$ and $X(t_2)$ may be obtained either from (3.35) or from

$$\Pr\{X(t_1) = k_1, X(t_2) = k_2 \mid X(0) = k_0\} = \Pr\{X(t_1) = k_1 \mid X(0) = k_0\}\Pr\{X(t_2) = k_2 \mid X(t_1) = k_1\}. \tag{3.37}$$

With either approach, we have

$$\Pr\{X(t_1) = k_1, X(t_2) = k_2 \mid X(0) = k_0\} = \prod_{i=0}^{1} \binom{k_{i+1} - 1}{k_{i+1} - k_i} \exp\left\{-k_i \int_{t_i}^{t_{i+1}} \lambda(\tau)\, d\tau\right\} \times \left[1 - \exp\left\{-\int_{t_i}^{t_{i+1}} \lambda(\tau)\, d\tau\right\}\right]^{k_{i+1}-k_i}. \tag{3.38}$$

If we are concerned with a sequence of random variables $\{X(t_i)\}$ for $0 < t_1 < t_2 < \ldots$, we shall have a chain of probability distributions which have the same form as (3.38), except that the upper limit of $(i + 1)$ will correspond to the last random variable of the sequence. ▶

4. THE POLYA PROCESS

In our original formulation of the pure birth process, the probability that a birth occurs in the time interval $(t, t + \Delta)$ was assumed to be $\lambda_k\Delta + o(\Delta)$. Suppose now that λ_k is a function of both k and t such that

$$\lambda_k(t) = \frac{\lambda + \lambda a k}{1 + \lambda a t} \tag{4.1}$$

where λ and a are nonnegative constants; this assumption defines the Polya process. With $X(0) = k_0$, the differential equations for the probability distribution are

$$\frac{d}{dt} p_{k_0}(t) = -\frac{\lambda + \lambda a k_0}{1 + \lambda a t} p_{k_0}(t) \tag{4.2a}$$

$$\frac{d}{dt} p_k(t) = -\frac{\lambda + \lambda a k}{1 + \lambda a t} p_k(t) + \frac{\lambda + \lambda a(k-1)}{1 + \lambda a t} p_{k-1}(t), \qquad k > k_0. \tag{4.2b}$$

We shall once again demonstrate the use of the p.g.f., although the above equations can also be solved successively. Let

$$G_X(s; t) = \sum_{k=k_0}^{\infty} p_k(t) s^k \tag{4.3}$$

with

$$G_X(s; 0) = s^{k_0}. \tag{4.4}$$

We take the derivative of $G_X(s; t)$ with respect to t and use the system (4.2) to obtain the partial differential equation

$$(1 + \lambda a t) \frac{\partial}{\partial t} G_X(s; t) + \lambda a s(1 - s) \frac{\partial}{\partial s} G_X(s; t) = -\lambda(1 - s) G_X(s; t). \tag{4.5}$$

The auxiliary equations are

$$\frac{dt}{1 + \lambda a t} = \frac{ds}{\lambda a s(1 - s)} = \frac{dG_X(s; t)}{-\lambda(1 - s) G_X(s; t)}. \tag{4.6}$$

The first equation may be written as

$$d \log (1 + \lambda a t) = d \log \frac{s}{1 - s} \tag{4.7}$$

with the solution

$$\frac{1 - s}{s}(1 + \lambda a t) = \text{const.} \tag{4.8}$$

From the second auxiliary equation we have

$$\frac{1}{a} d \log s = -d \log G_X(s; t) \tag{4.9}$$

so that

$$G_X(s; t) s^{1/a} = \text{const.} \tag{4.10}$$

Hence the general solution of (4.5) is

$$G_X(s; t) = s^{-1/a} \Phi\left\{\left(\frac{1 - s}{s}\right)(1 + \lambda a t)\right\} \tag{4.11}$$

where Φ is an arbitrary differentiable function.

For the particular solution corresponding to the initial condition (4.4), we set $t = 0$ in (4.11) and use (4.4) to write

$$s^{-1/a}\Phi\left\{\frac{1-s}{s}\right\} = s^{k_0}. \tag{4.12}$$

Equation (4.12) holds true for all s such that $|s| < 1$; hence, for any θ so that $|1/(1+\theta)| < 1$,

$$\Phi(\theta) = (1+\theta)^{-(k_0+1/a)}. \tag{4.13}$$

Now letting

$$\theta = \frac{1-s}{s}(1+\lambda at)$$

we have, from (4.12), the required solution for the p.g.f.

$$G_X(s;t) = s^{-1/a}\left[1 + \frac{1-s}{s}(1+\lambda at)\right]^{-(k_0+1/a)}$$

$$= s^{k_0}\left[\frac{1/(1+\lambda at)}{1 - s[\lambda at/(1+\lambda at)]}\right]^{k_0+1/a}. \tag{4.14}$$

Hence $X(t)$ is, except for the additive constant k_0, a negative binomial random variable [see equation (4.18), Chapter 2] with parameters

$$r = k_0 + \frac{1}{a} \quad \text{and} \quad p = \frac{1}{1+\lambda at}. \quad ▶$$

If we rewrite (4.14) as

$$G_X(s;t) = \left\{\frac{s}{1+\lambda at - \lambda ats}\right\}^{k_0}[1 + \lambda at - \lambda ats]^{-1/a} \tag{4.15}$$

it is clear that

$$\lim_{a\to 0} G_X(s;t) = s^{k_0}e^{-(1-s)\lambda t}. \tag{4.16}$$

Thus we see that as $a \to 0$ the Polya process (except for the additive constant k_0) approaches the Poisson process. This is expected in view of (4.1).

The situation is more surprising if we compare the Polya process with the weighted Poisson process in Section 2.2. In fact, the two processes are identical in the sense that, except for the additive constant k_0, they are both negative binomial distributions where the parameter p is the inverse of a linear function of t. (We recall that the Yule process also produced a negative binomial random variable, but with p an exponential function of time.) In the Polya case, the risk of an event was allowed to depend both on time and on the past history of the process; in the Poisson case the risk was assumed to be independent of both. Thus we have two very different mechanisms on the "micro" level which give rise to the same behavior of the system on the "macro" level. This means that the two underlying mechanisms cannot be distinguished by observations of the random variables $X(t)$.

5. PURE DEATH PROCESS

The pure death process is exactly analogous to the pure birth process, except that in the pure death process $X(t)$ is decreased rather than increased by the occurrence of an "event." Wc specialize immediately to the case in which each of a group of individuals alive at $t = 0$ is independently subject to the same risk of dying. In the usual applications this assumption is reasonable if t is interpreted as age (in this case the group is known as a *cohort*). Let $\mu(t)\Delta + o(\Delta)$ be the probability that an individual alive at age t will die in the interval $(t, t + \Delta)$; $\mu(t)$ is known as the *force of mortality*, *intensity of risk of dying*, or *failure rate*. If at age t, the $X(t) = k$ individuals are independently subject to the force of mortality $\mu(t)$, then the probability of one death occurring in $(t, t + \Delta)$ is $k\mu(t)\Delta + o(\Delta)$, the probability of two or more deaths occurring in $(t, t + \Delta)$ is $o(\Delta)$, and the probability of no change is $1 - k\mu(t)\Delta + o(\Delta)$. Following the steps in Section 3, we can formulate the differential equations for the probabilities of $X(t)$ and, from these, arrive at the solution

$$\Pr\{X(t) = k \mid X(0) = k_0\} = p_k(t) = \binom{k_0}{k} \exp\left\{-k\int_0^t \mu(\tau)\, d\tau\right\} \times \left(1 - \exp\left\{-\int_0^t \mu(\tau)\, d\tau\right\}\right)^{k_0-k}. \tag{5.1}$$

Rather than repeat the same procedure, we shall introduce a slightly different approach.

Since the k_0 individuals are assumed to be independently subject to the same force of mortality, we may start with $k_0 = 1$ and then compute the k_0-fold convolution to obtain the required solution for $k_0 > 1$. Let the continuous random variable T be the life span, so that the distribution function

$$F_T(t) = \Pr\{T \le t\} \tag{5.2}$$

is the probability that the individual will die prior to (or at) age t. Consider now the interval $(0, t + \Delta)$ and the corresponding distribution function $F_T(t + \Delta)$. For an individual to die prior to $t + \Delta$ he must die prior to t or else he must survive to t and die during the interval $(t, t + \Delta)$. Therefore the corresponding probabilities have the relation

$$F_T(t+\Delta) = F_T(t) + [1 - F_T(t)][\mu(t)\Delta + o(\Delta)] \tag{5.3}$$

or

$$\frac{F_T(t+\Delta) - F_T(t)}{\Delta} = [1 - F_T(t)]\mu(t) + \frac{o(\Delta)}{\Delta}. \tag{5.4}$$

Taking the limits of both sides of (5.4) as $\Delta \to 0$, we have the differential equation

$$\frac{d}{dt} F_T(t) = [1 - F_T(t)]\mu(t) \tag{5.5}$$

with the initial condition

$$F_T(0) = 0. \tag{5.6}$$

Integrating (5.5) and using (5.6) yield the solution

$$1 - F_T(t) = \exp\left\{-\int_0^t \mu(\tau)\, d\tau\right\}. \tag{5.7}$$

Equation (5.7) gives the probability that one individual alive at age 0 will survive to age t. If there are k_0 individuals alive at age 0, the number $X(t)$ of survivors to age t is clearly a binomial random variable with the probability of success $1 - F_T(t)$. Consequently $X(t)$ has the probability distribution (5.1), and the solution is completed. ▶

The survival probability in (5.7) has been known to life table students for more than two hundred years.[3] Unfortunately it has not been given due recognition by investigators in statistics, although different forms of this function have appeared in various areas of research. We shall mention a few below in terms of the probability density function of X,

$$\begin{aligned} f_T(t) = \frac{dF_T(t)}{dt} &= \mu(t) \exp\left\{-\int_0^t \mu(\tau)\, d\tau\right\} & t \geq 0 \\ &= 0 & t < 0. \end{aligned} \tag{5.8}$$

(i) Gompertz Distribution. In a celebrated paper on the law of human mortality, Benjamin Gompertz [1825] attributed death to either of two causes: chance or deterioration of the power to withstand destruction. In deriving his law of mortality, however, he considered only deterioration and assumed that man's power to resist death decreases at a rate proportional to the power itself. Since the force of mortality $\mu(t)$ is a measure of man's susceptibility to death, Gompertz used the reciprocal $1/\mu(t)$ as a

[3] E. Halley's famous table for the City of Breslau was published in the year 1693.

measure of man's resistance to death and thus arrived at the formula

$$\frac{d}{dt}\left(\frac{1}{\mu(t)}\right) = -h\,\frac{1}{\mu(t)}, \tag{5.9}$$

where h is a positive constant. Integrating (5.9) gives

$$\log\left(\frac{1}{\mu(t)}\right) = -ht + k \tag{5.10}$$

which when rearranged becomes the Gompertz law of mortality

$$\mu(t) = Bc^t. \tag{5.11}$$

The corresponding density function is

$$f(t) = Bc^t e^{-Bc^t/\log c}. \tag{5.12}$$

W. M. Makeham [1860] suggested the modification

$$\mu(t) = A + Bc^t, \tag{5.13}$$

which describes quite satisfactorily the mortality experience of human populations from about age 20 to the end of life. ▶

(ii) Weibull Distribution. When the force of mortality (or failure rate) is assumed to be an exponential function of t, $\mu(t) = \mu a t^{a-1}$, we have

$$f(t) = \mu a t^{a-1} e^{-\mu t^a}. \tag{5.14}$$

This distribution, recommended by W. Weibull [1939] for studies of the life span of materials, is used extensively in reliability theory. ▶

(iii) Exponential Distribution. If $\mu(t) = \mu$ is a constant, then

$$f(t) = \mu e^{-\mu t}, \tag{5.15}$$

a formula that plays a central role in the problem of life testing (Epstein and Sobel [1953]). ▶

6. BIRTH-DEATH PROCESSES

We now consider processes that permit a population to grow as well as to decline. These are more relevant to biological populations, in which both births and deaths occur, than are the processes considered in the preceding sections. We shall discuss two models in detail and briefly mention a few others of practical interest in the exercises of this chapter.

Again we let $X(t)$ denote the size of a population at time t for $0 \leq t < \infty$, with the initial size $X(0) = k_0$ and $p_k(t) = \Pr\{X(t) = k\}$. Given

$X(t) = k$, we assume that (i) the probability of exactly one birth occurring in interval $(t, t + \Delta)$ is $\lambda_k(t)\Delta + o(\Delta)$; (ii) the probability of exactly one death is $\mu_k(t)\Delta + o(\Delta)$; and (iii) the probability of more than one change is $o(\Delta)$. Therefore the probability of no change in $(t, t + \Delta)$ is

$$1 - \lambda_k(t)\Delta - \mu_k(t)\Delta - o(\Delta).$$

Consequently the probability $p_k(t + \Delta)$ at time $t + \Delta$ may be expressed in the difference equation

$$p_k(t + \Delta) = p_k(t)[1 - \lambda_k(t)\Delta - \mu_k(t)\Delta] + p_{k-1}(t)\lambda_{k-1}(t)\Delta + p_{k+1}(t)\mu_{k+1}(t)\Delta + o(\Delta), \tag{6.1}$$

from which we obtain the corresponding system of differential equations

$$\frac{d}{dt} p_0(t) = -[\lambda_0(t) + \mu_0(t)]p_0(t) + \mu_1(t)p_1(t)$$

$$\frac{d}{dt} p_k(t) = -[\lambda_k(t) + \mu_k(t)]p_k(t) + \lambda_{k-1}(t)p_{k-1}(t) + \mu_{k+1}(t)p_{k+1}(t), \qquad k \geq 1. \tag{6.2}$$

System (6.2), under the initial condition

$$p_{k_0}(0) = 1, \qquad p_k(0) = 0 \qquad \text{for} \qquad k \neq k_0, \tag{6.3}$$

completely determines the probability distribution $p_k(t)$. Appropriate assumptions may be made regarding the functions $\lambda_k(t)$ and $\mu_k(t)$ to obtain stochastic processes corresponding to empirical phenomena. We shall consider a few below.

6.1. Linear Growth

Suppose that both $\lambda_k(t)$ and $\mu_k(t)$ are independent of time but proportional to k,

$$\lambda_k(t) = k\lambda \qquad \text{and} \qquad \mu_k(t) = k\mu \tag{6.4}$$

where λ and μ are constant. Then the system (6.2) becomes

$$\frac{d}{dt} p_0(t) = \mu p_1(t)$$

$$\frac{d}{dt} p_k(t) = -k(\lambda + \mu)p_k(t) + (k - 1)\lambda p_{k-1}(t) + (k + 1)\mu p_{k+1}(t). \tag{6.5}$$

Since the equations (6.5) cannot be solved successively, we shall use the method of p.g.f.'s. Let

$$G_X(s; t) = \sum_{k=0}^{\infty} s^k p_k(t). \tag{6.6}$$

It is easily deduced from (6.5) that the p.g.f. satisfies the homogeneous partial differential equation

$$\frac{\partial}{\partial t} G_X(s; t) + (1 - s)(\lambda s - \mu) \frac{\partial}{\partial s} G_X(s; t) = 0 \tag{6.7}$$

with the initial condition at $t = 0$

$$G_X(s; 0) = s^{k_0}. \tag{6.8}$$

The auxiliary equations are

$$\frac{dt}{1} = \frac{ds}{(1 - s)(\lambda s - \mu)} \qquad \text{and} \qquad dG_X(s; t) = 0. \tag{6.9}$$

In the case $\lambda \neq \mu$ we may use partial fractions to rewrite the first auxiliary equation as

$$dt = \frac{\lambda}{(\lambda - \mu)(\lambda s - \mu)} ds + \frac{1}{(\lambda - \mu)(1 - s)} ds \tag{6.10}$$

or

$$(\lambda - \mu)\, dt = d \log \left\{\frac{\lambda s - \mu}{1 - s}\right\}. \tag{6.11}$$

Integrating both sides of (6.11) gives

$$\frac{1 - s}{\lambda s - \mu} e^{(\lambda - \mu)t} = \text{const.} \tag{6.12}$$

The second auxiliary equation in (6.9) implies

$$G_X(s; t) = \text{const.} \tag{6.13}$$

Therefore the general solution of (6.7) is

$$G_X(s; t) = \Phi\left\{\frac{1 - s}{\lambda s - \mu} e^{(\lambda - \mu)t}\right\} \tag{6.14}$$

where Φ is an arbitrary differentiable function. Using the initial condition (6.8), we see from (6.14) that at $t = 0$

$$\Phi\left\{\frac{1 - s}{\lambda s - \mu}\right\} = s^{k_0} \tag{6.15}$$

at least for all s with $|s| < 1$. Hence, for all θ such that $|1 + \theta\mu| < |1 + \theta\lambda|$,

$$\Phi\{\theta\} = \left\{\frac{1 + \theta\mu}{1 + \theta\lambda}\right\}^{k_0}. \tag{6.16}$$

Letting

$$\theta = \frac{1-s}{\lambda s - \mu} e^{(\lambda-\mu)t} \tag{6.17}$$

we have the particular solution for the case $\mu \neq \lambda$

$$G_X(s;t) = \left\{\frac{(\lambda s - \mu) + \mu(1-s)e^{(\lambda-\mu)t}}{(\lambda s - \mu) + \lambda(1-s)e^{(\lambda-\mu)t}}\right\}^{k_0}. \tag{6.18}$$

Letting

$$\alpha(t) = \mu \frac{1 - e^{(\lambda-\mu)t}}{\mu - \lambda e^{(\lambda-\mu)t}}, \qquad \beta(t) = \frac{\lambda}{\mu}\alpha(t) \tag{6.19}$$

(6.18) may be rewritten as

$$G_X(s;t) = \left\{\frac{\alpha(t) + [1-\alpha(t)-\beta(t)]s}{1-\beta(t)s}\right\}^{k_0}. \tag{6.20}$$

We now expand the p.g.f. $G_X(s;t)$ to obtain the probability distribution $\{p_k(t)\}$. We have

$$\{\alpha(t) + [1-\alpha(t)-\beta(t)]s\}^{k_0} = \sum_{j=0}^{k_0} \binom{k_0}{j} [\alpha(t)]^{k_0-j}[1-\alpha(t)-\beta(t)]^j s^j, \tag{6.21}$$

and, since clearly $|\beta(t)s| < 1$,

$$\{1-\beta(t)s\}^{-k_0} = \sum_{i=0}^{\infty} \binom{-k_0}{i} (-1)^i [\beta(t)]^i s^i = \sum_{i=0}^{\infty} \binom{k_0+i-1}{i} [\beta(t)]^i s^i. \tag{6.22}$$

Hence the probability

$$p_k(t) = \sum_{j=0}^{\min[k_0,k]} \binom{k_0}{j}\binom{k_0+k-j-1}{k-j} [\alpha(t)]^{k_0-j}[\beta(t)]^{k-j}[1-\alpha(t)-\beta(t)]^j \qquad k \geq 1,$$

$$p_0(t) = [\alpha(t)]^{k_0}. \tag{6.23}$$

When the initial population size is $k_0 = 1$,

$$G_X(s;t) = \frac{\alpha(t) + [1-\alpha(t)-\beta(t)s]}{1-\beta(t)s},$$

and

$$p_k(t) = [1-\alpha(t)][1-\beta(t)][\beta(t)]^{k-1} \qquad k \geq 1,$$

$$p_0(t) = \alpha(t). \tag{6.24}$$

By differentiating the p.g.f. in (6.20), we find the following expectation and variance for $X(t)$

$$E[X(t)] = k_0 \frac{1 - \alpha(t)}{1 - \beta(t)} = k_0 e^{(\lambda-\mu)t} \tag{6.25}$$

and

$$\sigma^2_{X(t)} = k_0 \frac{[1 - \alpha(t)][\alpha(t) + \beta(t)]}{[1 - \beta(t)]^2} = k_0 \left(\frac{\lambda + \mu}{\lambda - \mu}\right) e^{(\lambda-\mu)t}[e^{(\lambda-\mu)t} - 1]. \tag{6.26}$$

We now return to the case $\mu = \lambda$. The differential equation (6.7) becomes

$$\frac{\partial}{\partial t} G_X(s; t) - \lambda(1 - s)^2 \frac{\partial}{\partial s} G_X(s; t) = 0 \tag{6.27}$$

with the solution

$$G_X(s; t) = \left\{\frac{\alpha^*(t) + [1 - 2\alpha^*(t)]s}{1 - \alpha^*(t)s}\right\}^{k_0} \tag{6.28}$$

where $\alpha^*(t) = \lambda t/(1 + \lambda t)$ and the probabilities

$$p_k(t) = \sum_{j=0}^{\min[k_0,k]} \binom{k_0}{j} \binom{k_0 + k - j - 1}{k - j} [\alpha^*(t)]^{(k_0+k-2j)}[1 - 2\alpha^*(t)]^j \qquad k \geq 1$$

$$p_0(t) = \left\{\frac{\lambda t}{1 + \lambda t}\right\}^{k_0}. \tag{6.29}$$

Formulas (6.28) and (6.29) can also be obtained directly from (6.20) and (6.23) by letting $\mu \to \lambda$. The expected value and variance of $X(t)$ are

$$E[X(t)] = k_0 \qquad \text{and} \qquad \sigma^2_{X(t)} = 2k_0\lambda t; \tag{6.30}$$

thus when the birth rate equals the death rate the population size has a constant expectation but an increasing variance.

The limiting behavior of the birth-death process is analogous to that encountered in branching processes (see Section 8, Chapter 2). The probability that the population becomes extinct at time t is $p_0(t) = G_X(0; t)$. From (6.20) and (6.28) we see that, for $|s| < 1$,

$$\begin{aligned} \lim_{t\to\infty} G_X(s; t) &= \left(\frac{\mu}{\lambda}\right)^{k_0} \qquad \mu < \lambda \\ &= 1 \qquad \mu \geq \lambda \end{aligned} \tag{6.31}$$

Thus, if $\mu \geq \lambda$, the probability of extinction tends to unity as $t \to \infty$, and the population is certain to die out eventually. On the other hand, if $\mu < \lambda$, the probability of ultimate extinction is $(\mu/\lambda)^{k_0}$; furthermore, since the limiting "p.g.f." is constant, the probability that the population will increase without bound is $1 - (\mu/\lambda)^{k_0}$.

The relative magnitude of λ and μ also influences the asymptotic value of the expectation and variance of $X(t)$. From (6.25), (6.26), and (6.30), we see that

$$\begin{aligned} \lim_{t\to\infty} E[X(t)] &= 0 \qquad \text{if} \quad \mu > \lambda \\ &= k_0 \qquad \text{if} \quad \mu = \lambda \\ &= \infty \qquad \text{if} \quad \mu < \lambda \end{aligned} \tag{6.32}$$

and

$$\begin{aligned} \lim_{t\to\infty} \sigma^2_{X(t)} &= 0 \qquad \text{if} \quad \mu > \lambda \\ &= \infty \qquad \text{if} \quad \mu \le \lambda. \end{aligned} \tag{6.33}$$

When $\lambda = \mu$ we have an interesting case in which the probability of extinction tends to unity and yet the expected population size tends to k_0. These seemingly contradictory facts may be intuitively explained by the large value of the variance. Although most populations will eventually become extinct, a few others will attain huge sizes, so that the average size is k_0.

6.2. A Time-Dependent General Birth-Death Process

We may generalize the preceding model by letting both $\lambda = \lambda(t)$ and $\mu = \mu(t)$ be functions of time. The mathematics involved in deriving the probability distribution $p_k(t)$ is somewhat more complex than for linear growth, but we may follow the same procedure until we reach the partial differential equation analogous to (6.7)

$$\frac{\partial}{\partial t} G_X(s; t) + (1 - s)[s\lambda(t) - \mu(t)] \frac{\partial}{\partial s} G_X(s; t) = 0 \tag{6.34}$$

and the corresponding auxiliary equations

$$\frac{dt}{1} = \frac{ds}{(1 - s)[s\lambda(t) - \mu(t)]} \qquad \text{and} \qquad dG_X(s; t) = 0. \tag{6.35}$$

Since $\lambda(t)$ and $\mu(t)$ are functions of t, the method of partial fractions will not work here. We introduce a new variable

$$z = (1 - s)^{-1} \tag{6.36}$$

and rewrite the first equation in (6.35) as

$$\frac{dz}{dt} - [\lambda(t) - \mu(t)]z + \lambda(t) = 0 \tag{6.37}$$

which can be solved in the ordinary way. Letting

$$\gamma(t) = -\int_0^t [\lambda(\tau) - \mu(\tau)]\, d\tau \tag{6.38}$$

and multiplying (6.37) through by $e^{\gamma(t)}$, we have

$$e^{\gamma(t)} \frac{dz}{dt} - [\lambda(t) - \mu(t)] z e^{\gamma(t)} + \lambda(t) e^{\gamma(t)} = 0$$

or

$$\frac{d}{dt} [z e^{\gamma(t)}] + \lambda(t) e^{\gamma(t)} = 0. \tag{6.39}$$

Integrating (6.39) and replacing z by $(1 - s)^{-1}$ yield the equation

$$\frac{1}{1-s} e^{\gamma(t)} + \int_0^t \lambda(\tau) e^{\gamma(\tau)}\, d\tau = \text{const.} \tag{6.40}$$

Using (6.40) and the second auxiliary equation in (6.35) we obtain the general solution

$$G_X(s; t) = \Phi\left\{\frac{1}{1-s} e^{\gamma(t)} + \int_0^t \lambda(\tau) e^{\gamma(\tau)}\, d\tau\right\}. \tag{6.41}$$

If the initial population size is $X(0) = k_0$, then

$$G_X(s; 0) = \Phi\left\{\frac{1}{1-s}\right\} = s^{k_0} \tag{6.42}$$

at least for $|s| < 1$; hence, for θ such that $\left|1 - \frac{1}{\theta}\right| < 1$,

$$\Phi(\theta) = \left\{1 - \frac{1}{\theta}\right\}^{k_0}. \tag{6.43}$$

Therefore the required formula for the p.g.f. is

$$G_X(s; t) = \left\{1 - \frac{1}{\dfrac{e^{\gamma(t)}}{1-s} + \displaystyle\int_0^t \lambda(\tau) e^{\gamma(\tau)} d\tau}\right\}^{k_0} \tag{6.44}$$

or

$$G_X(s; t) = \left\{1 - \frac{1-s}{e^{\gamma(t)} + \displaystyle\int_0^t \lambda(\tau) e^{\gamma(\tau)}\, d\tau - s\int_0^t \lambda(\tau) e^{\gamma(\tau)}\, d\tau}\right\}^{k_0}. \tag{6.45}$$

Letting

$$\alpha(t) = 1 - \frac{1}{e^{\gamma(t)} + \displaystyle\int_0^t \lambda(\tau) e^{\gamma(\tau)}\, d\tau} \tag{6.46}$$

and

$$\beta(t) = 1 - e^{\gamma(t)}[1 - \alpha(t)] \tag{6.47}$$

we may rewrite (6.45) as

$$G_X(s; t) = \left\{\frac{\alpha(t) + [1 - \alpha(t) - \beta(t)]s}{1 - \beta(t)s}\right\}^{k_0}. \tag{6.48}$$

Except for the definitions of $\alpha(t)$ and $\beta(t)$, the p.g.f. (6.48) is of the same form as the one in (6.20); therefore the probability distribution $\{p_k(t)\}$ is given by (6.23) with $\alpha(t)$ and $\beta(t)$ as defined in (6.46) and (6.47).

The probability of extinction at time t may be obtained directly from (6.46) and (6.48)

$$p_0(t) = \left\{1 - \frac{1}{e^{\gamma(t)} + \int_0^t \lambda(\tau)e^{\gamma(\tau)}\, d\tau}\right\}^{k_0}. \tag{6.49}$$

Using the definition of $\gamma(t)$ in (6.38), we can write

$$e^{\gamma(t)} + \int_0^t \lambda(\tau)e^{\gamma(\tau)}\, d\tau = 1 + \int_0^t \mu(\tau)e^{\gamma(\tau)}\, d\tau \tag{6.50}$$

and

$$p_0(t) = \left\{\frac{\int_0^t \mu(\tau)e^{\gamma(\tau)}\, d\tau}{1 + \int_0^t \mu(\tau)e^{\gamma(\tau)}\, d\tau}\right\}^{k_0}. \tag{6.51}$$

We see that in this case $p_0(t) \to 1$ as $t \to \infty$ if, and only if, the integral in the above expression diverges as $t \to \infty$.

PROBLEMS

1. *Poisson process.* Let $\{X(t)\}$ be the simple Poisson process with the probability distribution given in (2.8). Find the covariance between $X(t)$ and $X(t + \tau)$.

2. *Renewal process.* Suppose the occurrence of events (renewals, for example, replacements of automobile tires) follows the simple Poisson process. Define T_n as the length of time up to the occurrence of the nth event. Derive the distribution function

$$F_n(t) = \Pr\{T_n \le t\}$$

and the expectation $E(T_n)$. [*Hint.* Establish a relation between $X(t)$ the number of events occurring up to t and T_n.]

3. *Continuation.* Derive the distribution function for the difference

$$T_{n+1} - T_n.$$

4. *Continuation.* Given $X(t) = n$, derive the joint distribution of $T_1, \ldots, T_n$.

5. Show that the expectation and variance of $X(t)$ with probability distribution

$$p_k(t) = \frac{\Gamma(k + \alpha)}{k!\,\Gamma(\alpha)} \beta^\alpha t^k (\beta + t)^{-(k+\alpha)}, \qquad k = 0, 1, \ldots, \tag{2.23}$$

are $E[X(t)] = \alpha t/\beta$ and

$$\sigma_X{}^2(t) = \alpha \frac{t}{\beta}\left(1 + \frac{t}{\beta}\right).$$

6. The Canadian Department of National Health and Welfare and the Dominion Bureau of Statistics conducted a sickness survey in 1950–1951 on a sample of approximately 10,000 households. Part of the published data has been fitted in a weighted Poisson distribution. The material in Table P6 was taken from Chiang [1965].

Let the random variable X be the number of doctors' calls or clinic visits for one person in a year, and suppose that X has the probability distribution

$$\Pr\{X = k\} = \frac{\Gamma(k + \alpha)}{k!\,\Gamma(\alpha)} \beta^\alpha (\beta + 1)^{-(k+\alpha)}, \qquad k = 0, 1, \ldots.$$

(a) Find $E(X)$ and $\sigma_X{}^2$.

Table P6. Observed number of persons under 15 years of age by number of doctor's calls or clinic visits

Number of Doctors' Calls or Clinic Visits k	Observed Number of Persons f_k
0	2367
1	749
2	350
3	222
4	136
5	95
6	64
7	41
8	25
9	14
10	12
11	11
12	9
13	8
14	5
15 or more	8
Total	4116

(b) Compute the sample mean $\bar{X}$ and sample variance S_X^2 from the frequency table above, and use these values to estimate α and β in the formula.

(c) Compute the expected number of persons F_k for each k and determine

$$\chi_0^2 = \sum \frac{(f_k - F_k)^2}{F_k}.$$

7. *Yule process.* From (3.9), derive the probability distribution

$$p_k(t) = \binom{k-1}{k-k_0} e^{-k_0 \lambda t}(1 - e^{-\lambda t})^{k-k_0}. \tag{3.10}$$

8. *Continuation.* Solve successively the system of differential equations in (3.16) in the time-dependent Yule process.

9. *Continuation.* Show that p.g.f. (3.27) satisfies the partial differential equation (3.20).

10. *Continuation.* Derive the covariance between $X(t_1)$ and $X(t_2)$ in (3.36) (i) from the p.g.f. (3.35), and (ii) by using the identity $E[X(t_1)X(t_2) \mid X(0)] = E[X(t_1)E\{X(t_2) \mid X(t_1)\} \mid X(0)]$.

11. *Polya process.* Solve the system of differential equations in (4.2).

12. *Pure death process.* Derive and solve the differential equations for the probability distribution in (5.1).

13. *Continuation.* Use the method of p.g.f. to derive the probability distribution in (5.1).

14. *Linear growth.* Show that the probability function (6.23) satisfies the system of differential equations in (6.5) and that

$$\sum_{k=0}^{\infty} p_k(t) = 1.$$

15. *Continuation.* Derive the p.g.f. in (6.28) by solving the differential equation in (6.7).

16. *Continuation.* Derive the p.g.f. in (6.28) from (6.18) by taking the limit as $\mu \to \lambda$.

17. *Migration process.* Suppose that the change in a population size results only from immigration and emigration so that

$$\Pr\{X(t+\Delta) = k+1 \mid X(t) = k\} = \eta\Delta + o(\Delta)$$

and

$$\Pr\{X(t+\Delta) = k-1 \mid X(t) = k\} = k\mu\Delta + o(\Delta)$$

where η and μ are constant. Derive the probability distribution $\Pr\{X(t) = k \mid X(0) = k_0\}$ and the corresponding p.g.f.

18. *Continuation.* Suppose that the increase in a population size results from "birth" and "immigration" so that

$$\Pr\{X(t+\Delta) = k+1 \mid X(t) = k\} = k\lambda\Delta + \eta\Delta + o(\Delta)$$

and

$$\Pr\{X(t + \Delta) = k - 1 \mid X(t) = k\} = k\mu\Delta + o(\Delta).$$

Derive the probability distribution for $X(t)$ given $X(0) = k_0$ and the corresponding p.g.f.

19. *General birth-death process.* Show that the p.g.f. (6.44) satisfies the partial differential equation (6.34).

20. *Telephone exchange.* In a telephone exchange with N available channels a connection (or conversation) is realized if the incoming call finds an idle channel; as soon as conversation is finished, the channel being utilized becomes immediately available for a new call. An incoming call is lost if all channels are busy. Suppose the probability of an incoming call during the interval $(t, t + \Delta)$ is $\lambda\Delta + o(\Delta)$ and the probability of a conversation ending during the interval $(t, t + \Delta)$ is $\mu\Delta + o(\Delta)$. Let $X(t)$ be the number of channels occupied at time t, with $X(0) = k_0$. Derive the appropriate differential equations for the probabilities $p_k(t) = \Pr\{X(t) = k\}$.

21. *Continuation.* Suppose that, as $t \to \infty$, $p_k(t) \to p_k$ so that the derivatives of $p_k(t)$ with respect to t vanish. Derive the limiting probabilities p_k.

22. *Continuation.* Explicit formulas can be obtained for the probabilities $p_k(t)$ when the number N of channels is infinity. Derive the p.g.f. and the probabilities $p_k(t)$ in this case.

23. *Queueing.* A counter has s stations for serving customers. If all s stations are occupied, newly arriving customers must form a *queue* (or line) and wait until service is available. A queueing system is determined by (i) the *input process*, (ii) the *queue discipline*, and (iii) the *service mechanism*. The queueing theory has applications in a large number of problems. Many writers have contributed to the development of these applications; see, for example, Forster [1953], Homma [1955], Kendall [1952, 1954], and others. If (i) customers arrive according to a Poisson process with intensity λ, (ii) the queue discipline is governed by the "first come, first served" rule, and (iii) the service time has the exponential distribution with a constant intensity μ for each customer and all s stations [see (5.7) in the pure-death process with $\mu(\tau) = \mu$]. Let $X(t)$ be the total number of customers being served or waiting in line at time t so that $X(t) - s$ is the length of the waiting line, and let $p_k(t) = \Pr\{X(t) = k\}$ be the probability distribution with $p_0(0) = 1$, $p_k(0) = 0$ for $k > 0$. Derive appropriate differential equations for the probabilities $p_k(t)$.

24. *Continuation.* It may be shown that as $t \to \infty$, the derivative $dp_k(t)/dt$ vanishes whatever may be $k = 0, 1, \ldots$. Find the limiting probabilities $p_k(t)$ as $t \to \infty$.

25. *Continuation.* Let Q be the length of the waiting line in the limiting case in Problem 24. Derive the expectation and variance of Q.

26. *Continuation.* Let W be the length of time that a customer has to wait in line for service. Find the expected waiting time $E(W)$ for the limiting case as $t \to \infty$.

CHAPTER 4

A Simple Illness-Death Process

1. INTRODUCTION

In the pure death process and in the birth-death processes individuals were assumed to have the same probability of dying within a given time interval. This assumption does not hold true in the human population. All individuals are not equally healthy, and the chance of dying varies from one person to another. Since death is usually preceded by an illness, a mortality study is incomplete unless illness is taken into consideration. Illness and death, however, are distinct and different types of events. Illnesses are potentially concurrent, repetitive, and reversible, whereas death is an irreversible or absorbing state. The study of illness adds a new dimension and a new complexity to the general problem of mortality. A further complexity is introduced by competition among various risks of death for the life of each individual so that an individual's probability of dying from one risk is influenced by the presence of competition from other risks. In a mortality study we must consider competing risks and we must be concerned with the probabilities of dying from a specific risk with and without competing risks. The study of mortality and competing risks of death is discussed in Chapters 9 to 12. Illness-death processes are dealt with here because of their close association with the general approach of stochastic processes.

A stochastic model of the illness-death process seems to have been suggested first by Fix and Neyman in their study of the probabilities of relapse, recovery, and death for cancer patients. Generally, in the study of illness-death processes, a population is divided into a number of *illness states* and *death states*. An individual is considered to be in a particular illness state if he contracts the corresponding disease, and he is considered to be in a particular death state if he dies of the corresponding cause. He may leave a state of illness at any time through recovery, death, or by contracting other diseases. Thus, illness states are transient whereas death states are absorbing states, since once an individual has arrived at a death state, he will remain in the state forever. In order to introduce basic concepts of the illness-death process, a simple model with two illness states and a finite number of death states is presented in this chapter and in Chapter 5. A general model for any finite number of illness states is given in Chapter 7.

Example 1. In a stochastic model of recovery, relapse, death, and loss of cancer patients, Fix and Neyman [1951] considered two illness states: the state of being under treatment of cancer, and the statc of "leading normal life"; and two death states: death from cancer or operative death, and death from other causes or cases lost because of tracing difficulties. The state of "leading normal life" includes patients who have "recovered" from cancer, whether the recovery is real or only clinically apparent. A patient "leading normal life" can leave that state by having a recurrence of cancer or by dying from causes not connected with cancer or being lost from observation, whereas a patient may leave the cancer state only through death from cancer or by recovery. Thus this model does not allow for the death of a cancer patient from other causes. ▶

Example 2. A California mental hospital has different types of wards for patients with varying degrees of illness, ranging from very mild to extremely severe. The ward to which a new patient is assigned depends upon the severity of his case. A patient may be transferred from one ward to another when his condition improves or worsens; he may be discharged from the hospital, transferred to another hospital, or he may die. A record is kept for each patient from the time of admission regarding treatments received, transferrals between wards, the time of transferring, and his duration of stay in each ward. Analysis of data of this kind can be helpful in determining the progress of patients, effectiveness of treatments, incidence and prevalence of mental illnesses in the general population, the need for hospital beds and other facilities, the size of hospital personnel, the state's budget for mental care, and other matters. ▶

Illness-death processes can be applied to other types of problems; for example, instead of illness states we may study occupations or geographic locations. Or, instead of the illness of a human being, we may study any repairable disorder of a mechanical object. We shall, however, take the human population as an example and use illness and death as criteria of classification.

2. ILLNESS TRANSITION PROBABILITY, $P_{\alpha\beta}(t)$ AND DEATH TRANSITION PROBABILITY, $Q_{\alpha\delta}(t)$

Imagine a population having two illness states, S_1 and S_2, and r death states, $R_1, \ldots, R_r$, whatever may be the finite positive integer r. Let (τ, t) be a time interval, with $0 \leq \tau \leq t < \infty$. At time τ let an individual be in state S_α. During the interval (τ, t) he may travel continually between S_α and S_β, for $\alpha, \beta = 1, 2$, and he may reach a state of death. His transitions from one state to another are governed by intensities of risk of illness ($\nu_{\alpha\beta}$) and of death ($\mu_{\alpha\delta}$), which are defined as follows:

$$\nu_{\alpha\beta}\Delta + o(\Delta) = \Pr\{\text{an individual in state } S_\alpha \text{ at time } \xi \text{ will be in state } S_\beta \text{ at time } \xi + \Delta\}, \quad (2.1)$$

$$\mu_{\alpha\delta}\Delta + o(\Delta) = \Pr\{\text{an individual in state } S_\alpha \text{ at time } \xi \text{ will be in state } R_\delta \text{ at time } \xi + \Delta\}, \quad (2.2)$$

$$\alpha \neq \beta;\ \alpha, \beta = 1, 2;\ \delta = 1, \ldots, r,$$

for each ξ, $\tau \leq \xi \leq t$. The intensities $\nu_{\alpha\beta}$ and $\mu_{\alpha\delta}$ are assumed to be independent of ξ for $\tau \leq \xi \leq t$. For notational convenience, we define

$$\nu_{\alpha\alpha} = -\left(\nu_{\alpha\beta} + \sum_{\delta=1}^{r} \mu_{\alpha\delta}\right), \quad (2.3)$$

so that

$$1 + \nu_{\alpha\alpha}\Delta + o(\Delta) = \Pr\{\text{an individual in state } S_\alpha \text{ at time } \xi \text{ will remain in } S_\alpha \text{ during the interval } (\xi, \xi + \Delta)\}. \quad (2.4)$$

It is clear that $\nu_{\alpha\alpha} \leq 0$. If $\nu_{\alpha\alpha} = 0$, then the state S_α is absorbing, and should be treated as a death state. Hence we shall assume that

$$\nu_{\alpha\alpha} < 0 \qquad \alpha = 1, 2. \quad (2.5)$$

The state S_1, for example, may be regarded as a health state and S_2 as an illness state, so that a transition from S_1 to S_2 means an illness or a relapse, and a transition from S_2 to S_1 means recovery. If it is assumed that an individual dies only through the illness state, then $\mu_{1\delta} = 0$, $\delta = 1, \ldots, r$; however, for convenience of presentation, we shall treat health as a special case of illness without explicit deletion of any of the $\mu_{\alpha\delta}$, and assume that

$$\sum_{\delta=1}^{r} \mu_{1\delta} > 0 \qquad \text{or} \qquad \sum_{\delta=1}^{r} \mu_{2\delta} > 0. \quad (2.6)$$

The case in which all $\mu_{\alpha\delta}$ vanish corresponds to a process with no absorbing states.

An individual in state S_α at time τ will be either in one of the illness states S_β or in one of the death states R_δ at time t. (*Note*. In the following discussion, α may be equal to β; clarification will be made whenever necessary.) The corresponding probabilities will be called the *illness transition probabilities*,

$$P_{\alpha\beta}(\tau, t) = \Pr\{\text{an individual in state } S_\alpha \text{ at time } \tau \text{ will be in state } S_\beta \text{ at time } t\}, \quad \alpha, \beta = 1, 2, \tag{2.7}$$

and the *death transition probabilities*,

$$Q_{\alpha\delta}(\tau, t) = \Pr\{\text{an individual in state } S_\alpha \text{ at time } \tau \text{ will be in state } R_\delta \text{ at time } t\}, \quad \alpha = 1, 2;\ \delta = 1, \ldots, r. \tag{2.8}$$

They satisfy the initial conditions

$$\begin{aligned} P_{\alpha\alpha}(\tau, \tau) &= 1, & \alpha &= 1, 2, \\ P_{\alpha\beta}(\tau, \tau) &= 0, & \alpha &\neq \beta; \alpha, \beta = 1, 2, \\ Q_{\alpha\delta}(\tau, \tau) &= 0, & \alpha &= 1, 2; \delta = 1, \ldots, r. \end{aligned} \tag{2.9}$$

We shall first derive expressions for the illness transition probabilities and then use these to obtain the death transition probabilities.

Consider a time interval (τ, t), a fixed time ξ between τ and t, and an individual who is in state S_α at time τ. We assume that the future transitions of the individual are independent of his transitions in the past. It follows that the probability that the individual is in a particular state S_β at time ξ and is in S_γ at time t is

$$P_{\alpha\beta}(\tau, \xi)P_{\beta\gamma}(\xi, t). \tag{2.10}$$

Since the events corresponding to (2.10) for different β are mutually exclusive, we have

$$P_{\alpha\gamma}(\tau, t) = \sum_{\beta=1}^{2} P_{\alpha\beta}(\tau, \xi)P_{\beta\gamma}(\xi, t). \tag{2.11}$$

Equation (2.11) is a version of the Chapman-Kolmogorov equation, which we shall encounter in more general form in Chapter 6.

The differential equations for $P_{\alpha\beta}(\tau, t)$ are obtained by considering two contiguous time intervals, (τ, t) and $(t, t + \Delta)$, and the probabilities $P_{\alpha\beta}(\tau, t + \Delta)$. Using (2.11), (2.1), and (2.4) gives

$$\begin{aligned} P_{\alpha\alpha}(\tau, t + \Delta) &= P_{\alpha\alpha}(\tau, t)[1 + \nu_{\alpha\alpha}\Delta] + P_{\alpha\beta}(\tau, t)\nu_{\beta\alpha}\Delta + o(\Delta) \\ P_{\alpha\beta}(\tau, t + \Delta) &= P_{\alpha\alpha}(\tau, t)\nu_{\alpha\beta}\Delta + P_{\alpha\beta}(\tau, t)[1 + \nu_{\beta\beta}\Delta] + o(\Delta). \end{aligned} \tag{2.12}$$

By rearranging (2.12) to form difference quotients and taking the limit as $\Delta \to 0$, we obtain the differential equations

$$\begin{aligned} \frac{\partial}{\partial t} P_{\alpha\alpha}(\tau, t) &= P_{\alpha\alpha}(\tau, t)\nu_{\alpha\alpha} + P_{\alpha\beta}(\tau, t)\nu_{\beta\alpha} \\ \frac{\partial}{\partial t} P_{\alpha\beta}(\tau, t) &= P_{\alpha\alpha}(\tau, t)\nu_{\alpha\beta} + P_{\alpha\beta}(\tau, t)\nu_{\beta\beta}, \qquad \alpha \neq \beta; \alpha, \beta = 1, 2. \end{aligned} \tag{2.13}$$

This is a system of linear homogeneous first-order differential equations with constant coefficients (notice that the resemblance to partial differential equations is only formal). As is usual for such systems, we look for solutions of the form

$$P_{\alpha\alpha}(\tau, t) = c_{\alpha\alpha}e^{\rho t}, \qquad P_{\alpha\beta}(\tau, t) = c_{\alpha\beta}e^{\rho t}. \tag{2.14}$$

Substituting (2.14) in (2.13) gives

$$\begin{aligned} \rho c_{\alpha\alpha}e^{\rho t} &= c_{\alpha\alpha}e^{\rho t}\nu_{\alpha\alpha} + c_{\alpha\beta}e^{\rho t}\nu_{\beta\alpha} \\ \rho c_{\alpha\beta}e^{\rho t} &= c_{\alpha\alpha}e^{\rho t}\nu_{\alpha\beta} + c_{\alpha\beta}e^{\rho t}\nu_{\beta\beta}, \end{aligned}$$

which can be rewritten as

$$\begin{aligned} (\rho - \nu_{\alpha\alpha})c_{\alpha\alpha} - \nu_{\beta\alpha}c_{\alpha\beta} &= 0 \\ -\nu_{\alpha\beta}c_{\alpha\alpha} + (\rho - \nu_{\beta\beta})c_{\alpha\beta} &= 0. \end{aligned} \tag{2.15}$$

Since (2.15) is a linear homogeneous system, nontrivial solutions for $c_{\alpha\alpha}$ and $c_{\alpha\beta}$ can exist only if the determinant

$$\begin{vmatrix} \rho - \nu_{\alpha\alpha} & -\nu_{\beta\alpha} \\ -\nu_{\alpha\beta} & \rho - \nu_{\beta\beta} \end{vmatrix} = \rho^2 - (\nu_{\alpha\alpha} + \nu_{\beta\beta})\rho + (\nu_{\alpha\alpha}\nu_{\beta\beta} - \nu_{\alpha\beta}\nu_{\beta\alpha}) = 0. \tag{2.16}$$

Equation (2.16) is called the *characteristic equation* of the differential equations (2.13). The roots

$$\begin{aligned} \rho_1 &= \tfrac{1}{2}[\nu_{\alpha\alpha} + \nu_{\beta\beta} + \sqrt{(\nu_{\alpha\alpha} - \nu_{\beta\beta})^2 + 4\nu_{\alpha\beta}\nu_{\beta\alpha}}] \\ \rho_2 &= \tfrac{1}{2}[\nu_{\alpha\alpha} + \nu_{\beta\beta} - \sqrt{(\nu_{\alpha\alpha} - \nu_{\beta\beta})^2 + 4\nu_{\alpha\beta}\nu_{\beta\alpha}}] \end{aligned} \tag{2.17}$$

of (2.16) are thus the only values of ρ for which the expressions (2.14) can be solutions of (2.13). Because the discriminant

$$(\nu_{\alpha\alpha} - \nu_{\beta\beta})^2 + 4\nu_{\alpha\beta}\nu_{\beta\alpha}$$

cannot be negative, the roots ρ_i are always real; furthermore, if

$$\nu_{\alpha\beta} > 0 \qquad \text{and} \qquad \nu_{\beta\alpha} > 0 \tag{2.18}$$

the discriminant is strictly positive and the roots are therefore distinct. We shall assume that (2.18) holds true, that is, that each state can be reached from the other state, and hence we shall consider only the case of distinct roots. From (2.5) and (2.17), we see that ρ_2 is strictly negative and ρ_1 is nonpositive. Closer examination of (2.17) in the light of (2.3), (2.5), and (2.6) shows that ρ_1 is also strictly negative.

We see from equations (2.15) that the coefficients $c_{\alpha\alpha i}$ and $c_{\alpha\beta i}$ corresponding to each root ρ_i are related as follows:

$$\frac{c_{\alpha\alpha i}}{c_{\alpha\beta i}} = \frac{\nu_{\beta\alpha}}{\rho_i - \nu_{\alpha\alpha}} = \frac{\rho_i - \nu_{\beta\beta}}{\nu_{\alpha\beta}}. \tag{2.19}$$

[Notice that the second equality is identical to equation (2.16).] Setting

$$k_i = \frac{c_{\alpha\alpha i}}{\rho_i - \nu_{\beta\beta}} = \frac{c_{\alpha\beta i}}{\nu_{\alpha\beta}} \tag{2.20}$$

we have, for each ρ_i, a pair of solutions

$$\begin{aligned} P_{\alpha\alpha}(\tau, t) &= k_i(\rho_i - \nu_{\beta\beta})e^{\rho_i t} \\ P_{\alpha\beta}(\tau, t) &= k_i \nu_{\alpha\beta} e^{\rho_i t} \end{aligned} \tag{2.21}$$

of the original differential equations (2.13). Because the two roots ρ_i are real and distinct, the general solution is

$$\begin{aligned} P_{\alpha\alpha}(\tau, t) &= \sum_{i=1}^{2} k_i(\rho_i - \nu_{\beta\beta})e^{\rho_i t} \\ P_{\alpha\beta}(\tau, t) &= \sum_{i=1}^{2} k_i \nu_{\alpha\beta} e^{\rho_i t}, \qquad \alpha \neq \beta;\, \alpha, \beta = 1, 2. \end{aligned} \tag{2.22}$$

In order to determine the constants k_i, we set $t = \tau$ in equation (2.22) and use the initial conditions (2.9) to write

$$\begin{aligned} P_{\alpha\alpha}(\tau, \tau) &= \sum_{i=1}^{2} k_i(\rho_i - \nu_{\beta\beta})e^{\rho_i \tau} = 1 \\ P_{\alpha\beta}(\tau, \tau) &= \sum_{i=1}^{2} k_i \nu_{\alpha\beta} e^{\rho_i \tau} = 0. \end{aligned} \tag{2.23}$$

Solving (2.23) for k_i gives

$$k_i = \frac{1}{\rho_i - \rho_j} e^{-\rho_i \tau}, \qquad i \neq j;\, i, j = 1, 2. \tag{2.24}$$

Substituting (2.24) in (2.22), we obtain the required formulas for the illness transition probabilities:

$$P_{\alpha\alpha}(\tau, t) = \sum_{i=1}^{2} \frac{\rho_i - \nu_{\beta\beta}}{\rho_i - \rho_j} e^{\rho_i(t-\tau)}$$
$$P_{\alpha\beta}(\tau, t) = \sum_{i=1}^{2} \frac{\nu_{\alpha\beta}}{\rho_i - \rho_j} e^{\rho_i(t-\tau)}. \tag{2.25}$$

The probabilities in (2.25) depend only on the difference $t - \tau$ but not on τ and t separately; thus the process is *homogeneous with respect to time.* We shall therefore let $\tau = 0$ and

$$t = t - \tau \tag{2.26}$$

be the interval length, and write

$$P_{\alpha\alpha}(\tau, t) = P_{\alpha\alpha}(t) = \sum_{i=1}^{2} \frac{\rho_i - \nu_{\beta\beta}}{\rho_i - \rho_j} e^{\rho_i t}$$
$$P_{\alpha\beta}(\tau, t) = P_{\alpha\beta}(t) = \sum_{i=1}^{2} \frac{\nu_{\alpha\beta}}{\rho_i - \rho_j} e^{\rho_i t}, \qquad j \neq i; \alpha \neq \beta; \alpha, \beta = 1, 2. \tag{2.27}$$

The *death transition probability*, which may now be written as $Q_{\alpha\delta}(t)$, has a definite relation to the illness transition probabilities. The relation can be established as follows: an individual in illness state S_α may reach the death state R_δ directly from S_α or by way of S_β, $\beta \neq \alpha$. Since an individual in R_δ at time t may have reached that state at any time prior to t, let us consider an infinitesimal time interval $(\tau, \tau + d\tau)$ for a fixed τ, $0 < \tau \leq t$. The probability that an individual in state S_α at time 0 will reach the state R_δ in the interval $(\tau, \tau + d\tau)$ is

$$P_{\alpha\alpha}(\tau)\mu_{\alpha\delta}\, d\tau + P_{\alpha\beta}(\tau)\mu_{\beta\delta}\, d\tau. \tag{2.28}$$

As τ varies over the interval $(0, t)$ the corresponding events, whose probabilities are given in (2.28), are mutually exclusive. Hence

$$Q_{\alpha\delta}(t) = \int_0^t P_{\alpha\alpha}(\tau)\mu_{\alpha\delta}\, d\tau + \int_0^t P_{\alpha\beta}(\tau)\mu_{\beta\delta}\, d\tau. \tag{2.29}$$

Now we substitute (2.27) in (2.29) and integrate the resulting expression to obtain the formula for the death transition probability,

$$Q_{\alpha\delta}(t) = \sum_{i=1}^{2} \frac{e^{\rho_i t} - 1}{\rho_i(\rho_i - \rho_j)} [(\rho_i - \nu_{\beta\beta})\mu_{\alpha\delta} + \nu_{\alpha\beta}\mu_{\beta\delta}]$$
$$j \neq i; \alpha \neq \beta; \alpha, \beta = 1, 2; \delta = 1, \ldots, r. \tag{2.30}$$

3. CHAPMAN-KOLMOGOROV EQUATIONS

The illness-death process, described in this chapter, is a Markov process in the sense that the future transitions of an individual are independent of transitions made in the past (see Definition 1 in Chapter 6). An important consequence of this property is the Chapman-Kolmogorov equation in (2.11). Since the process is homogeneous with respect to time, equation (2.11) may be rewritten as

$$P_{\alpha\alpha}(t) = P_{\alpha\alpha}(\tau)P_{\alpha\alpha}(t-\tau) + P_{\alpha\beta}(\tau)P_{\beta\alpha}(t-\tau) \tag{3.1a}$$

$$P_{\alpha\beta}(t) = P_{\alpha\alpha}(\tau)P_{\alpha\beta}(t-\tau) + P_{\alpha\beta}(\tau)P_{\beta\beta}(t-\tau) \tag{3.1b}$$

for $0 \le \tau \le t$, $\alpha \ne \beta$; $\alpha, \beta = 1, 2$. We now verify that the transition probabilities in (2.27) satisfy equation (3.1*a*), and leave the verification for (3.1*b*) to the reader. Substituting (2.27) in (3.1*a*) yields

$$\sum_{i=1}^{2} \frac{\rho_i - \nu_{\beta\beta}}{\rho_i - \rho_j} e^{\rho_i t} = \left(\sum_{i=1}^{2} \frac{\rho_i - \nu_{\beta\beta}}{\rho_i - \rho_j} e^{\rho_i \tau}\right)\left(\sum_{i=1}^{2} \frac{\rho_i - \nu_{\beta\beta}}{\rho_i - \rho_j} e^{\rho_i (t-\tau)}\right) + \left(\sum_{i=1}^{2} \frac{\nu_{\alpha\beta}}{\rho_i - \rho_j} e^{\rho_i \tau}\right)\left(\sum_{i=1}^{2} \frac{\nu_{\beta\alpha}}{\rho_i - \rho_j} e^{\rho_i (t-\tau)}\right). \tag{3.2}$$

The right side of (3.2) simplifies to

$$\sum_{i=1}^{2} \frac{(\rho_i - \nu_{\beta\beta})^2 + \nu_{\alpha\beta}\nu_{\beta\alpha}}{(\rho_i - \rho_j)^2} e^{\rho_i t} - \frac{(\rho_1 - \nu_{\beta\beta})(\rho_2 - \nu_{\beta\beta}) + \nu_{\alpha\beta}\nu_{\beta\alpha}}{(\rho_1 - \rho_2)^2}\left(\sum_{\substack{i=1 \\ j \ne i}}^{2} e^{\rho_i \tau + \rho_j (t-\tau)}\right). \tag{3.3}$$

The numerator of the first term in (3.3) may be rewritten

$$(\rho_i - \nu_{\beta\beta})^2 + \nu_{\alpha\beta}\nu_{\beta\alpha} = (\rho_i - \nu_{\beta\beta})(\rho_i - \rho_j), \tag{3.4}$$

whereas the second term vanishes, since

$$(\rho_1 - \nu_{\beta\beta})(\rho_2 - \nu_{\beta\beta}) + \nu_{\alpha\beta}\nu_{\beta\alpha} = 0. \tag{3.5}$$

When (3.4) and (3.5) are substituted in (3.3), we recover the left side of (3.2), verifying equation (3.2) and hence also equation (3.1*a*).

Chapman-Kolmogorov type equations may be established also for the transition probabilities leading to death. Consider an individual in illness state S_α at time 0 and the probability $Q_{\alpha\delta}(t)$ that he will be in death state R_δ at time t, and let τ be a fixed point in the interval $(0, t)$. The individual may reach state R_δ either prior to τ or after τ. The probability of the former event is $Q_{\alpha\delta}(\tau)$. In the latter case, he must be in either illness state

S_α or S_β at time τ and enter R_δ during the interval (τ, t); the corresponding probabilities are $P_{\alpha\alpha}(\tau)Q_{\alpha\delta}(t-\tau)$ and $P_{\alpha\beta}(\tau)Q_{\beta\delta}(t-\tau)$, respectively. Taking these possibilities together, we have the equation

$$Q_{\alpha\delta}(t) = Q_{\alpha\delta}(\tau) + P_{\alpha\alpha}(\tau)Q_{\alpha\delta}(t-\tau) + P_{\alpha\beta}(\tau)Q_{\beta\delta}(t-\tau), \quad \alpha \neq \beta;\ \alpha, \beta = 1, 2. \tag{3.6}$$

When expressed in terms of the transition probabilities in (2.27) and (2.30), equation (3.6) can be verified by direct computation.

4. EXPECTED DURATION IN ILLNESS AND DEATH

Another important concept is the average duration that an individual is expected to stay in each of the states $S_1, S_2, R_1, \ldots, R_r$, within a time period of length t. This duration depends, of course, on the initial state. For an individual in state S_α at time 0, let

$$e_{\alpha\beta}(t) = \text{the expected duration of stay in } S_\beta \text{ in the interval } (0, t), \qquad \beta = 1, 2, \tag{4.1}$$

$$\varepsilon_{\alpha\delta}(t) = \text{the expected duration of stay in } R_\delta \text{ in the interval } (0, t), \qquad \delta = 1, \ldots, r. \tag{4.2}$$

To obtain expressions for $e_{\alpha\beta}(t)$ and $\varepsilon_{\alpha\delta}(t)$, let an individual be in S_α at time 0 and, for each τ, $0 \le \tau \le t$, define indicator functions $I_{\alpha\beta}(\tau)$ and $J_{\alpha\delta}(\tau)$ so that

$$I_{\alpha\beta}(\tau) = \begin{cases} 1 & \text{if the individual is in } S_\beta \text{ at time } \tau \\ 0 & \text{otherwise} \end{cases} \tag{4.3}$$

and

$$J_{\alpha\delta}(\tau) = \begin{cases} 1 & \text{if the individual is in } R_\delta \text{ at time } \tau \\ 0 & \text{otherwise} \end{cases} \tag{4.4}$$

with the expectations

$$E[I_{\alpha\beta}(\tau)] = P_{\alpha\beta}(\tau) \tag{4.5}$$

$$E[J_{\alpha\delta}(\tau)] = Q_{\alpha\delta}(\tau). \tag{4.6}$$

It is easy to see that

$$e_{\alpha\beta}(t) = E\int_0^t I_{\alpha\beta}(\tau)\, d\tau \tag{4.7}$$

$$\varepsilon_{\alpha\delta}(t) = E\int_0^t J_{\alpha\delta}(\tau)\, d\tau. \tag{4.8}$$

Interchanging the expectation and the integral signs gives

$$e_{\alpha\beta}(t) = \int_0^t P_{\alpha\beta}(\tau)\,d\tau \tag{4.9}$$

and

$$\varepsilon_{\alpha\delta}(t) = \int_0^t Q_{\alpha\delta}(\tau)\,d\tau. \tag{4.10}$$

Using formulas (2.27) and (2.30) for the transition probabilities $P_{\alpha\beta}(\tau)$ and $Q_{\alpha\delta}(\tau)$, we obtain the explicit forms

$$e_{\alpha\alpha}(t) = \int_0^t \sum_{i=1}^{2} \frac{\rho_i - \nu_{\beta\beta}}{\rho_i - \rho_j} e^{\rho_i\tau}\,d\tau = \sum_{i=1}^{2} \frac{\rho_i - \nu_{\beta\beta}}{\rho_i(\rho_i - \rho_j)} (e^{\rho_i t} - 1) \tag{4.11}$$

$$e_{\alpha\beta}(t) = \int_0^t \sum_{i=1}^{2} \frac{\nu_{\alpha\beta}}{\rho_i - \rho_j} e^{\rho_i\tau}\,d\tau = \sum_{i=1}^{2} \frac{\nu_{\alpha\beta}}{\rho_i(\rho_i - \rho_j)} (e^{\rho_i t} - 1) \tag{4.12}$$

and

$$\varepsilon_{\alpha\delta}(t) = \int_0^t \sum_{i=1}^{2} \frac{e^{\rho_i\tau} - 1}{\rho_i(\rho_i - \rho_j)} [(\rho_i - \nu_{\beta\beta})\mu_{\alpha\delta} + \nu_{\alpha\beta}\mu_{\beta\delta}]\,d\tau$$

$$= \sum_{i=1}^{2} \left[\frac{1}{\rho_i}(e^{\rho_i t} - 1) - t\right] \frac{(\rho_i - \nu_{\beta\beta})\mu_{\alpha\delta} + \nu_{\alpha\beta}\mu_{\beta\delta}}{\rho_i(\rho_i - \rho_j)}. \tag{4.13}$$

The sum of the expected durations of stay over all states is equal to the entire length of the interval,

$$e_{\alpha 1}(t) + e_{\alpha 2}(t) + \varepsilon_{\alpha 1}(t) + \cdots + \varepsilon_{\alpha r}(t) = t,\ \alpha = 1, 2. \tag{4.14}$$

Proof of (4.14) is left to the reader.

5. POPULATION SIZES IN ILLNESS STATES AND DEATH STATES

An individual in S_α at time 0 must be either in one of the illness states or in one of the death states at time t; consequently the corresponding transition probabilities add to one, or

$$\sum_{\beta=1}^{2} P_{\alpha\beta}(t) + \sum_{\delta=1}^{r} Q_{\alpha\delta}(t) = 1. \tag{5.1}$$

Having derived explicit forms for the probabilities $P_{\alpha\beta}(t)$ and $Q_{\alpha\delta}(t)$, we can now verify equation (5.1) directly. Using (2.30) and (2.3), we have

$$\sum_{\delta=1}^{r} Q_{\alpha\delta}(t) = \sum_{i=1}^{2} \frac{e^{\rho_i t} - 1}{\rho_i(\rho_i - \rho_j)} \left[(\rho_i - \nu_{\beta\beta}) \sum_{\delta=1}^{r} \mu_{\alpha\delta} + \nu_{\alpha\beta} \sum_{\delta=1}^{r} \mu_{\beta\delta} \right]$$

$$= \sum_{i=1}^{2} \frac{e^{\rho_i t} - 1}{\rho_i(\rho_i - \rho_j)} [-(\rho_i - \nu_{\beta\beta})(\nu_{\alpha\alpha} + \nu_{\alpha\beta}) - \nu_{\alpha\beta}(\nu_{\beta\alpha} + \nu_{\beta\beta})]$$

$$= \sum_{i=1}^{2} \frac{e^{\rho_i t} - 1}{\rho_i(\rho_i - \rho_j)} [-\rho_i^2 + \rho_i \nu_{\beta\beta} - \rho_i \nu_{\alpha\beta}] \tag{5.2}$$

since ρ_i is a root of the characteristic equation (2.16). The last expression in (5.2) can be simplified to give

$$\sum_{\delta=1}^{r} Q_{\alpha\delta}(t) = 1 - \sum_{i=1}^{2} \frac{e^{\rho_i t}}{\rho_i - \rho_j} [(\rho_i - \nu_{\beta\beta}) + \nu_{\alpha\beta}]. \tag{5.3}$$

Combining (5.3) with (2.27), we recover equation (5.1).

Equation (5.1) may be used to derive the probability distribution of population sizes in states S_β and R_δ at any time t. At time $t = 0$, let there be $x_1(0)$ individuals in state S_1 and $x_2(0)$ individuals in state S_2, and let the sum

$$x(0) = x_1(0) + x_2(0) \tag{5.4}$$

be the initial size of the population. Suppose that the $x(0)$ individuals travel independently from one state to another, and that at the end of the interval $(0, t)$, $X_\beta(t)$ individuals are in illness state S_β and $Y_\delta(t)$ individuals are in state R_δ. Obviously

$$x(0) = X_1(t) + X_2(t) + Y_1(t) + \cdots + Y_r(t). \tag{5.5}$$

Each of the random variables on the right side of (5.5) is composed of two parts,

$$X_\beta(t) = X_{1\beta}(t) + X_{2\beta}(t), \qquad \beta = 1, 2, \tag{5.6}$$

where $X_{\alpha\beta}(t)$ is the number of those people in state S_β at time t who were in state S_α at time 0, and

$$Y_\delta(t) = Y_{1\delta}(t) + Y_{2\delta}(t), \qquad \delta = 1, \ldots, r, \tag{5.7}$$

where $Y_{\alpha\delta}(t)$ is the number of those in state R_δ at time t who were in state S_α at time 0. On the other hand, each of the $x_\alpha(0)$ people in S_α at time 0 must be in one of the illness states or in one of the death states at time t; therefore we have

$$x_\alpha(0) = X_{\alpha 1}(t) + X_{\alpha 2}(t) + Y_{\alpha 1}(t) + \cdots + Y_{\alpha r}(t). \tag{5.8}$$

When the summation in (5.8) is performed over $\alpha = 1, 2$, we get (5.5).

Table 1. Distribution of individuals in illness states S_β and death states R_δ at time t according to initial state S_α at time 0

State at Time 0	State at Time t			Row Totals (initial population sizes)
	S_1	S_2	$R_1 \cdots R_r$	
S_1	$X_{11}(t)$	$X_{12}(t)$	$Y_{11}(t) \cdots Y_{1r}(t)$	$x_1(0)$
S_2	$X_{21}(t)$	$X_{22}(t)$	$Y_{21}(t) \cdots Y_{2r}(t)$	$x_2(0)$
Column totals (population sizes at time t)	$X_1(t)$	$X_2(t)$	$Y_1(t) \cdots Y_r(t)$	$x(0)$

The population sizes of the states at time 0 and at time t are summarized in Table 1. Because of equation (5.1), for each α the random variables on the right side of (5.8) have a multinomial distribution with p.g.f. given by [see Equation (6.14), Chapter 2]:

$$E[s_1^{X_{\alpha 1}(t)} s_2^{X_{\alpha 2}(t)} z_1^{Y_{\alpha 1}(t)} \cdots z_r^{Y_{\alpha r}(t)} \mid x_\alpha(0)]$$
$$= [P_{\alpha 1}(t)s_1 + P_{\alpha 2}(t)s_2 + Q_{\alpha 1}(t)z_1 + \cdots + Q_{\alpha r}(t)z_r]^{x_\alpha(0)} \quad (5.9)$$

Therefore the p.g.f. of the joint probability distribution for the population sizes of all the states at time t is

$$E[s_1^{X_1(t)} s_2^{X_2(t)} z_1^{Y_1(t)} \cdots z_r^{Y_r(t)} \mid x_1(0), x_2(0)]$$
$$= \prod_{\alpha=1}^{2} [P_{\alpha 1}(t)s_1 + P_{\alpha 2}(t)s_2 + Q_{\alpha 1}(t)z_1 + \cdots + Q_{\alpha r}(t)z_r]^{x_\alpha(0)} \quad (5.10)$$

and the joint probabilities are

$$\Pr\{X_1(t) = x_1, X_2(t) = x_2, Y_1(t) = y_1, \ldots, Y_r(t) = y_r \mid x_1(0), x_2(0)\}$$
$$= \sum \prod_{\alpha=1}^{2} \frac{x_\alpha(0)!}{x_{\alpha 1}!\, x_{\alpha 2}!\, y_{\alpha 1}! \cdots y_{\alpha r}!} P_{\alpha 1}(t)^{x_{\alpha 1}} P_{\alpha 2}(t)^{x_{\alpha 2}} Q_{\alpha 1}(t)^{y_{\alpha 1}} \cdots Q_{\alpha r}(t)^{y_{\alpha r}}, \quad (5.11)$$

where the summation is taken over all possible values of $x_{\alpha\beta}$ and $y_{\alpha\delta}$, $\alpha, \beta = 1, 2$; $\delta = 1, \ldots, r$, so that

$$x_{1\beta} + x_{2\beta} = x_\beta, \qquad \beta = 1, 2,$$

and

$$y_{1\delta} + y_{2\delta} = y_\delta, \qquad \delta = 1, \ldots, r.$$

The expected number of individuals in each state at time t can be computed directly from (5.10),

$$E[X_\beta(t) \mid x_1(0), x_2(0)] = x_1(0)P_{1\beta}(t) + x_2(0)P_{2\beta}(t), \quad (5.12)$$

and

$$E[Y_\delta(t) \mid x_1(0), x_2(0)] = x_1(0)Q_{1\delta}(t) + x_2(0)Q_{2\delta}(t). \tag{5.13}$$

The corresponding variances are

$$\sigma^2_{X_\beta(t)} = \sum_{\alpha=1}^{2} x_\alpha(0)P_{\alpha\beta}(t)[1 - P_{\alpha\beta}(t)] \tag{5.14}$$

and

$$\sigma^2_{Y_\delta(t)} = \sum_{\alpha=1}^{2} x_\alpha(0)Q_{\alpha\delta}(t)[1 - Q_{\alpha\delta}(t)]. \tag{5.15}$$

The population becomes extinct at time t if $X_1(t) = 0$ and $X_2(t) = 0$. The probability of extinction at time t may be obtained from (5.10) by setting $s_1 = s_2 = 0$ and $z_1 = \cdots = z_r = 1$; then, by using equation (5.3), we have

$$\begin{aligned}\Pr\{X_1(t) = 0, X_2(t) = 0 \mid x_1(0), x_2(0)\} \\ &= \prod_{\alpha=1}^{2} [Q_{\alpha 1}(t) + \cdots + Q_{\alpha r}(t)]^{x_\alpha(0)} \\ &= \prod_{\alpha=1}^{2} \left\{1 - \sum_{i=1}^{2} \frac{e^{\rho_i t}}{\rho_i - \rho_j} [(\rho_i - \nu_{\beta\beta}) + \nu_{\alpha\beta}]\right\}^{x_\alpha(0)}. \end{aligned} \tag{5.16}$$

5.1. The Limiting Distribution

The roots ρ_i are negative; therefore, as the length t of the time interval approaches infinity, each of the illness transition probabilities in equation (2.27) approaches zero:

$$P_{\alpha\alpha}(\infty) = \lim_{t\to\infty} P_{\alpha\alpha}(t) = \lim_{t\to\infty} \sum_{i=1}^{2} \frac{\rho_i - \nu_{\beta\beta}}{\rho_i - \rho_j} e^{\rho_i t} = 0, \tag{5.17}$$

$$P_{\alpha\beta}(\infty) = \lim_{t\to\infty} P_{\alpha\beta}(t) = \lim_{t\to\infty} \sum_{i=1}^{2} \frac{\nu_{\alpha\beta}}{\rho_i - \rho_j} e^{\rho_i t} = 0,$$

$$\alpha \neq \beta;\ \alpha, \beta = 1, 2, \tag{5.18}$$

and each of the death transition probabilities in (2.30) approaches a constant:

$$\begin{aligned} Q_{\alpha\delta}(\infty) &= \lim_{t\to\infty} Q_{\alpha\delta}(t) = \lim_{t\to\infty} \sum_{i=1}^{2} \frac{e^{\rho_i t} - 1}{\rho_i(\rho_i - \rho_j)} [(\rho_i - \nu_{\beta\beta})\mu_{\alpha\delta} + \nu_{\alpha\beta}\mu_{\beta\delta}] \\ &= \sum_{i=1}^{2} \frac{-[(\rho_i - \nu_{\beta\beta})\mu_{\alpha\delta} + \nu_{\alpha\beta}\mu_{\beta\delta}]}{\rho_i(\rho_i - \rho_j)}, \qquad \alpha = 1, 2;\ \delta = 1, \ldots, r. \end{aligned} \tag{5.19}$$

It is easily seen from (5.2) and (5.3) that

$$\sum_{\delta=1}^{r} Q_{\alpha\delta}(\infty) = \sum_{\delta=1}^{r} \sum_{i=1}^{2} \frac{-[(\rho_i - \nu_{\beta\beta})\mu_{\alpha\delta} + \nu_{\alpha\beta}\mu_{\beta\delta}]}{\rho_i(\rho_i - \rho_j)} = 1. \tag{5.20}$$

The limiting form of the p.g.f. (5.10) as $t \to \infty$ is

$$\prod_{\alpha=1}^{2} [z_1 Q_{\alpha 1}(\infty) + \cdots + z_r Q_{\alpha r}(\infty)]^{x_\alpha(0)}. \tag{5.21}$$

Thus as $t \to \infty$, the initial populations $x_1(0)$ and $x_2(0)$ die out, the random variables $X_1(\infty)$ and $X_2(\infty)$ degenerate to the value 0, and the limiting distribution is the convolution of two independent multinomial distributions with probabilities $Q_{\alpha\delta}(\infty)$.

PROBLEMS

1. In a stochastic model for follow-up study of cancer patients, Fix and Neyman [1951] considered two illness states: S_1, the state of "leading normal life," and S_2, the state of being under treatment for cancer; and two death states, R_1, death from other causes, and R_2, death from cancer. For illustration, they suggested the following two sets of numerical values for illness and death intensities.

(a)	$\nu_{11} = -0.4$,	$\nu_{12} = 0.2$,	$\mu_{11} = 0.2$,	$\mu_{12} = 0$,
	$\nu_{21} = 2.7$,	$\nu_{22} = -3.7$,	$\mu_{21} = 0$,	$\mu_{22} = 1.0$;
(b)	$\nu_{11} = -0.5$,	$\nu_{12} = 0.4$,	$\mu_{11} = 0.1$,	$\mu_{12} = 0$,
	$\nu_{21} = 0.2$,	$\nu_{22} = -0.4$,	$\mu_{21} = 0$,	$\mu_{22} = 0.2$.

For each set of data, compute the transition probabilities $P_{\alpha\beta}(t)$ and $Q_{\alpha\delta}(t)$, and the expected durations of stay $e_{\alpha\beta}(t)$ and $\varepsilon_{\alpha\delta}(t)$, for $\alpha, \beta, \delta = 1, 2$ and $t = 1$. Explain briefly the meaning of these numerical results.

2. *Transition probabilities.* If, instead of (2.20), we use

$$k_i = \frac{c_{\alpha\alpha i}}{\nu_{\beta\alpha}} = \frac{c_{\alpha\beta i}}{\rho_i - \nu_{\alpha\alpha}} \tag{2.20a}$$

then the general solution becomes

$$\begin{aligned} P_{\alpha\alpha}(\tau, t) &= \sum_{i=1}^{2} k_i \nu_{\beta\alpha} e^{\rho_i t} \\ P_{\alpha\beta}(\tau, t) &= \sum_{i=1}^{2} k_i(\rho_i - \nu_{\alpha\alpha}) e^{\rho_i t} \end{aligned} \tag{2.22a}$$

Derive from (2.22a) the particular solution for the transition probabilities corresponding to the initial conditions (2.9).

3. *Continuation.* By taking the derivatives of the formulas on the right, verify that solution (2.25) satisfies the differential equations (2.13) and the initial condition (2.9).

4. *Chapman-Kolmogorov equation.* Verify that solution (2.27) satisfies the Chapman-Kolmogorov equation (3.1b).

5. Verify equations (3.4) and (3.5).

6. Substitute (2.27) and (2.30) for the transition probabilities $P_{\alpha\beta}(t)$ and $Q_{\alpha\delta}(t)$ in (3.6) and verify the resulting equation.

7. *An alternative model.* Consider an illness-death process whose structure is similar to the one described in this chapter except that transition of an individual is possible only from illness state S_1 to S_2 but not from S_2 to S_1, so that $\nu_{12} > 0$ and $\nu_{21} = 0$. Derive the transition probabilities $P_{12}(t)$, $Q_{1\delta}(t)$, and $Q_{2\delta}(t)$.

8. *Expected duration of stay.* Use (4.11), (4.12), and (4.13) to show that the sum of the expected durations of stay in the illness and death states during a time interval $(0, t)$ is equal to the length of the interval t.

9. *A two dimensional process.* The probability distribution of the population sizes of the illness states at a particular time t, as discussed in Section 5, can be derived with a different approach. Given the initial population sizes $x_1(0)$ and $x_2(0)$ at time 0, let $X_1(t)$ and $X_2(t)$ be the population sizes in S_1 and S_2 at time t with the probability distribution denoted by $P(x_1, x_2; t)$,

$$P(x_1, x_2; t) = \Pr\{X_1(t) = x_1, X_2(t) = x_2 \mid x_1(0), x_2(0)\}.$$

Verify the following system of differential equations.

$$\begin{aligned}\frac{d}{dt}P(x_1, x_2; t) &= P(x_1, x_2; t)(x_1\nu_{11} + x_2\nu_{22}) \\ &\quad + P(x_1 + 1, x_2 - 1; t)(x_1 + 1)\nu_{12} + P(x_1 - 1, x_2 + 1; t)(x_2 + 1)\nu_{21} \\ &\quad + P(x_1 + 1, x_2; t)(x_1 + 1)\mu_1 + P(x_1, x_2 + 1; t)(x_2 + 1)\mu_2\end{aligned}$$

where $\mu_\alpha = \mu_{\alpha 1} + \cdots + \mu_{\alpha r}$.

10. *Continuation.* Let

$$G_X(s_1, s_2; t) = E[s_1^{X_1(t)} s_2^{X_2(t)} \mid x_1(0), x_2(0)]$$

be the p.g.f. of the probability distribution in Problem 9 with the initial condition at $t = 0$

$$G_X(s_1, s_2; 0) = s_1^{x_1(0)} s_2^{x_2(0)}.$$

Establish the partial differential equation for the p.g.f. and derive the particular solution corresponding to the above initial condition.

11. *Continuation.* From the p.g.f. in Problem 10, derive formulas for the probability distribution $P(x_1, x_2; t)$, the expectations, variance, and covariance of $X_1(t)$ and $X_2(t)$. Compare your results with those given in (5.11), (5.12), and (5.14).

12. *Smoking Habit.* A cigarette smoker wants to give up smoking. When he has an urge to smoke during a time interval $(t, t + \Delta)$, the probability of his refraining is $\nu_{12}\Delta + 0(\Delta)$ and the probability of smoking is $1 + \nu_{11}\Delta + 0(\Delta)$, with $\nu_{11} = -\nu_{12}$. After having given up the habit, he has the desire to smoke again; the probability of his resuming the habit in a time interval $(t, t + \Delta)$ is $\nu_{21}\Delta + 0(\Delta)$ and of remaining a nonsmoker is $1 + \nu_{22}\Delta + 0(\Delta)$, with $\nu_{22} = -\nu_{21}$. The transition probability $P_{12}(t)$ then is the probability that a person who smokes at the beginning of time interval $(0, t)$ will be a nonsmoker at time t. Derive the formulas for the transition probabilities $P_{\alpha\beta}(t)$, for $\alpha, \beta = 1, 2$. (The difference should be noted between a nonsmoker and a smoker who is not smoking at a particular moment.)

13. *Continuation.* Show that $P_{\alpha 1}(t) + P_{\alpha 2}(t) = 1$, for $\alpha = 1, 2$.

14. *Continuation.* Given a person who smokes at the beginning of a time interval, find the expected durations of time that he will be a smoker and a nonsmoker during the interval $(0, t)$.

CHAPTER 5

Multiple Transitions in the Simple Illness-Death Process

1. INTRODUCTION

The transition probabilities $P_{\alpha\beta}(t)$ and $Q_{\alpha\delta}(t)$ presented in Chapter 4 describe the end results of the movement of an individual in a given time interval. For example, if we consider an individual in the healthy state S_1 at the beginning of the interval $(0, t)$, then $P_{11}(t)$ is the probability that he will be healthy at the end of the interval, but this tells us nothing about his health *during* the interval. He may become ill and recover a number of times during $(0, t)$, or he may remain healthy throughout the interval; the probability $P_{11}(t)$ does not enable us to distinguish one case from the other. The probabilities $P_{\alpha\beta}(t)$ and $Q_{\alpha\delta}(t)$ are also insufficient for describing an individual's health during an interval. In order to present a more detailed picture of the health condition of an individual throughout the entire interval, we need to study *multiple transitions:* the number of times that an individual leaves the initial state and the number of times that he enters another state. The first part of this chapter is devoted to deriving explicit formulas for the *multiple transition probabilities* and various related problems.

In the formulation of multiple transition probabilities, we are concerned with the number of transitions made in a given time interval. Alternatively, one may study the time required for making a given number of transitions, for example, how long it takes a patient to return to good health, or how long will it be before the occurrence of a second illness. Clearly in such investigations, time is a random variable. In the latter part of this chapter we shall present the probability density functions for the *multiple transition times* and problems associated with the multiple transition times.

2. MULTIPLE EXIT TRANSITION PROBABILITIES, $P_{\alpha\beta}^{(m)}(t)$

Consider again the time interval (τ, t) and an individual in state S_α at τ. We are now interested not only in which state S_β the individual occupies at t, but also in the number of transitions he undergoes to get to S_β from S_α during (τ, t). The *multiple exit transition probability* $P_{\alpha\beta}^{(m)}(\tau, t)$ is defined by

$$
\begin{aligned}
P_{\alpha\beta}^{(m)}(\tau, t) &= \Pr\{\text{an individual in } S_\alpha \text{ at time } \tau \text{ will } \textit{leave } S_\alpha \\
&\qquad m \text{ times during } (\tau, t) \text{ and will be in } S_\beta \text{ at time } t\} \\
&= \Pr\{M_{\alpha\beta}(\tau, t) = m\}, \qquad \alpha,\beta = 1, 2 \\
&\qquad\qquad\qquad\qquad\qquad m = 0, 1, \ldots. \qquad (2.1)
\end{aligned}
$$

The random variable $M_{\alpha\beta}(\tau, t)$, corresponding to the probabilities $P_{\alpha\beta}^{(m)}(\tau, t)$, is thus the number of times the individual *leaves* S_α for S_β before he reaches S_β at time t. Since the individual may not be in S_β at t, $M_{\alpha\beta}(\tau, t)$ is an improper random variable, as is demonstrated later in this section. We now derive an explicit solution for $P_{\alpha\beta}^{(m)}(\tau, t)$.

We first derive systems of differential equations for $P_{\alpha\beta}^{(m)}(\tau, t)$ by considering two contiguous time intervals (τ, t) and $(t, t + \Delta)$, and the probability that $M_{\alpha\beta}(\tau, t + \Delta) = m$. Direct enumeration yields the following system of differential equations:

$$
\begin{aligned}
\frac{\partial}{\partial t} P_{\alpha\alpha}^{(m)}(\tau, t) &= P_{\alpha\alpha}^{(m)}(\tau, t)\nu_{\alpha\alpha} + P_{\alpha\beta}^{(m)}(\tau, t)\nu_{\beta\alpha} \\
\frac{\partial}{\partial t} P_{\alpha\beta}^{(m)}(\tau, t) &= P_{\alpha\alpha}^{(m-1)}(\tau, t)\nu_{\alpha\beta} + P_{\alpha\beta}^{(m)}(\tau, t)\nu_{\beta\beta}, \qquad \alpha \neq \beta \\
&\qquad\qquad\qquad\qquad \alpha, \beta = 1, 2.
\end{aligned} \tag{2.2}
$$

Let us define a system of p.g.f.'s

$$
g_{\alpha\beta}(s; \tau, t) = \sum_{m=0}^{\infty} s^m P_{\alpha\beta}^{(m)}(\tau, t), \qquad \alpha, \beta = 1, 2. \tag{2.3}
$$

Taking derivatives of $g_{\alpha\beta}(s; \tau, t)$ in (2.3) with respect to t and substituting (2.2) in the resulting expressions, we obtain the differential equations for the p.g.f.'s

$$
\begin{aligned}
\frac{\partial}{\partial t} g_{\alpha\alpha}(s; \tau, t) &= g_{\alpha\alpha}(s; \tau, t)\nu_{\alpha\alpha} + g_{\alpha\beta}(s; \tau, t)\nu_{\beta\alpha} \\
\frac{\partial}{\partial t} g_{\alpha\beta}(s; \tau, t) &= g_{\alpha\alpha}(s; \tau, t)s\nu_{\alpha\beta} + g_{\alpha\beta}(s; \tau, t)\nu_{\beta\beta}, \qquad \alpha \neq \beta; \alpha, \beta = 1, 2.
\end{aligned} \tag{2.4}
$$

Since $P_{\alpha\alpha}^{(0)}(\tau, \tau) = 1$ and $P_{\alpha\beta}^{(m)}(\tau, \tau) = 0$ for $\alpha \neq \beta$ and all $m \geq 0$, the initial conditions of (2.4) are

$$g_{\alpha\alpha}(s; \tau, \tau) = 1, \qquad g_{\alpha\beta}(s; \tau, \tau) = 0, \qquad \alpha \neq \beta; \quad \alpha, \beta = 1, 2. \tag{2.5}$$

The differential equations in (2.4) can be solved in exactly the same way as those in (2.13), Chapter 4. Here the characteristic equation is

$$\begin{vmatrix} (\rho - \nu_{\alpha\alpha}) & -\nu_{\beta\alpha} \\ -s\nu_{\alpha\beta} & (\rho - \nu_{\beta\beta}) \end{vmatrix} = \rho^2 - (\nu_{\alpha\alpha} + \nu_{\beta\beta})\rho + (\nu_{\alpha\alpha}\nu_{\beta\beta} - \nu_{\alpha\beta}\nu_{\beta\alpha}s) = 0 \tag{2.6}$$

with the roots

$$\rho_1 = \tfrac{1}{2}[(\nu_{\alpha\alpha} + \nu_{\beta\beta}) + \sqrt{(\nu_{\alpha\alpha} - \nu_{\beta\beta})^2 + 4\nu_{\alpha\beta}\nu_{\beta\alpha}s}]$$

and

$$\rho_2 = \tfrac{1}{2}[(\nu_{\alpha\alpha} + \nu_{\beta\beta}) - \sqrt{(\nu_{\alpha\alpha} - \nu_{\beta\beta})^2 + 4\nu_{\alpha\beta}\nu_{\beta\alpha}s}]. \tag{2.7}$$

Notice that, except for the appearance of s, (2.6) and (2.7) are identical to the corresponding expressions in (2.16) and (2.17) in Chapter 4. The particular solution of (2.4) corresponding to the initial condition (2.5) turns out to be

$$g_{\alpha\alpha}(s; \tau, t) = \sum_{i=1}^{2} \frac{(\rho_i - \nu_{\beta\beta})}{\rho_i - \rho_j} e^{\rho_i(t-\tau)}$$

$$g_{\alpha\beta}(s; \tau, t) = \sum_{i=1}^{2} \frac{s\nu_{\alpha\beta}}{\rho_i - \rho_j} e^{\rho_i(t-\tau)}, \qquad j \neq i; \quad \alpha \neq \beta; \alpha, \beta = 1, 2. \tag{2.8}$$

When $s = 1$,

$$g_{\alpha\alpha}(1; \tau, t) = P_{\alpha\alpha}(\tau, t),$$

and

$$g_{\alpha\beta}(1; \tau, t) = P_{\alpha\beta}(\tau, t), \qquad \alpha \neq \beta; \alpha, \beta = 1, 2. \tag{2.9}$$

The p.g.f.'s in (2.8) depend only on the difference $(t - \tau)$ and thus the process is again homogeneous with respect to time, as we would anticipate from the transition probabilities $P_{\alpha\beta}(t)$. Once again we may let $\tau = 0$, and let $t = t - \tau$ be the length of the time interval in which the number of transitions from S_α to S_β is

$$M_{\alpha\beta}(t) = M_{\alpha\beta}(0, t) \tag{2.10}$$

with the multiple transition probabilities

$$P_{\alpha\beta}^{(m)}(t) = P_{\alpha\beta}^{(m)}(0, t) \tag{2.11}$$

and the corresponding p.g.f.'s

$$g_{\alpha\beta}(s; t) = g_{\alpha\beta}(s; 0, t). \tag{2.12}$$

The equations (2.8) then reduce to

$$g_{\alpha\alpha}(s; t) = \sum_{i=1}^{2} \frac{(\rho_i - \nu_{\beta\beta})}{\rho_i - \rho_j} e^{\rho_i t},$$
$$g_{\alpha\beta}(s; t) = \sum_{i=1}^{2} \frac{s\nu_{\alpha\beta}}{\rho_i - \rho_j} e^{\rho_i t}, \qquad j \neq i;\ \alpha \neq \beta;\ \alpha, \beta = 1, 2. \tag{2.13}$$

The p.g.f.'s in (2.13) have s under all the square root signs in ρ_i; therefore it is not practical to derive the multiple transition probabilities by taking derivatives,

$$P_{\alpha\beta}^{(m)}(t) = \frac{1}{m!} \frac{\partial^m}{\partial s^m} g_{\alpha\beta}(s; t)\bigg|_{s=0}. \tag{2.14}$$

A simpler way to obtain the probabilities $P_{\alpha\beta}^{(m)}(t)$ is to expand the right side of each of the equations (2.13) as a power series in s. We let

$$\sqrt{\cdot} = \sqrt{(\nu_{\alpha\alpha} - \nu_{\beta\beta})^2 + 4\nu_{\alpha\beta}\nu_{\beta\alpha}s}, \tag{2.15}$$

and use (2.7) to rewrite the first equation of (2.13) as

$$g_{\alpha\alpha}(s; t) = e^{(t/2)(\nu_{\alpha\alpha}+\nu_{\beta\beta})}\left[\frac{\nu_{\alpha\alpha} - \nu_{\beta\beta}}{2\sqrt{\cdot}}\left(e^{(t/2)\sqrt{\cdot}} - e^{-(t/2)\sqrt{\cdot}}\right) + \tfrac{1}{2}\left(e^{(t/2)\sqrt{\cdot}} + e^{-(t/2)\sqrt{\cdot}}\right)\right]. \tag{2.16}$$

Expanding the exponentials in power series, we find that

$$\frac{\nu_{\alpha\alpha} - \nu_{\beta\beta}}{2\sqrt{\cdot}}\left(e^{(t/2)\sqrt{\cdot}} - e^{-(t/2)\sqrt{\cdot}}\right) = \sum_{k=0}^{\infty} \frac{\nu_{\alpha\alpha} - \nu_{\beta\beta}}{(2k+1)!}\left(\frac{t}{2}\right)^{2k+1} (\sqrt{\cdot})^{2k} \tag{2.17}$$

and

$$\tfrac{1}{2}\left(e^{(t/2)\sqrt{\cdot}} + e^{-(t/2)\sqrt{\cdot}}\right) = \sum_{k=0}^{\infty} \frac{1}{(2k)!}\left(\frac{t}{2}\right)^{2k} (\sqrt{\cdot})^{2k}. \tag{2.18}$$

Using (2.15) and substituting (2.17) and (2.18) in (2.16), we have

$$g_{\alpha\alpha}(s; t) = e^{(t/2)(\nu_{\alpha\alpha}+\nu_{\beta\beta})} \sum_{k=0}^{\infty} \left[\frac{\nu_{\alpha\alpha} - \nu_{\beta\beta}}{(2k+1)!}\left(\frac{t}{2}\right)^{2k+1} + \frac{1}{(2k)!}\left(\frac{t}{2}\right)^{2k}\right] \times [(\nu_{\alpha\alpha} - \nu_{\beta\beta})^2 + 4\nu_{\alpha\beta}\nu_{\beta\alpha}s]^k. \tag{2.19}$$

Expanding the last binomial function in (2.19) and interchanging summation signs yield

$$g_{\alpha\alpha}(s;t) = \sum_{m=0}^{\infty} (4\nu_{\alpha\beta}\nu_{\beta\alpha})^m e^{(t/2)(\nu_{\alpha\alpha}+\nu_{\beta\beta})}$$

$$\times \left\{\sum_{k=m}^{\infty} \binom{k}{m} (\nu_{\alpha\alpha} - \nu_{\beta\beta})^{2(k-m)} \left[\frac{\nu_{\alpha\alpha} - \nu_{\beta\beta}}{(2k+1)!}\left(\frac{t}{2}\right)^{2k+1} + \frac{1}{(2k)!}\left(\frac{t}{2}\right)^{2k}\right]\right\} s^m,$$

$$\alpha \neq \beta;\ \beta = 1, 2. \quad (2.20)$$

The probability $P_{\alpha\alpha}^{(m)}(t)$ is the coefficient of s^m in (2.20),

$$P_{\alpha\alpha}^{(m)}(t) = (4\nu_{\alpha\beta}\nu_{\beta\alpha})^m e^{(t/2)(\nu_{\alpha\alpha}+\nu_{\beta\beta})}$$

$$\left\{\sum_{k=m}^{\infty} \binom{k}{m} (\nu_{\alpha\alpha} - \nu_{\beta\beta})^{2(k-m)} \times \left[\frac{\nu_{\alpha\alpha} - \nu_{\beta\beta}}{(2k+1)!}\left(\frac{t}{2}\right)^{2k+1} + \frac{1}{(2k)!}\left(\frac{t}{2}\right)^{2k}\right]\right\},$$

$$m = 0, 1, \ldots;\ \alpha \neq \beta;\ \alpha, \beta = 1, 2. \quad (2.21)$$

For $m = 0$, (2.21) is the probability that an individual in S_α will remain there for a time period of length t,

$$P_{\alpha\alpha}^{(0)}(t) = e^{(t/2)(\nu_{\alpha\alpha}+\nu_{\beta\beta})} \sum_{k=0}^{\infty} (\nu_{\alpha\alpha} - \nu_{\beta\beta})^{2k} \left[\frac{\nu_{\alpha\alpha} - \nu_{\beta\beta}}{(2k+1)!}\left(\frac{t}{2}\right)^{2k+1} + \frac{1}{(2k)!}\left(\frac{t}{2}\right)^{2k}\right]$$

$$= e^{\nu_{\alpha\alpha}t}. \quad (2.22)$$

This expression can also be derived directly from (2.13) by evaluating $g_{\alpha\alpha}(s;t)$ at $s = 0$. When $m = 1$, we have *the probability of first return* to the original state within a time interval of length t,

$$P_{\alpha\alpha}^{(1)}(t) = \frac{\nu_{\alpha\beta}\nu_{\beta\alpha}}{\nu_{\alpha\alpha} - \nu_{\beta\beta}}\left[te^{\nu_{\alpha\alpha}t} - \frac{1}{\nu_{\alpha\alpha} - \nu_{\beta\beta}}(e^{\nu_{\alpha\alpha}t} - e^{\nu_{\beta\beta}t})\right]. \quad (2.23)$$

The second equation of (2.13) also can be expressed in a power series of s. Using the same argument as in deriving (2.20), we obtain an explicit solution for the probability

$$P_{\alpha\beta}^{(m)}(t) = 2\nu_{\alpha\beta}(4\nu_{\alpha\beta}\nu_{\beta\alpha})^{m-1} e^{(t/2)(\nu_{\alpha\alpha}+\nu_{\beta\beta})}$$

$$\times \left[\sum_{k=m-1}^{\infty} \binom{k}{m-1} (\nu_{\alpha\alpha} - \nu_{\beta\beta})^{2(k-m+1)} \frac{1}{(2k+1)!}\left(\frac{t}{2}\right)^{2k+1}\right],$$

$$m = 1, 2, \ldots;\ \alpha \neq \beta;\ \alpha, \beta = 1, 2. \quad (2.24)$$

For $m = 1$ we have the *first passage probability* that exactly one transition will take place from S_α to S_β during a time period of length t,

$$P_{\alpha\beta}^{(1)}(t) = 2\nu_{\alpha\beta}e^{(t/2)(\nu_{\alpha\alpha}+\nu_{\beta\beta})}\sum_{k=0}^{\infty}(\nu_{\alpha\alpha} - \nu_{\beta\beta})^{2k}\frac{1}{(2k+1)!}\left(\frac{t}{2}\right)^{2k+1}$$

$$= \frac{\nu_{\alpha\beta}}{\nu_{\alpha\alpha} - \nu_{\beta\beta}}(e^{\nu_{\alpha\alpha}t} - e^{\nu_{\beta\beta}t}). \tag{2.25}$$

It should be noted that, for fixed α, β and t, the sum of the probabilities $P_{\alpha\beta}^{(m)}(t)$ over all possible values of m is less than one. In fact, using (2.17) and (2.27) in Chapter 4 and (2.21) above, the sum

$$\sum_{m=0}^{\infty} P_{\alpha\alpha}^{(m)}(t) = \tfrac{1}{2}e^{(t/2)(\nu_{\alpha\alpha}+\nu_{\beta\beta})}\left[\left(1 + \frac{\nu_{\alpha\alpha} - \nu_{\beta\beta}}{\sqrt{}}\right)e^{(t/2)\sqrt{}} + \left(1 - \frac{\nu_{\alpha\alpha} - \nu_{\beta\beta}}{\sqrt{}}\right)e^{-(t/2)\sqrt{}}\right]$$

$$= P_{\alpha\alpha}(t), \qquad \alpha \neq \beta;\ \alpha, \beta = 1, 2 \tag{2.26}$$

with

$$\sqrt{} = \sqrt{(\nu_{\alpha\alpha} - \nu_{\beta\beta})^2 + 4\nu_{\alpha\beta}\nu_{\beta\alpha}}, \tag{2.27}$$

is the probability that an individual in S_α at $t = 0$ will be in S_α at t. Similarly, the sum

$$\sum_{m=1}^{\infty} P_{\alpha\beta}^{(m)}(t) = \frac{\nu_{\alpha\beta}}{\sqrt{}}e^{(t/2)(\nu_{\alpha\alpha}+\nu_{\beta\beta})}[e^{(t/2)\sqrt{}} - e^{-(t/2)\sqrt{}}] = P_{\alpha\beta}(t),$$

$$\alpha \neq \beta;\ \alpha, \beta = 1, 2. \tag{2.28}$$

is the probability that an individual in S_α at time 0 will be in S_β at time t. Because of (5.1) in Chapter 4, both of these probabilities are less than one. Thus there is a positive probability that the random variables $M_{\alpha\beta}(t)$ will not take on any value at all and hence, as noted earlier, they are improper random variables. Since an individual in state S_α at time 0 may be in S_α or in S_β at time t, or may even have left the population through death prior to time t, it is not surprising that the probabilities (2.26) and (2.28) (and their sum) are less than one.

2.1. Conditional Distribution of the Number of Transitions

If we consider only those journeys having a specified ending state as well as a specified starting state, then the corresponding numbers of transitions are proper random variables. Let $M_{\alpha\beta}^*(t)$ be the number of transitions that a person makes from S_α to S_β during $(0, t)$, given that he starts in S_α at time 0 and ends in S_β at time t, and let

$$\Pr\{M_{\alpha\beta}^*(t) = m\} = P_{\alpha\beta}^{*(m)}(t), \qquad m = 0, 1, \ldots \tag{2.29}$$

be the corresponding probability. Because of the assumption that the individual is in S_β at time t, (2.29) is given by the conditional probability

$$P_{\alpha\beta}^{*(m)}(t) = \frac{P_{\alpha\beta}^{(m)}(t)}{P_{\alpha\beta}(t)} \tag{2.30}$$

with

$$\sum_{m=0}^{\infty} P_{\alpha\beta}^{*(m)}(t) = 1. \tag{2.31}$$

Thus $M_{\alpha\beta}^*(t)$ is a proper random variable.

The probability $P_{\alpha\beta}^{*(m)}(t)$ is the theoretical equivalent of the proportion of people starting in S_α at 0 and ending in S_β at t who make m transitions from S_α to S_β. The sum of the proportions for all possible values of m must be equal to one, as indicated in (2.31). It is meaningful here to determine the expected value and variance of $M_{\alpha\beta}^*(t)$. They are most easily obtained from the corresponding p.g.f.'s in (2.13). For example, the expected numbers are

$$\begin{aligned} E[M_{\alpha\alpha}^*(t)] &= \frac{1}{P_{\alpha\alpha}(t)} \frac{\partial}{\partial s} g_{\alpha\alpha}(s; t)\bigg|_{s=1} \\ &= \frac{\nu_{\alpha\beta}\nu_{\beta\alpha} t}{\rho_1 - \rho_2} + \frac{\nu_{\alpha\beta}\nu_{\beta\alpha}(\nu_{\alpha\alpha} - \nu_{\beta\beta})}{P_{\alpha\alpha}(t)(\rho_1 - \rho_2)^2} \\ &\quad \times \left\{\left[1 - \frac{\rho_1 - \rho_2}{(\nu_{\alpha\alpha} - \nu_{\beta\beta})}\right] t e^{\rho_2 t} - \frac{e^{\rho_1 t} - e^{\rho_2 t}}{\rho_1 - \rho_2}\right\} \end{aligned} \tag{2.32}$$

and

$$E[M_{\alpha\beta}^*(t)] = 1 + \frac{\nu_{\alpha\beta}\nu_{\beta\alpha} t}{\rho_1 - \rho_2}\left(\frac{e^{\rho_1 t} + e^{\rho_2 t}}{e^{\rho_1 t} - e^{\rho_2 t}}\right) - \frac{2\nu_{\alpha\beta}\nu_{\beta\alpha}}{(\rho_1 - \rho_2)^2} \tag{2.33}$$

where ρ_1 and ρ_2 are given in (2.7) with $s = 1$, and $P_{\alpha\alpha}(t)$ in (2.26).

3. MULTIPLE TRANSITION PROBABILITIES LEADING TO DEATH, $Q_{\alpha\delta}^{(m)}(t)$

Since the processes we are considering are homogeneous with respect to time, we shall simply consider a time interval $(0, t)$ and the transitions during that interval by an individual who starts in S_α at time 0. The probabilities for multiple transitions leading to death are defined by

$$Q_{\alpha\delta}^{(m)}(t) = \Pr\{\text{an individual in } S_\alpha \text{ at time 0 will leave } S_\alpha \; m \text{ times during } (0, t) \text{ and will be in } R_\delta \text{ at } t\}. \tag{3.1}$$

We shall denote by $D_{\alpha\delta}(t)$ the corresponding random variable, which is the number of times the individual leaves S_α before reaching R_δ sometime during $(0, t)$, so that

$$Q_{\alpha\delta}^{(m)}(t) = \Pr\{D_{\alpha\delta}(t) = m\}. \tag{3.2}$$

An explicit expression for $Q_{\alpha\delta}^{(m)}(t)$ can be obtained from the multiple illness transition probabilities presented in Section 2. Using the same argument as in deriving $Q_{\alpha\delta}(t)$, we have

$$\begin{aligned} Q_{\alpha\delta}^{(m)}(t) &= \int_0^t P_{\alpha\alpha}^{(m-1)}(\tau)\mu_{\alpha\delta}\,d\tau + \int_0^t P_{\alpha\beta}^{(m)}(\tau)\mu_{\beta\delta}\,d\tau \\ &= Q_{\alpha\delta.\alpha}^{(m)}(t) + Q_{\alpha\delta.\beta}^{(m)}(t), \\ &\qquad m = 1, 2, \ldots;\ \alpha \neq \beta;\ \alpha, \beta = 1, 2;\ \delta = 1, \ldots, r. \end{aligned} \tag{3.3}$$

The first term on the right side is the probability of the multiple transition leading to R_δ from the original state S_α, and the second term is the probability of the transition to R_δ by way of another state S_β. Substituting (2.21) in the first term, we have

$$\begin{aligned} Q_{\alpha\delta.\alpha}^{(m)}(t) &= \int_0^t P_{\alpha\alpha}^{(m-1)}(\tau)\mu_{\alpha\delta}\,d\tau \\ &= (4\nu_{\alpha\beta}\nu_{\beta\alpha})^{m-1}\mu_{\alpha\delta}\sum_{k=m-1}^{\infty}\binom{k}{m-1}(\nu_{\alpha\alpha}-\nu_{\beta\beta})^{2(k-m+1)} \\ &\quad \times \left[\frac{\nu_{\alpha\alpha}-\nu_{\beta\beta}}{(2k+1)!}\int_0^t\left(\frac{\tau}{2}\right)^{2k+1} e^{(\tau/2)(\nu_{\alpha\alpha}+\nu_{\beta\beta})}\,d\tau \right. \\ &\qquad \left. + \frac{1}{(2k)!}\int_0^t \left(\frac{\tau}{2}\right)^{2k} e^{(\tau/2)(\nu_{\alpha\alpha}+\nu_{\beta\beta})}\,d\tau\right]. \end{aligned} \tag{3.4}$$

Here the integrals, which are incomplete gamma functions, may be evaluated directly:

$$\int_0^t \left(\frac{\tau}{2}\right)^k e^{(\tau/2)(\nu_{\alpha\alpha}+\nu_{\beta\beta})}\,d\tau = (-1)^{k+1}\frac{2}{(\nu_{\alpha\alpha}+\nu_{\beta\beta})^{k+1}}\,k!\left(1-\sum_{l=0}^{k}\frac{\theta^l}{l!}e^{-\theta}\right), \tag{3.5}$$

where

$$\theta = -\frac{t}{2}(\nu_{\alpha\alpha}+\nu_{\beta\beta}). \tag{3.6}$$

Therefore

$$\begin{aligned} Q_{\alpha\delta.\alpha}^{(m)}(t) &= 2(4\nu_{\alpha\beta}\nu_{\beta\alpha})^{m-1}\sum_{k=m-1}^{\infty}\binom{k}{m-1}\frac{(\nu_{\alpha\alpha}-\nu_{\beta\beta})^{2(k-m+1)}}{(\nu_{\alpha\alpha}+\nu_{\beta\beta})^{2(k+1)}} \\ &\quad \times\left[-2\nu_{\beta\beta}\mu_{\alpha\delta}\left(1-\sum_{l=0}^{2k+1}\frac{\theta^l}{l!}e^{-\theta}\right) - (\nu_{\alpha\alpha}+\nu_{\beta\beta})\mu_{\alpha\delta}\frac{\theta^{2k+1}}{(2k+1)!}e^{-\theta}\right]. \end{aligned} \tag{3.7}$$

Similarly, substituting formula (2.24) in the last term in (3.3) and integrating give

$$Q_{\alpha\delta.\beta}^{(m)}(t) = \int_0^t P_{\alpha\beta}^{(m)}(\tau)\mu_{\beta\delta}\, d\tau$$

$$= 2(4\nu_{\alpha\beta}\nu_{\beta\alpha})^{m-1} \sum_{k=m-1}^{\infty} \binom{k}{m-1} \frac{(\nu_{\alpha\alpha} - \nu_{\beta\beta})^{2(k-m+1)}}{(\nu_{\alpha\alpha} + \nu_{\beta\beta})^{2(k+1)}} 2\nu_{\alpha\beta}\mu_{\beta\delta}$$

$$\times \left(1 - \sum_{l=0}^{2k+1} \frac{\theta^l}{l!} e^{-\theta}\right). \tag{3.8}$$

The probability $Q_{\alpha\delta}^{(m)}(t)$ is obtained by substituting (3.7) and (3.8) in (3.3)

$$Q_{\alpha\delta}^{(m)}(t) = 2(4\nu_{\alpha\beta}\nu_{\beta\alpha})^{m-1} \sum_{k=m-1}^{\infty} \binom{k}{m-1} \frac{(\nu_{\alpha\alpha} - \nu_{\beta\beta})^{2(k-m+1)}}{(\nu_{\alpha\alpha} + \nu_{\beta\beta})^{2(k+1)}}$$

$$\times \left[2(\nu_{\alpha\beta}\mu_{\beta\delta} - \nu_{\beta\beta}\mu_{\alpha\delta}) \left(1 - \sum_{l=0}^{2k+1} \frac{\theta^l}{l!} e^{-\theta}\right) \right.$$

$$\left. - (\nu_{\alpha\alpha} + \nu_{\beta\beta})\mu_{\alpha\delta} \frac{\theta^{2k+1}}{(2k+1)!} e^{-\theta} \right],$$

$$m = 1, 2, \ldots ; \alpha \neq \beta; \alpha, \beta = 1, 2; \delta = 1, \ldots, r. \tag{3.9}$$

It should be noted that the random variables $D_{\alpha\delta}(t)$ also are improper and that

$$\sum_{m=1}^{\infty} Q_{\alpha\delta}^{(m)}(t) = Q_{\alpha\delta}(t) < 1. \tag{3.10}$$

Although the formulas for the probabilities $Q_{\alpha\delta.\alpha}^{(m)}(t)$, $Q_{\alpha\delta.\beta}^{(m)}(t)$, and $Q_{\alpha\delta}^{(m)}(t)$ are complicated, the corresponding p.g.f.'s take on simple forms. Let us consider the p.g.f.'s

$$h_{\alpha\delta.\alpha}(s; t) = \sum_{m=1}^{\infty} s^m Q_{\alpha\delta.\alpha}^{(m)}(t) \tag{3.11}$$

$$h_{\alpha\delta.\beta}(s; t) = \sum_{m=1}^{\infty} s^m Q_{\alpha\delta.\beta}^{(m)}(t) \tag{3.12}$$

and

$$h_{\alpha\delta}(t) = h_{\alpha\delta.\alpha}(s; t) + h_{\alpha\delta.\beta}(s; t). \tag{3.13}$$

From (3.4) we see that

$$h_{\alpha\delta.\alpha}(s;t) = \sum_{m=1}^{\infty} s^m \int_0^t P_{\alpha\alpha}^{(m-1)}(\tau)\mu_{\alpha\delta}\, d\tau$$

$$= \int_0^t \sum_{m=1}^{\infty} s^m P_{\alpha\alpha}^{(m-1)}(\tau)\mu_{\alpha\delta}\, d\tau$$

$$= \int_0^t g_{\alpha\alpha}(s;\tau)s\mu_{\alpha\delta}\, d\tau; \tag{3.14}$$

and similarly

$$h_{\alpha\delta.\beta}(s;t) = \int_0^t g_{\alpha\beta}(s;\tau)\mu_{\beta\delta}\, d\tau. \tag{3.15}$$

Using the explicit forms of the p.g.f's $g_{\alpha\beta}(s;t)$ in (2.13), (3.14) and (3.15) become

$$h_{\alpha\delta.\alpha}(s;t) = s\int_0^t \sum_{i=1}^{2} \frac{\rho_i - \nu_{\beta\beta}}{\rho_i - \rho_j} e^{\rho_i\tau}\mu_{\alpha\delta}\, d\tau$$

$$= \sum_{i=1}^{2} \frac{s(\rho_i - \nu_{\beta\beta})}{\rho_i(\rho_i - \rho_j)}(e^{\rho_i t} - 1)\mu_{\alpha\delta} \tag{3.16}$$

and

$$h_{\alpha\delta.\beta}(s;t) = \int_0^t \sum_{i=1}^{2} \frac{s\nu_{\alpha\beta}}{\rho_i - \rho_j} e^{\rho_i\tau}\mu_{\beta\delta}\, d\tau$$

$$= \sum_{i=1}^{2} \frac{s\nu_{\alpha\beta}}{\rho_i(\rho_i - \rho_j)}(e^{\rho_i t} - 1)\mu_{\beta\delta}, \tag{3.17}$$

where the formulas for ρ_i are given in (2.7). Substituting (3.16) and (3.17) in (3.13)

$$h_{\alpha\delta}(s;t) = \sum_{i=1}^{2} \frac{s(\rho_i - \nu_{\beta\beta})}{\rho_i(\rho_i - \rho_j)}(e^{\rho_i t} - 1)\mu_{\alpha\delta} + \sum_{i=1}^{2} \frac{s\nu_{\alpha\beta}}{\rho_i(\rho_i - \rho_j)}(e^{\rho_i t} - 1)\mu_{\beta\delta},$$

$$i \neq j;\ \alpha \neq \beta;\ i, j, \alpha, \beta = 1, 2;\ \delta = 1, \ldots, r. \tag{3.18}$$

Formulas (3.16), (3.17), and (3.18) can be used to compute the expectations and variances of the corresponding "properized" random variables. We illustrate this with an example. Let $D^*_{\alpha\delta.\beta}(t)$ be the number of times that a person leaves state S_α, given that he is in S_α at time 0 and enters R_δ by way of S_β during the time interval $(0, t)$. If S_α denotes the state of health and S_β the state of illness, then $D_{\alpha\delta.\beta}(t)$ is the number of times a person becomes ill before dying from an illness in $(0, t)$. The corresponding probability is

$$\Pr\{D^*_{\alpha\delta.\beta}(t) = m\} = \frac{Q^{(m)}_{\alpha\delta.\beta}(t)}{Q_{\alpha\delta.\beta}(t)}, \qquad m = 1, 2, \ldots, \tag{3.19}$$

where

$$Q_{\alpha\delta.\beta}(t) = \sum_{m=1}^{\infty} Q_{\alpha\delta.\beta}^{(m)}(t) = h_{\alpha\delta.\beta}(1, t)$$

$$= \sum_{i=1}^{2} \frac{\nu_{\alpha\beta}}{\rho_i(\rho_i - \rho_j)} (e^{\rho_i t} - 1)\mu_{\beta\delta} \tag{3.20}$$

with ρ_i as given in (2.7) for $s = 1$. The expectation of $D_{\alpha\delta.\beta}^*(t)$ can be computed as follows:

$$E[D_{\alpha\delta.\beta}^*(t)] = \frac{1}{Q_{\alpha\delta.\beta}(t)} \frac{\partial}{\partial s} h_{\alpha\delta.\beta}(t) \bigg|_{s=1}$$

$$= \frac{\mu_{\beta\delta}}{Q_{\alpha\delta.\beta}(t)} \sum_{i=1}^{2} \left\{ \frac{\nu_{\alpha\beta}}{\rho_i(\rho_i - \rho_j)^3} [(\rho_i - \rho_j)^2 - 2\nu_{\alpha\beta}\nu_{\beta\alpha}](e^{\rho_i t} - 1) \right.$$

$$\left. + \frac{\nu_{\alpha\beta}^2\, \nu_{\beta\alpha}}{(\rho_1 - \rho_2)^2} \left[\frac{t}{\rho_i} e^{\rho_i t} - \frac{1}{\rho_i^2} (e^{\rho_i t} - 1) \right] \right\}. \tag{3.21}$$

4. CHAPMAN-KOLMOGOROV EQUATIONS

In deriving differential equations (2.2) for the multiple transition probabilities, we have implicitly assumed that the future transitions of an individual are stochastically independent of his past transitions. Since the process is homogeneous with respect to time, this assumption implies that, for $0 \le \tau \le t$,

$$P_{\alpha\alpha}^{(m)}(t) = \sum_{l=0}^{m} P_{\alpha\alpha}^{(l)}(\tau) P_{\alpha\alpha}^{(m-l)}(t - \tau) + \sum_{l=1}^{m} P_{\alpha\beta}^{(l)}(\tau) P_{\beta\alpha}^{(m-l+1)}(t - \tau), \tag{4.1}$$

and

$$P_{\alpha\beta}^{(m)}(t) = \sum_{l=0}^{m-1} P_{\alpha\alpha}^{(l)}(\tau) P_{\alpha\beta}^{(m-l)}(t - \tau) + \sum_{l=1}^{m} P_{\alpha\beta}^{(l)}(\tau) P_{\beta\beta}^{(m-l)}(t - \tau) \tag{4.2}$$

for $\beta \ne \alpha$; $\alpha, \beta = 1, 2$; $m = 1, 2, \ldots$. We shall now verify that the solutions $P_{\alpha\beta}^{(m)}(t)$, obtained in Section 2, actually satisfy (4.1) and (4.2).

To verify (4.1), we notice that the sequence in the first term

$$\sum_{l=0}^{m} P_{\alpha\alpha}^{(l)}(\tau) P_{\alpha\alpha}^{(m-l)}(t - \tau) \tag{4.3}$$

on the right side of the equality sign is the convolution of the two sequences $\{P_{\alpha\alpha}^{(m)}(\tau)\}$ and $\{P_{\alpha\alpha}^{(m)}(t - \tau)\}$; and thus the p.g.f. of (4.3) is the product of the p.g.f.'s of the two component sequences $\{P_{\alpha\alpha}^{(m)}(\tau)\}$ and $\{P_{\alpha\alpha}^{(m)}(t - \tau)\}$, or

$$g_{\alpha\alpha}(s; \tau) g_{\alpha\alpha}(s; t - \tau). \tag{4.4}$$

Similarly, the p.g.f. of the sequence in the second term

$$\sum_{l=1}^{m} P_{\alpha\beta}^{(l)}(\tau)P_{\beta\alpha}^{(m-l+1)}(t-\tau) \tag{4.5}$$

is the product of the p.g.f.'s of the two component sequences $\{P_{\alpha\beta}^{(m)}(\tau)\}$ and $\{P_{\beta\alpha}^{(m)}(t-\tau)\}$. Making adjustment for the power of s, the p.g.f. of (4.5) is equal to

$$\frac{1}{s}\, g_{\alpha\beta}(s;\tau)g_{\beta\alpha}(s;t-\tau). \tag{4.6}$$

In order for equation (4.1) to hold true for all $m = 0, 1, \ldots$, it is necessary and sufficient that the p.g.f. of the sequence $\{P_{\alpha\alpha}^{(m)}(t)\}$ on the left side of (4.1) be equal to the sum of the products in (4.4) and (4.6), or

$$g_{\alpha\alpha}(s;t) = g_{\alpha\alpha}(s;\tau)g_{\alpha\alpha}(s;t-\tau) + \frac{1}{s}\, g_{\alpha\beta}(s;\tau)g_{\beta\alpha}(s;t-\tau). \tag{4.7}$$

Substituting (2.13) for the p.g.f.'s in (4.7) gives

$$\sum_{i=1}^{2}\frac{\rho_i - \nu_{\beta\beta}}{\rho_i - \rho_j}e^{\rho_i t} = \left(\sum_{i=1}^{2}\frac{\rho_i - \nu_{\beta\beta}}{\rho_i - \rho_j}e^{\rho_i\tau}\right)\left[\sum_{i=1}^{2}\frac{\rho_i - \nu_{\beta\beta}}{\rho_i - \rho_j}e^{\rho_i(t-\tau)}\right]$$
$$+\left(\sum_{i=1}^{2}\frac{\nu_{\alpha\beta}}{\rho_i - \rho_j}e^{\rho_i\tau}\right)\left[\sum_{i=1}^{2}\frac{s\nu_{\beta\alpha}}{\rho_i - \rho_j}e^{\rho_i(t-\tau)}\right]. \tag{4.8}$$

Equation (4.8) can be verified in the same way as was equation (3.2) in Chapter 4; this shows that the multiple transition probabilities in (2.21) and (2.24) satisfy equation (4.1).

Using the same argument we see that for equation (4.2) to hold true for all $m = 1, 2, \ldots$, it is necessary and sufficient that the corresponding p.g.f.'s satisfy the following equation

$$g_{\alpha\beta}(s;t) = g_{\alpha\alpha}(s;\tau)g_{\alpha\beta}(s;t-\tau) + g_{\alpha\beta}(s;\tau)g_{\beta\beta}(s;t-\tau). \tag{4.9}$$

Verification of (4.9) is similar to that of (4.7) and is left to the reader.

When $s = 1$, equations (4.7) and (4.9) become identically equal to equations (3.1*a*) and (3.1*b*) in Chapter 4.

Chapman-Kolmogorov equations can also be derived for the probabilities of multiple transitions leading to death. The reasoning is analogous to that used in the illness transition probabilities, but the consideration of death states as well as illness states introduces some additional complexities.

Consider again an individual who starts in S_α at time 0, leaves S_α m times during $(0, t)$ and reaches R_δ at or before t. At an arbitrary but fixed time τ between 0 and t, the individual either (i) has already reached R_δ after leaving S_α m times, or (ii) has left S_α l times and is now in S_α, or (iii) has left S_α l times and is now in S_β. These possibilities are mutually exclusive, so that

$$Q_{\alpha\delta}^{(m)}(t) = Q_{\alpha\delta}^{(m)}(\tau) + \sum_{l=0}^{m-1} P_{\alpha\alpha}^{(l)}(\tau)Q_{\alpha\delta}^{(m-l)}(t-\tau) + \sum_{l=1}^{m} P_{\alpha\beta}^{(l)}(\tau)Q_{\beta\delta}^{*(m-l)}(t-\tau), \tag{4.10}$$

where $Q_{\beta\delta}^{*(m-l)}(t-\tau)$ is the probability that the individual, starting at τ in S_β, leaves S_α $m-l$ times during (τ, t) and is in R_δ by t. In other words

$$\begin{aligned} Q_{\beta\delta}^{*(m-l)}(t-\tau) &= \int_\tau^t P_{\beta\beta}^{(m-l)}(\xi-\tau)\mu_{\beta\delta}\,d\xi + \int_\tau^t P_{\beta\alpha}^{(m-l)}(\xi-\tau)\mu_{\alpha\delta}\,d\xi \\ &= Q_{\beta\delta.\beta}^{(m-l+1)}(t-\tau) + Q_{\beta\delta.\alpha}^{(m-l)}(t-\tau). \end{aligned} \tag{4.11}$$

Consequently the required equation is

$$\begin{aligned} Q_{\alpha\delta}^{(m)}(t) &= Q_{\alpha\delta}^{(m)}(\tau) + \sum_{l=0}^{m-1} P_{\alpha\alpha}^{(l)}(\tau)Q_{\alpha\delta}^{(m-l)}(t-\tau) \\ &\quad + \sum_{l=1}^{m} P_{\alpha\beta}^{(l)}(\tau)[Q_{\beta\delta.\beta}^{(m-l+1)}(t-\tau) + Q_{\beta\delta.\alpha}^{(m-l)}(t-\tau)], \\ &\qquad \alpha \neq \beta;\ \alpha, \beta = 1, 2;\ \delta = 1, \ldots, r;\ m = 1, 2, \ldots. \end{aligned} \tag{4.12}$$

Equation (4.12) can be verified also by using p.g.f.'s. It is clear that equation (4.12) holds true for every positive integer m if, and only if, the corresponding p.g.f.'s satisfy the equation

$$\begin{aligned} h_{\alpha\delta}(s; t) &= h_{\alpha\delta}(s; \tau) + g_{\alpha\alpha}(s; \tau)h_{\alpha\delta}(s; t-\tau) \\ &\quad + g_{\alpha\beta}(s; \tau)\left[\frac{1}{s} h_{\beta\delta.\beta}(s; t-\tau) + h_{\beta\delta.\alpha}(s; t-\tau)\right]. \end{aligned} \tag{4.13}$$

Explicit forms of $g_{\alpha\beta}(t)$, $h_{\alpha\delta}(t)$, $h_{\beta\delta.\alpha}(s; t)$, and $h_{\beta\delta.\beta}(s; t)$ are given in (2.13), (3.18), (3.17), and (3.16), respectively. After substitutions of these explicit forms, (4.13) can be verified by direct computations. When $s = 1$, (4.13) becomes identical to equation (3.6) in Chapter 4.

5. MORE IDENTITIES FOR MULTIPLE TRANSITION PROBABILITIES

The Chapman-Kolmogorov equations for the multiple transition probabilities hold true for each fixed time τ in the interval $(0, t)$. Alternatively we may keep the number l of transitions fixed and vary the time

τ. To illustrate, consider that an individual in S_α at time 0 will leave S_α m times and be in S_β at time t, and let l be a fixed number with $0 \le l < m$. The probability that the $(l+1)$st exit transition from S_α occurs at a particular time τ [or, more precisely, during the infinitesimal time interval $(\tau, \tau + d\tau)$] and the remaining $m - l - 1$ transitions during (τ, t) is

$$P_{\alpha\alpha}^{(l)}(\tau)[\nu_{\alpha\beta}\, d\tau]P_{\beta\beta}^{(m-l-1)}(t-\tau). \tag{5.1}$$

Since the events corresponding to (5.1) for different τ are mutually exclusive, we may integrate (5.1) to obtain the identity

$$P_{\alpha\beta}^{(m)}(t) = \int_0^t P_{\alpha\alpha}^{(l)}(\tau)\nu_{\alpha\beta}P_{\beta\beta}^{(m-l-1)}(t-\tau)\, d\tau, \qquad l = 0, \ldots, m-1. \tag{5.2}$$

Another identity is obtained if we let τ be the time of the first transition from S_β to S_α after the individual has already left S_α l times; this approach leads to

$$P_{\alpha\beta}^{(m)}(t) = \int_0^t P_{\alpha\beta}^{(l)}(\tau)\nu_{\beta\alpha}P_{\alpha\beta}^{(m-l)}(t-\tau)\, d\tau, \qquad l = 1, \ldots, m-1. \tag{5.3}$$

Using a similar reasoning, we have the two identities for $P_{\alpha\alpha}^{(m)}(t)$,

$$P_{\alpha\alpha}^{(m)}(t) = \int_0^t P_{\alpha\beta}^{(l)}(\tau)\nu_{\beta\alpha}P_{\alpha\alpha}^{(m-l)}(t-\tau)\, d\tau, \qquad l = 1, \ldots, m \tag{5.4}$$

and

$$P_{\alpha\alpha}^{(m)}(t) = \int_0^t P_{\alpha\alpha}^{(l)}(\tau)\nu_{\alpha\beta}P_{\beta\alpha}^{(m-l)}(t-\tau)\, d\tau, \qquad l = 0, \ldots, m-1. \tag{5.5}$$

Verification of identities (5.2) to (5.5) for any arbitrary value of m requires the following three equalities. First,

$$\int_0^t \left(\frac{\tau}{2}\right)^{2i+1}\left(\frac{t-\tau}{2}\right)^{2j+1} d\tau = 2\left(\frac{t}{2}\right)^{2k+1}\frac{(2i+1)!\,(2j+1)!}{(2k+1)!} \tag{5.6}$$

where $k = i + j + 1$. To prove the equality, we let $y = \tau/t$ so that the integral is transformed to a complete beta function for which the integral is well known. Thus we have

$$2\left(\frac{t}{2}\right)^{2k+1}\int_0^1 y^{2i+1}(1-y)^{2j+1}\, dy = 2\left(\frac{t}{2}\right)^{2k+1}\frac{(2i+1)!\,(2j+1)!}{(2k+1)!} \tag{5.7}$$

which demonstrates the identity (5.6).

Second, if $\{a_{ij}\}$ is an arbitrary sequence of numbers dependent upon two indices, then

$$\sum_{i=l-1}^{\infty} \sum_{j=m-l-1}^{\infty} a_{ij} = \sum_{k=m-1}^{\infty} \sum_{j=m-l-1}^{k-l} a_{k-j-1,j}. \tag{5.8}$$

Equality (5.8) can be most easily seen from the following picture of an infinite grid of the points (i, j). Summation over the points may be taken in different ways. The left side of (5.8) represents the summation taken along the vertical $(j + 1)$ and the horizontal (i) axes, whereas the right side of (5.8) represents the summation taken first along the diagonal lines $k = i + j + 1 = \text{const.}$, and then added over all the diagonals indicated. Equality (5.8) thus becomes obvious.

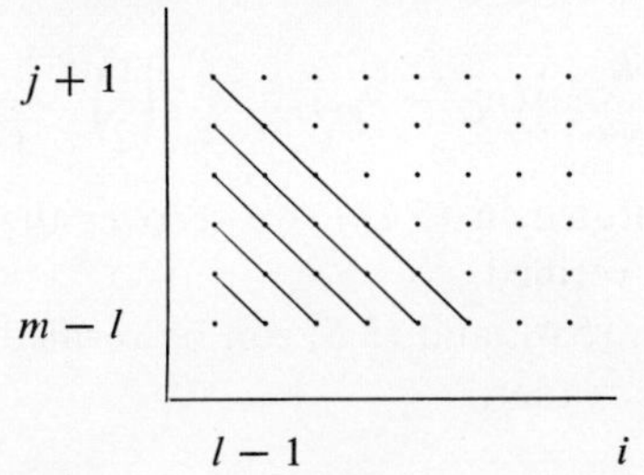

The third equality is the combinatorial equality,

$$\sum_{j=m-l-1}^{k-l} \binom{k-j-1}{l-1} \binom{j}{m-l-1} = \binom{k}{m-1}. \tag{5.9}$$

The proof of (5.9) lies in the expansion of the identity

$$(1 - c)^{-m} = (1 - c)^{-l}(1 - c)^{-(m-l)}, \tag{5.10}$$

where $|c| < 1$. Using Newton's binomial formula to expand both sides of (5.10) and equating the coefficients of c^n yield

$$(-1)^n \binom{-m}{n} = \sum_{i=0}^{n} (-1)^n \binom{-l}{n-i} \binom{-(m-l)}{i} \tag{5.11}$$

which may be rewritten as

$$\binom{m+n-1}{m-1} = \sum_{i=0}^{n} \binom{l+n-i-1}{l-1} \binom{m-l+i-1}{m-l-1}. \tag{5.12}$$

Making the substitutions $m + n - 1 = k$ and $m - l + i - 1 = j$, we recover equality (5.9).

Using the three equalities we can verify equation (5.3). Substituting formula (2.24) for the transition probability $P_{\alpha\beta}^{(m)}(t)$ in the right side of

(5.3) and using (5.6) to integrate the resulting expression, we obtain

$$\int_0^t P_{\alpha\beta}^{(l)}(\tau)\nu_{\beta\alpha}P_{\alpha\beta}^{(m-l)}(t-\tau)\,d\tau = 2\nu_{\alpha\beta}(4\nu_{\alpha\beta}\nu_{\beta\alpha})^{m-1}e^{(t/2)(\nu_{\alpha\alpha}+\nu_{\beta\beta})}$$
$$\times\left[\sum_{i=l-1}^{\infty}\sum_{j=m-l-1}^{\infty}\binom{i}{l-1}\binom{j}{m-l-1}\right.$$
$$\left.\times(\nu_{\alpha\alpha}-\nu_{\beta\beta})^{2(k-m+1)}\left(\frac{t}{2}\right)^{2k+1}\frac{1}{(2k+1)!}\right] \tag{5.13}$$

where $k = i + j + 1$. Now (5.8) and (5.9) are used successively and the quantity inside the brackets is reduced to

$$\left[\sum_{k=m-1}^{\infty}\binom{k}{m-1}(\nu_{\alpha\alpha}-\nu_{\beta\beta})^{2(k-m+1)}\left(\frac{t}{2}\right)^{2k+1}\frac{1}{(2k+1)!}\right]. \tag{5.14}$$

When (5.14) is substituted in (5.13), we recover the formula for $P_{\alpha\beta}^{(m)}(t)$, and equation (5.3) is verified.

The identities (5.2), (5.4), and (5.5) can be verified in the same manner. Details are left to the reader.

6. MULTIPLE ENTRANCE TRANSITION PROBABILITIES, $p_{\alpha\beta}^{(n)}(t)$

The random variable $M_{\alpha\beta}(\tau, t)$ associated with the transition probabilities $P_{\alpha\beta}^{(m)}(\tau, t)$ is the number of times an individual *leaves* S_α in the interval (τ, t). Alternatively, we may consider the number of times an individual *enters* S_β in going from S_α at τ to S_β at t. We denote this random variable by $N_{\alpha\beta}(\tau, t)$ and its probability distribution by

$$\begin{aligned} p_{\alpha\beta}^{(n)}(\tau, t) &= \Pr\{N_{\alpha\beta}(\tau, t) = n\} \\ &= \Pr\{\text{an individual in } S_\alpha \text{ at } \tau \text{ will } \textit{enter } S_\beta\ n \text{ times during} \\ &\quad (\tau, t) \text{ and will be in } S_\beta \text{ at } t\} \qquad \alpha, \beta = 1, 2; n = 0, 1, \ldots. \end{aligned} \tag{6.1}$$

Entrance transition into S_β is conceptually different from the exit transition from S_α and generally the random variable $N_{\alpha\beta}(\tau, t)$ has a different probability distribution from $M_{\alpha\beta}(\tau, t)$, as we shall see in Chapter 7. In the present case of two illness states, however, as soon as an individual exits from one illness state he must enter the other; therefore the number of exit transitions $M_{\alpha\beta}(\tau, t)$ and the number of entrance transitions $N_{\alpha\beta}(\tau, t)$ have the same probability distribution. In this section we derive the p.g.f.'s of $p_{\alpha\beta}^{(n)}(\tau, t)$ and show that they are the same as the p.g.f.'s of $P_{\alpha\beta}^{(m)}(\tau, t)$.

In Section 2 we obtained differential equations for the exit transition probabilities $P_{\alpha\beta}^{(m)}(\tau, t)$ by considering the contiguous intervals (τ, t) and $(t, t + \Delta)$. The same approach works here. However, we take a "backward" approach and consider the intervals $(\tau, \tau + \Delta)$ and $(\tau + \Delta, t + \Delta)$. Since $\nu_{\alpha\beta}$ are independent of time, the entrance transition probabilities $p_{\alpha\beta}^{(n)}(\tau, t)$ also are time homogeneous and therefore

$$p_{\alpha\beta}^{(n)}(\tau + \Delta, t + \Delta) = p_{\alpha\beta}^{(n)}(\tau, t) \tag{6.2}$$

whatever may be Δ. By a direct enumeration of the possible events during the intervals $(\tau, \tau + \Delta)$ and $(\tau + \Delta, t + \Delta)$, we find that

$$p_{\alpha\beta}^{(n)}(\tau, t + \Delta) = (1 + \nu_{\alpha\alpha}\Delta)p_{\alpha\beta}^{(n)}(\tau + \Delta, t + \Delta) + \nu_{\alpha\beta}\,\Delta p_{\beta\beta}^{(n-1)}(\tau + \Delta, t + \Delta)$$
$$p_{\beta\beta}^{(n)}(\tau, t + \Delta) = \nu_{\beta\alpha}\,\Delta p_{\alpha\beta}^{(n)}(\tau + \Delta, t + \Delta) + (1 + \nu_{\beta\beta}\Delta)p_{\beta\beta}^{(n)}(\tau + \Delta, t + \Delta),$$

which, in view of (6.2), may be rewritten as

$$\begin{aligned} p_{\alpha\beta}^{(n)}(\tau, t + \Delta) &= (1 + \nu_{\alpha\alpha}\Delta)p_{\alpha\beta}^{(n)}(\tau, t) + \nu_{\alpha\beta}\,\Delta p_{\beta\beta}^{(n-1)}(\tau, t) \\ p_{\beta\beta}^{(n)}(\tau, t + \Delta) &= \nu_{\beta\alpha}\,\Delta p_{\alpha\beta}^{(n)}(\tau, t) + (1 + \nu_{\beta\beta}\Delta)p_{\beta\beta}^{(n)}(\tau, t). \end{aligned} \tag{6.3}$$

Forming difference quotients and letting $\Delta \to 0$, we obtain the differential equations

$$\begin{aligned} \frac{\partial}{\partial t} p_{\alpha\beta}^{(n)}(\tau, t) &= \nu_{\alpha\alpha}p_{\alpha\beta}^{(n)}(\tau, t) + \nu_{\alpha\beta}p_{\beta\beta}^{(n-1)}(\tau, t) \\ \frac{\partial}{\partial t} p_{\beta\beta}^{(n)}(\tau, t) &= \nu_{\beta\alpha}p_{\alpha\beta}^{(n)}(\tau, t) + \nu_{\beta\beta}p_{\beta\beta}^{(n)}(\tau, t), \qquad \alpha \neq \beta;\ \alpha, \beta = 1, 2. \end{aligned} \tag{6.4}$$

It follows that the corresponding p.g.f.'s defined by

$$g_{\alpha\beta}(s; \tau, t) = \sum_{n=1}^{\infty} s^n p_{\alpha\beta}^{(n)}(\tau, t)$$

and

$$g_{\beta\beta}(s; \tau, t) = \sum_{n=0}^{\infty} s^n p_{\beta\beta}^{(n)}(\tau, t) \tag{6.5}$$

satisfy

$$\frac{\partial}{\partial t} g_{\alpha\beta}(s; \tau, t) = \nu_{\alpha\alpha}g_{\alpha\beta}(s; \tau, t) + \nu_{\alpha\beta}s g_{\beta\beta}(s; \tau, t)$$

and

$$\frac{\partial}{\partial t} g_{\beta\beta}(s; \tau, t) = \nu_{\beta\alpha}g_{\alpha\beta}(s; \tau, t) + \nu_{\beta\beta}g_{\beta\beta}(s; \tau, t) \tag{6.6}$$

with the initial conditions

$$\begin{aligned} g_{\alpha\beta}(s;\tau,\tau) &= 0 \\ g_{\beta\beta}(s;\tau,\tau) &= 1. \end{aligned} \tag{6.7}$$

Proceeding as we did to solve the system (2.4), we obtain the following solution of (6.6) corresponding to the initial condition (6.7):

$$\begin{aligned} g_{\alpha\beta}(s;\tau,t) &= \sum_{i=1}^{2} \frac{s\nu_{\alpha\beta}}{\rho_i - \rho_j} e^{\rho_i(t-\tau)} \\ g_{\beta\beta}(s;\tau,t) &= \sum_{i=1}^{2} \frac{\rho_i - \nu_{\alpha\alpha}}{\rho_i - \rho_j} e^{\rho_i(t-\tau)}, \qquad \alpha \neq \beta, \alpha, \beta = 1, 2, \end{aligned} \tag{6.8}$$

where ρ_1 and ρ_2 are given in (2.7). The formulas in (6.8) are identical with the corresponding p.g.f.'s of $P_{\alpha\beta}^{(m)}(\tau, t)$ in (2.8). Therefore the probability distribution $\{p_{\alpha\beta}^{(n)}(\tau, t)\}$ is the same as $\{P_{\alpha\beta}^{(m)}(\tau, t)\}$.

7. MULTIPLE TRANSITION TIME, $T_{\alpha\beta}^{(m)}$

Thus far in the discussion of multiple transitions, the number of transitions was treated as a random variable and the length of the time interval was taken to be constant. Our results acquire a new interpretation if we consider the length of time required for a given number of transitions as a random variable.

Consider an individual in state S_α and let a random variable $T_{\alpha\beta}^{(m)}$ be the length of time up to the mth transition from S_α to S_β. Clearly for each set of possible values of α, β, and m, $T_{\alpha\beta}^{(m)}$ takes on nonnegative real values. The simplest example is $m = 1$. If $\beta = \alpha$, $T_{\alpha\alpha}^{(1)}$ is the *first return time*; if $\beta \neq \alpha$, $T_{\alpha\beta}^{(1)}$ is the *first passage time*. If S_α stands for the state of health and S_β for the state of illness, then $T_{\alpha\beta}^{(1)}$ is the length of time a person enjoys his health before an illness occurs, and $T_{\beta\alpha}^{(1)}$ is the time needed to recover from the illness.

Let $f_{\alpha\beta}^{(m)}(t)$ be the probability density function of $T_{\alpha\beta}^{(m)}$, $\alpha, \beta = 1, 2$. An explicit expression for $f_{\alpha\beta}^{(m)}(t)$ can be derived from the results obtained in Section 2. For the case where $\beta = \alpha$ with the probability density function $f_{\alpha\alpha}^{(m)}(t)$, let us consider two contiguous time intervals $(0, t)$ and $(t, t + dt)$. Starting in state S_α at time 0, an individual will make m transitions from S_α to S_β in the interval $(0, t)$ with a probability $P_{\alpha\beta}^{(m)}(t)$; and he exits from S_β to S_α within the interval $(t, t + dt)$ with a probability $\nu_{\beta\alpha}\, dt$. Therefore the density function of $T_{\alpha\alpha}^{(m)}$ is given by

$$f_{\alpha\alpha}^{(m)}(t)\, dt = P_{\alpha\beta}^{(m)}(t)\nu_{\beta\alpha}\, dt, \qquad \alpha \neq \beta;\ \alpha, \beta = 1, 2. \qquad m = 1, 2, \ldots. \tag{7.1}$$

Substituting (2.24) for the multiple transition probability in (7.1) gives the explicit form

$$f_{\alpha\alpha}^{(m)}(t)\,dt = \tfrac{1}{2}(4\nu_{\alpha\beta}\nu_{\beta\alpha})^m e^{(t/2)(\nu_{\alpha\alpha}+\nu_{\beta\beta})} \times \left[\sum_{k=m-1}^{\infty}\binom{k}{m-1}(\nu_{\alpha\alpha}-\nu_{\beta\beta})^{2(k-m+1)}\frac{1}{(2k+1)!}\left(\frac{t}{2}\right)^{2k+1}\right]dt,$$
$$m = 0, 1, \ldots;\ \alpha \neq \beta;\ \alpha, \beta = 1, 2. \quad (7.2)$$

For $\beta \neq \alpha$, we use similar reasoning and formula (2.21) to obtain

$$f_{\alpha\beta}^{(m)}(t)\,dt = P_{\alpha\alpha}^{(m-1)}(t)\nu_{\alpha\beta}\,dt$$
$$= \nu_{\alpha\beta}(4\nu_{\alpha\beta}\nu_{\beta\alpha})^{m-1}e^{(t/2)(\nu_{\alpha\alpha}+\nu_{\beta\beta})}\sum_{k=m-1}^{\infty}\binom{k}{m-1}(\nu_{\alpha\alpha}-\nu_{\beta\beta})^{2(k-m+1)}$$
$$\times\left[\frac{\nu_{\alpha\alpha}-\nu_{\beta\beta}}{(2k+1)!}\left(\frac{t}{2}\right)^{2k+1}+\frac{1}{(2k)!}\left(\frac{t}{2}\right)^{2k}\right]dt,$$
$$m = 1, \ldots;\ \alpha \neq \beta;\ \alpha, \beta = 1, 2. \quad (7.3)$$

The corresponding distribution functions

$$F_{\alpha\beta}^{(m)}(t) = \int_0^t f_{\alpha\beta}^{(m)}(\tau)\,d\tau, \qquad \alpha, \beta = 1, 2, \quad (7.4)$$

can be obtained by direct integration. Substituting (7.2) and (7.3) in (7.4) and integrating the resulting expressions give

$$F_{\alpha\alpha}^{(m)}(t) = (4\nu_{\alpha\beta}\nu_{\beta\alpha})^m\sum_{k=m-1}^{\infty}\binom{k}{m-1}\frac{(\nu_{\alpha\alpha}-\nu_{\beta\beta})^{2(k-m+1)}}{(\nu_{\alpha\alpha}+\nu_{\beta\beta})^{2(k+1)}}\left(1-\sum_{l=0}^{2k+1}\frac{\theta^l}{l!}e^{-\theta}\right) \quad (7.5)$$

and

$$F_{\alpha\beta}^{(m)}(t) = 2\nu_{\alpha\beta}(4\nu_{\alpha\beta}\nu_{\beta\alpha})^{m-1}\sum_{k=m-1}^{\infty}\binom{k}{m-1}\frac{(\nu_{\alpha\alpha}-\nu_{\beta\beta})^{2(k-m+1)}}{(\nu_{\alpha\alpha}+\nu_{\beta\beta})^{2(k+1)}}$$
$$\times\left[-2\nu_{\beta\beta}\left(1-\sum_{l=0}^{2k+1}\frac{\theta^l}{l!}e^{-\theta}\right)-(\nu_{\alpha\alpha}+\nu_{\beta\beta})\frac{\theta^{2k+1}}{(2k+1)!}e^{-\theta}\right] \quad (7.6)$$

where

$$\theta = -\frac{t}{2}(\nu_{\alpha\alpha}+\nu_{\beta\beta}). \quad (7.7)$$

In particular, the *first return time* $T_{\alpha\alpha}^{(1)}$ has the probability density function

$$f_{\alpha\alpha}^{(1)}(t)\,dt = \frac{\nu_{\alpha\beta}\nu_{\beta\alpha}}{\nu_{\alpha\alpha}-\nu_{\beta\beta}}\,[e^{\nu_{\alpha\alpha}t}-e^{\nu_{\beta\beta}t}]\,dt \quad (7.8)$$

and the distribution function

$$F_{\alpha\alpha}^{(1)}(t) = \frac{\nu_{\alpha\beta}\nu_{\beta\alpha}}{\nu_{\alpha\alpha}\nu_{\beta\beta}} + \frac{\nu_{\alpha\beta}\nu_{\beta\alpha}}{\nu_{\alpha\alpha} - \nu_{\beta\beta}}\left(\frac{1}{\nu_{\alpha\alpha}} e^{\nu_{\alpha\alpha}t} - \frac{1}{\nu_{\beta\beta}} e^{\nu_{\beta\beta}t}\right); \tag{7.9}$$

the *first passage time* $T_{\alpha\beta}^{(1)}$ has the probability density function

$$f_{\alpha\beta}^{(1)}(t)\, dt = e^{\nu_{\alpha\alpha}t}\nu_{\alpha\beta}\, dt \tag{7.10}$$

and the distribution function

$$F_{\alpha\beta}^{(1)}(t) = \frac{\nu_{\alpha\beta}}{\nu_{\alpha\alpha}}(e^{\nu_{\alpha\alpha}t} - 1). \tag{7.11}$$

An individual in state S_α at time 0 may reach a death state directly without ever entering S_β or returning to S_α; therefore the random variables $T_{\alpha\beta}^{(m)}$ are improper random variables with

$$F_{\alpha\beta}^{(m)}(\infty) = \Pr\{T_{\alpha\beta}^{(m)} < \infty\} < 1, \qquad \alpha, \beta = 1, 2. \tag{7.12}$$

The difference $1 - \Pr\{T_{\alpha\beta}^{(m)} < \infty\}$ is the probability that an individual in S_α will never be in S_β in m transitions. The quantities in (7.12) can be obtained explicitly from (7.5) and (7.6); as $t \to \infty$, $e^{-\theta} \to 0$ and we have

$$F_{\alpha\alpha}^{(m)}(\infty) = (4\nu_{\alpha\beta}\nu_{\beta\alpha})^m \sum_{k=m-1}^{\infty} \binom{k}{m-1} \frac{(\nu_{\alpha\alpha} - \nu_{\beta\beta})^{2(k-m+1)}}{(\nu_{\alpha\alpha} + \nu_{\beta\beta})^{2(k+1)}} \tag{7.13}$$

and

$$F_{\alpha\beta}^{(m)}(\infty) = -4\nu_{\alpha\beta}\nu_{\beta\beta}(4\nu_{\alpha\beta}\nu_{\beta\alpha})^{m-1} \sum_{k=m-1}^{\infty} \binom{k}{m-1} \frac{(\nu_{\alpha\alpha} - \nu_{\beta\beta})^{2(k-m+1)}}{(\nu_{\alpha\alpha} + \nu_{\beta\beta})^{2(k+1)}}. \tag{7.14}$$

Since the $T_{\alpha\beta}^{(m)}$ are improper random variables, it is not meaningful to discuss their expectations and variances. However, we can create proper random variables $T_{\alpha\beta}^{(m)*}$ by imposing the condition that the individual will at some time reach S_β in m transitions, so that $T_{\alpha\beta}^{(m)*}$ will have the probability density function

$$f_{\alpha\beta}^{(m)*}(t)\, dt = \frac{f_{\alpha\beta}^{(m)}(t)}{F_{\alpha\beta}^{(m)}(\infty)}\, dt, \qquad \alpha, \beta = 1, 2, \tag{7.15}$$

with

$$\int_0^\infty f_{\alpha\beta}^{(m)*}(t)\, dt = 1, \qquad \alpha, \beta = 1, 2. \tag{7.16}$$

The expectation $E[T_{\alpha\beta}^{(m)*}]$ means the average length of time needed for m transitions from S_α to S_β among those individuals who reach S_β from S_α in m transitions. Exact formulas for the expectation and variance of

$T_{\alpha\beta}^{(m)*}$ can be computed using (7.15); the expectations are

$$E[T_{\alpha\alpha}^{(m)*}] = \int_0^\infty t\,\frac{f_{\alpha\alpha}^{(m)}(t)}{F_{\alpha\alpha}^{(m)}(\infty)}\,dt$$

$$= \frac{-4\sum_{k=m-1}^{\infty}(k+1)\binom{k}{m-1}(\nu_{\alpha\alpha}-\nu_{\beta\beta})^{2k}(\nu_{\alpha\alpha}+\nu_{\beta\beta})^{-(2k+1)}}{\sum_{k=m-1}^{\infty}\binom{k}{m-1}(\nu_{\alpha\alpha}-\nu_{\beta\beta})^{2k}(\nu_{\alpha\alpha}+\nu_{\beta\beta})^{-2k}} \tag{7.17}$$

and

$$E[T_{\alpha\beta}^{(m)*}]$$

$$= \frac{\sum_{k=m-1}^{\infty}[2(2k+1)\nu_{\beta\beta}-(\nu_{\alpha\alpha}-\nu_{\beta\beta})]\binom{k}{m-1}(\nu_{\alpha\alpha}-\nu_{\beta\beta})^{2k}(\nu_{\alpha\alpha}+\nu_{\beta\beta})^{-(2k+1)}}{-\nu_{\beta\beta}\sum_{k=m-1}^{\infty}\binom{k}{m-1}(\nu_{\alpha\alpha}-\nu_{\beta\beta})^{2k}(\nu_{\alpha\alpha}+\nu_{\beta\beta})^{-2k}}$$

$$\alpha \neq \beta;\ \alpha, \beta = 1, 2;\ m = 1, 2, \ldots. \tag{7.18}$$

In particular,

$$E[T_{\alpha\alpha}^{(1)*}] = -\left(\frac{1}{\nu_{\alpha\alpha}} + \frac{1}{\nu_{\beta\beta}}\right) \tag{7.19}$$

and

$$E[T_{\alpha\beta}^{(1)*}] = -\frac{1}{\nu_{\alpha\alpha}}. \tag{7.20}$$

7.1. Multiple Transition Time Leading to Death $\tau_{\alpha\delta}^{(m)}$

Given an individual in S_α, let $\tau_{\alpha\delta}^{(m)}$ be the length of time until he enters a death state R_δ after making m exit transitions from S_α. The probability density function $\phi_{\alpha\delta}^{(m)}(t)$ of $\tau_{\alpha\delta}^{(m)}$ can be derived from the transition probability $P_{\alpha\beta}^{(m)}(t)$. Clearly

$$\phi_{\alpha\delta}^{(m)}(t)\,dt = [P_{\alpha\alpha}^{(m-1)}(t)\mu_{\alpha\delta} + P_{\alpha\beta}^{(m)}(t)\mu_{\beta\delta}]\,dt \tag{7.21}$$

with the corresponding distribution function

$$\Phi_{\alpha\delta}^{(m)}(t) = \int_0^t \phi_{\alpha\delta}^{(m)}(\tau)\,d\tau$$

$$= \int_0^t [P_{\alpha\alpha}^{(m-1)}(\tau)\mu_{\alpha\delta} + P_{\alpha\beta}^{(m)}(\tau)\mu_{\beta\delta}]\,d\tau. \tag{7.22}$$

Substituting formulas (2.21) and (2.24) for the transition probabilities $P_{\alpha\alpha}^{(m)}(\tau)$ and $P_{\alpha\beta}^{(m)}(\tau)$ in (7.21) and (7.22) gives the explicit forms for $\phi_{\alpha\delta}^{(m)}(t)$

and $\Phi_{\alpha\delta}^{(m)}(t)$. For $m = 1$, for example, we have

$$\phi_{\alpha\delta}^{(1)}(t)\,dt = \left[e^{\nu_{\alpha\alpha}t}\mu_{\alpha\delta} + \frac{\nu_{\alpha\beta}}{\nu_{\alpha\alpha} - \nu_{\beta\beta}}(e^{\nu_{\alpha\alpha}t} - e^{\nu_{\beta\beta}t})\mu_{\beta\delta}\right] dt \tag{7.23}$$

and

$$\Phi_{\alpha\delta}^{(1)}(t) = \frac{-1}{\nu_{\alpha\alpha}}(1 - e^{\nu_{\alpha\alpha}t})\mu_{\alpha\delta} - \frac{\nu_{\alpha\beta}}{\nu_{\alpha\alpha} - \nu_{\beta\beta}}\left[\left(\frac{1}{\nu_{\alpha\alpha}} - \frac{1}{\nu_{\beta\beta}}\right) - \left(\frac{e^{\nu_{\alpha\alpha}t}}{\nu_{\alpha\alpha}} - \frac{e^{\nu_{\beta\beta}t}}{\nu_{\beta\beta}}\right)\right]\mu_{\beta\delta} \tag{7.24}$$

The transition times $\tau_{\alpha\delta}^{(m)}$ also are improper random variables with

$$\int_0^\infty \phi_{\alpha\delta}^{(m)}(\tau)\,d\tau < 1. \tag{7.25}$$

However, since the R_δ are absorbing states, starting with any state S_α an individual will sooner or later be in one of the death states R_δ. This means that

$$\sum_{\delta=1}^{r}\sum_{m=1}^{\infty}\Phi_{\alpha\delta}^{(m)}(\infty) = 1. \tag{7.26}$$

For verification of (7.26), see Problem 28.

7.2. Identities for Multiple Transition Time

The identities in this section follow in a straightforward manner from consideration of the multiple transition time. Consider, for example, $T_{\alpha\alpha}^{(m)}$, the length of time for an individual to make m transitions from S_α to S_α. For fixed l, this length of time may be divided into two periods: a period of length $T_{\alpha\alpha}^{(l)}$ for making l transitions from S_α to S_α, and a second period of length $T_{\alpha\alpha}^{(m-l)}$ for making the remaining $(m - l)$ transitions. Therefore

$$T_{\alpha\alpha}^{(m)} = T_{\alpha\alpha}^{(l)} + T_{\alpha\alpha}^{(m-l)}, \qquad l = 1, \ldots, m - 1. \tag{7.27}$$

Alternatively, we may divide the time into a period of $T_{\alpha\beta}^{(l)}$ for making l transitions from S_α to S_β and another period of $T_{\beta\alpha}^{(m-l+1)}$ for making $(m - l + 1)$ transitions from S_β to S_α, or

$$T_{\alpha\alpha}^{(m)} = T_{\alpha\beta}^{(l)} + T_{\beta\alpha}^{(m-l+1)}, \qquad l = 1, \ldots, m - 1. \tag{7.28}$$

In the same way, we have two identities for the multiple transition time $T_{\alpha\beta}^{(m)}$,

$$T_{\alpha\beta}^{(m)} = T_{\alpha\alpha}^{(l)} + T_{\alpha\beta}^{(m-l)}, \tag{7.29}$$

and

$$T_{\alpha\beta}^{(m)} = T_{\alpha\beta}^{(l)} + T_{\beta\beta}^{(m-l)}, \qquad l = 1, \ldots, m - 1. \tag{7.30}$$

These identities are equivalent to the identities in the transition probabilities $P_{\alpha\beta}^{(m)}(t)$ in Section 5. For example, because of the independence of future transitions from past transitions, (7.27) holds true if and only if the corresponding probability density functions have the relation

$$f_{\alpha\alpha}^{(m)}(t) = \int_0^t f_{\alpha\alpha}^{(l)}(\tau) f_{\alpha\alpha}^{(m-l)}(t-\tau)\, d\tau, \tag{7.31}$$

which, in view of (7.1), becomes

$$P_{\alpha\beta}^{(m)}(t)\nu_{\beta\alpha} = \int_0^t P_{\alpha\beta}^{(l)}(\tau)\nu_{\beta\alpha} P_{\alpha\beta}^{(m-l)}(t-\tau)\nu_{\beta\alpha}\, d\tau \tag{7.32}$$

or

$$P_{\alpha\beta}^{(m)}(t) = \int_0^t P_{\alpha\beta}^{(l)}(\tau)\nu_{\beta\alpha} P_{\alpha\beta}^{(m-l)}(t-\tau)\, d\tau, \tag{7.33}$$

which is the same as the identity (5.3).

These identities can be easily generalized. Thus we have

$$T_{\alpha\alpha}^{(m)} = \sum_{i=1}^{k} T_{\alpha\alpha}^{(m_i)}, \tag{7.34}$$

$$T_{\alpha\alpha}^{(m)} = \sum_{i=1}^{k} [T_{\alpha\beta}^{(l_i)} + T_{\beta\alpha}^{(m_i-l_i+1)}], \tag{7.35}$$

$$T_{\alpha\beta}^{(m)} = \sum_{i=1}^{k} [T_{\alpha\alpha}^{(l_i)} + T_{\alpha\beta}^{(m_i-l_i)}], \tag{7.36}$$

and

$$T_{\alpha\beta}^{(m)} = \sum_{i=1}^{k} [T_{\alpha\beta}^{(l_i)} + T_{\beta\beta}^{(m_i-l_i)}], \tag{7.37}$$

where $m_1 + \cdots + m_k = m$.

PROBLEMS

1. Differential equations for the multiple exit transition probabilities can be derived with a slightly different approach. By considering two contiguous intervals $(\tau - \Delta, \tau)$ and (τ, t), we find that

$$\frac{\partial}{\partial t} P_{\alpha\alpha}^{(m)}(\tau, t) = \nu_{\alpha\alpha} P_{\alpha\alpha}^{(m)}(\tau, t) + \nu_{\alpha\beta} P_{\beta\alpha}^{(m)}(\tau, t)$$

$$\frac{\partial}{\partial t} P_{\beta\alpha}^{(m)}(\tau, t) = \nu_{\beta\alpha} P_{\alpha\alpha}^{(m-1)}(\tau, t) + \nu_{\beta\beta} P_{\beta\alpha}^{(m)}(\tau, t)$$

for $m = 0, 1, \ldots$.

(a) Verify these differential equations.

(b) Derive partial differential equations for the p.g.f.'s $g_{\alpha\alpha}(s; \tau, t)$ and $g_{\beta\alpha}(s; \tau, t)$, and state initial conditions.

(c) Solve the differential equations in (b) for $g_{\alpha\alpha}(s;\tau,t)$ and $g_{\beta\alpha}(s;\tau,t)$.
(d) Obtain the multiple transition probabilities from the p.g.f.'s in (c).
(e) Compare your formulas with those in Section 2.

2. Show that solution (2.8) satisfies the partial differential equations (2.4) and the initial conditions (2.5).

3. By taking derivatives of the p.g.f. in (2.13), derive the multiple transition probabilities $P_{\alpha\beta}^{(m)}(t)$ for $m = 1, 2$.

4. Verify the second equation in (2.26).

5. Show that $P_{\alpha\alpha}^{(0)}(t)$, $P_{\alpha\alpha}^{(1)}(t)$, and $P_{\alpha\beta}^{(1)}(t)$ as given in (2.22), (2.23), and (2.25), respectively, satisfy the differential equations:

$$\frac{d}{dt}P_{\alpha\alpha}^{(1)}(t) = P_{\alpha\alpha}^{(1)}(t)\nu_{\alpha\alpha} + P_{\alpha\beta}^{(1)}(t)\nu_{\beta\alpha}$$

$$\frac{d}{dt}P_{\alpha\beta}^{(1)}(t) = P_{\alpha\alpha}^{(0)}(t)\nu_{\alpha\beta} + P_{\alpha\beta}^{(1)}(t)\nu_{\beta\beta}$$

and also

$$\frac{d}{dt}P_{\alpha\alpha}^{(1)}(t) = \nu_{\alpha\alpha}P_{\alpha\alpha}^{(1)}(t) + \nu_{\alpha\beta}P_{\beta\alpha}^{(1)}(t)$$

$$\frac{d}{dt}P_{\beta\alpha}^{(1)}(t) = \nu_{\beta\alpha}P_{\alpha\alpha}^{(0)}(t) + \nu_{\beta\beta}P_{\beta\alpha}^{(1)}(t).$$

6. *Probability of first return.* Verify Equation (2.23) for the probability of first return to the original state within a time interval of length t.

7. *Continuation.* Show that whatever may be $t \geq 0$, $\nu_{\alpha\alpha} < 0$, and $\nu_{\beta\beta} < 0$, the quantity on the right side of (2.23) is never negative.

8. Verify the second equation in (2.25) for the first passage probability $P_{\alpha\beta}^{(1)}(t)$.

9. Derive the expression in (2.26) for the transition probability $P_{\alpha\alpha}(t)$ by adding the probabilities $P_{\alpha\alpha}^{(m)}(t)$ for $m = 0, 1, \ldots,$ and reduce this expression to the one in (2.27) in Chapter 4.

10. Derive the expression in (2.28) for the transition probability $P_{\alpha\beta}(t)$ by adding the probabilities $P_{\alpha\beta}^{(m)}(t)$ for $m = 1, 2, \ldots,$ and reduce this expression to the one in (2.27) in Chapter 4.

11. Derive from the second equation in (2.13) an explicit formula for the p.g.f. $g_{\alpha\beta}^{(m)}(t)$ and, from this, the formula for $P_{\alpha\beta}^{(m)}(t)$ in (2.24).

12. Verify the expectations $E[M_{\alpha\alpha}^*(t)]$ in (2.32) and $E[M_{\alpha\beta}^*(t)]$ in (2.33).

13. Evaluate the incomplete gamma function $\int_0^t y^n e^{-y}\,dy$, and use the result to verify (3.5).

14. Verify the formulas for $Q_{\alpha\delta\cdot\alpha}^{(m)}(t)$ in (3.7) and $Q_{\alpha\delta\cdot\beta}^{(m)}(t)$ in (3.8).

15. Find $Q_{\alpha\delta\cdot\alpha}^{(1)}(t)$ and $Q_{\alpha\delta\cdot\beta}^{(1)}(t)$ from the corresponding p.g.f.'s in (3.16) and (3.17).

16. Justify and verify the equation

$$g_{\alpha\beta}(s;t) = g_{\alpha\alpha}(s;\tau)g_{\alpha\beta}(s;t-\tau) + g_{\alpha\beta}(s;\tau)g_{\beta\beta}(s;t-\tau). \qquad (4.9)$$

17. Substitute the explicit formulas for the p.g.f.'s $g_{\alpha\beta}(s;t)$, $h_{\alpha\delta}(s;t)$, $h_{\beta\delta\cdot\beta}(s;t)$, and $h_{\beta\delta\cdot\alpha}(s;t)$ in (4.13) and verify the resultant equation.

18. Integrate the beta function

$$\int_0^1 y^i(1-y)^j\,dy. \qquad (5.7)$$

19. Use Equations (5.6), (5.8), and (5.9) to verify the identities (5.2), (5.4), and (5.5).

20. Verify the distribution functions of the multiple transition time in (7.5) and (7.6).

21. Verify the density function and the distribution function of the first return time $T_{\alpha\alpha}^{(1)}$ in (7.8) and (7.9).

22. Verify the density function and the distribution function of the first passage time $T_{\alpha\beta}^{(1)}$ in (7.10) and (7.11).

23. Use the explicit form for the multiple transition probabilities $P_{\alpha\beta}^{(m)}(t)$ to prove the identities

$$T_{\alpha\alpha}^{(2)} = T_{\alpha\beta}^{(1)} + T_{\beta\alpha}^{(2)} \qquad (7.28a)$$

and

$$T_{\alpha\beta}^{(2)} = T_{\alpha\alpha}^{(1)} + T_{\alpha\beta}^{(1)}. \qquad (7.30a)$$

24. Justify the relation

$$P_{\alpha\alpha}^{(1)}(t) = P_{\alpha\alpha}^{(1)}(\tau)P_{\alpha\alpha}^{(0)}(t-\tau) + P_{\alpha\alpha}^{(0)}(\tau)P_{\alpha\alpha}^{(1)}(t-\tau) + P_{\alpha\beta}^{(1)}(\tau)P_{\beta\alpha}^{(1)}(t-\tau) \qquad (4.1a)$$

where $0 < \tau < t$, and verify it with the explicit formulas in (2.22), (2.23), and (2.25).

25. Justify the relation

$$P_{\alpha\beta}^{(1)}(t) = P_{\alpha\alpha}^{(0)}(\tau)P_{\alpha\beta}^{(1)}(t-\tau) + P_{\alpha\beta}^{(1)}(\tau)P_{\beta\beta}^{(0)}(t-\tau) \qquad (4.2a)$$

where $0 < \tau < t$, and verify it with the explicit formulas in (2.22), and (2.25).

26. Justify the relation

$$P_{\alpha\beta}^{(1)}(t) = \int_0^t P_{\alpha\alpha}^{(0)}(\tau)\nu_{\alpha\beta}P_{\beta\beta}^{(0)}(t-\tau)\,d\tau \qquad (5.2a)$$

and verify it with the explicit formulas in (2.22) and (2.25).

27. Justify the relation

$$P_{\alpha\alpha}^{(1)}(t) = \int_0^t P_{\alpha\beta}^{(1)}(\tau)\nu_{\beta\alpha}P_{\alpha\alpha}^{(0)}(t-\tau)\,d\tau \qquad (5.4a)$$

and verify it with the explicit formulas in (2.22), (2.23), and (2.25).

28. Let $\Phi_{\alpha\delta}^{(m)}(t)$ be the distribution function of the multiple transition time $\tau_{\alpha\delta}^{(m)}$ as given in (7.22). Show that

$$\sum_{\delta=1}^{r}\sum_{m=1}^{\infty}\Phi_{\alpha\delta}^{(m)}(\infty) = 1. \qquad (7.26)$$

CHAPTER 6

The Kolmogorov Differential Equations and Finite Markov Processes

1. MARKOV PROCESSES AND THE CHAPMAN-KOLMOGOROV EQUATION

Thus far we have been concerned with specific stochastic processes. We now present a general formulation that will include these processes as special cases. We consider a system that is capable of being in a finite or denumerable number of states, which we shall label by the integers $0, 1, \ldots$. For each time t, we define a discrete random variable $X(t)$ whose value indicates the state of the system; that is, the event "$X(t) = j$" is the same as the event "the system is in state j at time t." In the population growth models of Chapter 3, for example, the state of the system is the population size and $X(t)$ has an infinite number of possible values, whereas in the illness-death process of Chapter 4, the state of the system is an illness or death state of one individual and the system has $r + 2$ possible states.

In Chapter 3 we made a rather restricted assumption that, within a small time element $(t, t + \Delta)$, a population size may increase (or decrease) by only one with an appreciable probability. We now remove this restriction by allowing a system to change from any (nonabsorbing) state to any other state in the system. We denote by $P_{ij}(\tau, t)$ the (conditional) probability that the system is in state j at time t, given that it was in state i at time τ; that is,

$$P_{ij}(\tau, t) = \Pr\{X(t) = j \mid X(\tau) = i\}, \qquad i, j = 0, 1, \ldots. \tag{1.1}$$

The transition probability (1.1) indicates the stochastic dependence of $X(t)$ on $X(\tau)$ for arbitrary $\tau < t$; two forms of dependence are of particular importance and are introduced in the following definitions.

A discrete-valued stochastic process $\{X(t)\}$ is called a *Markov process* if, for any $t_0 < t_1 < \cdots < t_i < \cdots < t_j$, and any integers $k_0, k_1, \ldots, k_j$,

$$\begin{aligned}\Pr\{X(t_j) = k_j \mid X(t_0) = k_0, X(t_1) = k_1, \ldots, X(t_i) = k_i\} \\ = \Pr\{X(t_j) = k_j \mid X(t_i) = k_i\}.\end{aligned} \tag{1.2}$$

Thus, in a Markov process, given $X(t_i)$ (present), the conditional probability distribution of $X(t_j)$ (future) is independent of $X(t_0), \ldots, X(t_{i-1})$ (past). The discussion in this book is confined exclusively to Markov processes. ▶

A Markov process $\{X(t)\}$ is said to be *homogeneous with respect to time* if the transition probability (1.1) depends only on the difference $t - \tau$ but not on τ or t separately. In such cases, we may write

$$P_{ij}(t - \tau) = \Pr\{X(t) = j \mid X(\tau) = i\}, \qquad i, j = 0, 1, \ldots. \qquad ▶(1.3)$$

The simple Poisson process and the illness-death process, for example, are homogeneous, whereas the Polya process is nonhomogeneous.

The Markov property implies important relations among the transition probabilities $P_{ij}(\tau, t)$. Let ξ be a fixed point in the interval (τ, t) so that $\tau < \xi < t$, and let $X(\tau)$, $X(\xi)$, and $X(t)$ be the corresponding random variables. Because of equation (1. 2) we have

$$\begin{aligned}\Pr\{X(t) = k \mid X(\tau) = i, X(\xi) = j\} &= \Pr\{X(t) = k \mid X(\xi) = j\} \\ &= P_{jk}(\xi, t)\end{aligned} \tag{1.4}$$

so that

$$\Pr\{X(\xi) = j \text{ and } X(t) = k \mid X(\tau) = i\} = P_{ij}(\tau, \xi)P_{jk}(\xi, t). \tag{1.5}$$

The last expression is the probability of a passage from $X(\tau) = i$ to $X(t) = k$ by way of a particular state j at time ξ. Since the passages corresponding to the various possible states at time ξ are mutually exclusive, we have

$$P_{ik}(\tau, t) = \sum_j P_{ij}(\tau, \xi)P_{jk}(\xi, t), \qquad i, k = 0, 1, \ldots; \quad \tau < \xi < t. \tag{1.6}$$

Equation (1.6), which is known as the *Chapman-Kolmogorov equation*, is the starting point for the general discussion of Markov processes. In the case of a time-homogeneous process, (1.6) becomes

$$P_{ik}(t - \tau) = \sum_j P_{ij}(\xi - \tau)P_{jk}(t - \xi),$$

and, in general, whatever may be $\tau > 0$ and $t > 0$,

$$P_{ik}(\tau + t) = \sum_j P_{ij}(\tau)P_{jk}(t). \qquad ▶(1.7)$$

2. THE KOLMOGOROV DIFFERENTIAL EQUATIONS

The Kolmogorov differential equations play a central role in the treatment of Markov processes in continuous time. In order to derive the Kolmogorov differential equations from the Chapman-Kolmogorov equations, we require the following assumptions.

Regularity Assumptions

(i) For every integer i, there exists a continuous function $\nu_{ii}(\tau) \leq 0$ such that

$$\lim_{\Delta\to 0} \frac{1 - P_{ii}(\tau, \tau + \Delta)}{\Delta} = -\nu_{ii}(\tau) \tag{2.1}$$

(ii) For every pair of integers $i \neq j$, there exists a continuous function $\nu_{ij}(\tau) \geq 0$ such that

$$\lim_{\Delta\to 0} \frac{P_{ij}(\tau, \tau + \Delta)}{\Delta} = \nu_{ij}(\tau); \tag{2.2}$$

furthermore, for fixed j the passage in (2.2) is uniform with respect to i.▶

The functions $\nu_{ij}(\tau)$ are called *intensity functions*. If we agree to require that

$$P_{ij}(\tau, \tau) = \delta_{ij} \tag{2.3}$$

where δ_{ij} is the Kronecker delta,[1] then it is clear from (2.1) and (2.2) that

$$\nu_{ii}(\tau) = \frac{\partial}{\partial t} P_{ii}(\tau, t)\bigg|_{t=\tau} \tag{2.4}$$

$$\nu_{ij}(\tau) = \frac{\partial}{\partial t} P_{ij}(\tau, t)\bigg|_{t=\tau} \qquad i \neq j. \tag{2.5}$$

The probabilistic meaning of the intensity functions emerges if (2.1) and (2.2) are written as

$$P_{ii}(\tau, \tau + \Delta) = 1 + \nu_{ii}(\tau)\Delta + o(\Delta) \tag{2.6}$$

$$P_{ij}(\tau, \tau + \Delta) = \nu_{ij}(\tau)\Delta + o(\Delta). \tag{2.7}$$

We note that if

$$\sum_j P_{ij}(\tau, t) = 1 \tag{2.8}$$

whatever may be $\tau < t$, then

$$\nu_{ii}(\tau) = -\sum_{j\neq i} \nu_{ij}(\tau). \tag{2.9}$$

---

[1] The Kronecker delta δ_{ij} is defined as follows

$$\delta_{ii} = 1 \qquad \text{and} \qquad \delta_{ij} = 0 \qquad j \neq i.$$

We shall use this function frequently in the book.

2.1. Derivation of the Kolmogorov Differential Equations

We now use the regularity conditions to derive Kolmogorov's differential equations. Let us consider three points $\tau < t < t + \Delta$ on the time axis and use the Chapman-Kolmogorov equation (1.6) to write

$$P_{ik}(\tau, t + \Delta) = P_{ik}(\tau, t)P_{kk}(t, t + \Delta) + \sum_{j \neq k} P_{ij}(\tau, t)P_{jk}(t, t + \Delta). \quad (2.10)$$

Using (2.6), (2.10) may be rewritten as

$$\frac{P_{ik}(\tau, t + \Delta) - P_{ik}(\tau, t)}{\Delta} = P_{ik}(\tau, t)\nu_{kk}(t) + \sum_{j \neq k} P_{ij}(\tau, t)\frac{P_{jk}(t, t + \Delta)}{\Delta} + \frac{o(\Delta)}{\Delta}. \quad (2.11)$$

According to Assumption (ii) $P_{jk}(t, t + \Delta)/\Delta \to \nu_{jk}(t)$ uniform in j, the right side of (2.11) tends to a limit as $\Delta \to 0$; hence the left side also tends to a limit. Therefore the transition probability $P_{ik}(\tau, t)$ satisfies the differential equation

$$\frac{\partial}{\partial t} P_{ik}(\tau, t) = \sum_{j} P_{ij}(\tau, t)\nu_{jk}(t). \quad (2.12)$$

This is the system of *Kolmogorov forward differential equations* with the initial conditions (2.3). It should be noted that in spite of the appearance of a partial derivative sign, (2.12) is a system of ordinary differential equations, since τ may be regarded as fixed.

We now keep t fixed and consider the transition probability $P_{ik}(\tau, t)$ as a function of τ in order to derive the backward differential equations. We start with the Chapman-Kolmogorov equation in the form

$$P_{ik}(\tau - \Delta, t) = P_{ii}(\tau - \Delta, \tau)P_{ik}(\tau, t) + \sum_{j \neq i} P_{ij}(\tau - \Delta, \tau)P_{jk}(\tau, t) \quad (2.13)$$

for $\tau - \Delta < \tau < t$ and, from (2.6),

$$P_{ii}(\tau - \Delta, \tau) = 1 + \nu_{ii}(\tau - \Delta)\Delta + o(\Delta),$$

therefore

$$\frac{P_{ik}(\tau - \Delta, t) - P_{ik}(\tau, t)}{-\Delta} = -\nu_{ii}(\tau - \Delta)P_{ik}(\tau, t) + \sum_{j \neq i} \frac{P_{ij}(\tau - \Delta, \tau)}{-\Delta} P_{jk}(\tau, t) + \frac{o(\Delta)}{-\Delta} \cdot \quad (2.14)$$

Passing to the limit as $\Delta \to 0$ and using (2.2), we get the system of *Kolmogorov backward differential equations*

$$\frac{\partial}{\partial \tau} P_{ik}(\tau, t) = -\sum_{j} \nu_{ij}(\tau)P_{jk}(\tau, t). \quad (2.15)$$

Again we have a system of ordinary differential equations for the transition probabilities. In this case, t is fixed and the initial conditions are

$$P_{ik}(t, t) = \delta_{ik}. \tag{2.16}$$

The above results may be summarized in

Theorem 1. *Let* $P_{ik}(\tau, t)$ *be the transition probabilities of a Markov process defined in* (1.1), *and suppose that the regularity conditions* (i) *and* (ii) *hold true. Then the transition probabilities satisfy the system of forward differential equations* (2.12) *with the initial conditions* (2.3) *and the system of backward differential equations* (2.15) *with the initial conditions* (2.16). ▶

These two systems of differential equations were first derived by A. Kolmogorov [1931]; detailed discussions on their theoretical aspects may be found in Chung [1960] and Doob [1953]. Feller [1940] has shown that, if $\sum_k P_{ik}(\tau, t) = 1$, there always exists a unique solution $P_{ik}(\tau, t)$ satisfying both the forward and the backward differential equations as well as the Chapman-Kolmogorov equation.

In general the solution $P_{ik}(\tau, t)$ is nonhomogeneous with respect to time; however, it may be shown that if the intensity functions $\nu_{ij}(t)$ are equal to constant ν_{ij} independent of time, then the process is homogeneous with respect to time. In this case, the two systems of differential equations become

$$\frac{d}{dt} P_{ik}(t) = \sum_j P_{ij}(t)\nu_{jk} \tag{2.17}$$

and

$$\frac{d}{dt} P_{ik}(t) = \sum_j \nu_{ij} P_{jk}(t), \tag{2.18}$$

with the common initial conditions

$$P_{ik}(0) = \delta_{ik}. \tag{2.19}$$

Kolmogorov's differential equations may be more compactly presented in matrix notation. Let

$$\mathbf{P}(\tau, t) = \begin{pmatrix} P_{00}(\tau, t) & P_{01}(\tau, t) & \cdots \\ P_{10}(\tau, t) & P_{11}(\tau, t) & \cdots \\ \cdot & \cdot & \cdot \\ \cdot & \cdot & \cdot \\ \cdot & \cdot & \cdot \end{pmatrix} \tag{2.20}$$

be the *transition probability matrix* and

$$\mathbf{V}(t) = \begin{pmatrix} \nu_{00}(t) & \nu_{01}(t) & \cdots \\ \nu_{10}(t) & \nu_{11}(t) & \cdots \\ \cdot & \cdot & \cdot \\ \cdot & \cdot & \cdot \\ \cdot & \cdot & \cdot \end{pmatrix} \tag{2.21}$$

be the *intensity function matrix*. Then the Chapman-Kolmogorov equation (1.6) becomes

$$\mathbf{P}(\tau, t) = \mathbf{P}(\tau, \xi)\mathbf{P}(\xi, t). \tag{2.22}$$

The forward differential equations (2.12) can be written

$$\frac{\partial}{\partial t}\mathbf{P}(\tau, t) = \mathbf{P}(\tau, t)\mathbf{V}(t) \tag{2.23}$$

with the initial condition

$$\mathbf{P}(\tau, \tau) = \mathbf{I} \tag{2.24}$$

and the backward differential equations in (2.15) can be written

$$\frac{\partial}{\partial \tau}\mathbf{P}(\tau, t) = -\mathbf{V}(\tau)\mathbf{P}(\tau, t) \tag{2.25}$$

with the initial condition

$$\mathbf{P}(t, t) = \mathbf{I} \tag{2.26}$$

where **I** is an identity matrix.

We shall present two explicit solutions for the finite time-homogeneous case in Section 4. We conclude this section by restating some of the stochastic processes in this new formulation.

2.2. Examples

Example 1. In the *Poisson process* the intensity matrix is

$$\mathbf{V} = \begin{pmatrix} -\lambda & \lambda & 0 & 0 & \cdots \\ 0 & -\lambda & \lambda & 0 & \cdots \\ 0 & 0 & -\lambda & \lambda & \cdots \\ \cdots & \cdots & \cdots & \cdots & \cdots \end{pmatrix} \tag{2.27}$$

and the Kolmogorov equations are

$$\begin{aligned} \frac{\partial}{\partial t}P_{ij}(\tau, t) &= P_{ij}(\tau, t)(-\lambda) + P_{i,j-1}(\tau, t)\lambda \\ \frac{\partial}{\partial \tau}P_{ij}(\tau, t) &= \lambda P_{ij}(\tau, t) - \lambda P_{i+1,j}(\tau, t). \end{aligned} \tag{2.28}$$

►

Example 2. In the *pure death process* the intensity matrix is

$$\mathbf{V}(t) = \begin{pmatrix} 0 & 0 & 0 & 0 & \cdots \\ \mu(t) & -\mu(t) & 0 & 0 & \cdots \\ 0 & 2\mu(t) & -2\mu(t) & 0 & \cdots \\ 0 & 0 & 3\mu(t) & -3\mu(t) & \cdots \\ \cdots & \cdots & \cdots & \cdots & \cdots \end{pmatrix} \tag{2.29}$$

and the Kolmogorov differential equations are

$$\begin{aligned} \frac{\partial}{\partial t} P_{ij}(\tau, t) &= P_{ij}(\tau, t)[-j\mu(t)] + P_{i,j+1}(\tau, t)(j+1)\mu(t) \\ \frac{\partial}{\partial \tau} P_{ij}(\tau, t) &= i\mu(\tau)P_{ij}(\tau, t) - i\mu(\tau)P_{i-1,j}(\tau, t). \end{aligned} \tag{2.30}$$

▶

Example 3. The intensity matrix for the *general birth-death process* is

$$\mathbf{V}(t) = \begin{pmatrix} 0 & 0 & 0 & 0 & \cdots \\ \mu(t) & -[\lambda(t)+\mu(t)] & \lambda(t) & 0 & \cdots \\ 0 & 2\mu(t) & -2[\lambda(t)+\mu(t)] & 2\lambda(t) & \cdots \\ \cdots & \cdots & \cdots & \cdots & \cdots \\ \cdots & \cdots & \cdots & \cdots & \cdots \end{pmatrix}. \tag{2.31}$$

The two systems of equations are

$$\begin{aligned} \frac{\partial}{\partial t} P_{ij}(\tau, t) &= P_{i,j-1}(\tau, t)(j-1)\lambda(t) + P_{ij}(\tau, t)\{-j[\lambda(t)+\mu(t)]\} \\ &\quad + P_{i,j+1}(\tau, t)(j+1)\mu(t) \\ \frac{\partial}{\partial \tau} P_{ij}(\tau, t) &= -i\lambda(\tau)P_{i+1,j}(\tau, t) + i[\lambda(\tau)+\mu(\tau)]P_{ij}(\tau, t) \\ &\quad - i\mu(\tau)P_{i-1,j}(\tau, t). \end{aligned} \tag{2.32}$$

▶

3. MATRICES, EIGENVALUES, AND DIAGONALIZATION

Before proceeding to explicit solutions of the Kolmogorov differential equations, let us review briefly the necessary concepts from linear algebra. Reference may be made to Birkhoff and MacLane [1953] and Gantmacher [1960]. First we recall a few definitions.

Let

$$\mathbf{W} = \begin{pmatrix} w_{11} & w_{12} & \cdots & w_{1s} \\ w_{21} & w_{22} & \cdots & w_{2s} \\ \cdot & \cdot & \cdot & \cdot \\ \cdot & \cdot & \cdot & \cdot \\ \cdot & \cdot & \cdot & \cdot \\ w_{s1} & w_{s2} & \cdots & w_{ss} \end{pmatrix} \tag{3.1}$$

be any $s \times s$ matrix of elements w_{ij}; let $\mathbf{W}'$ be its transpose, and $|W|$ its determinant. A *minor* of order r of $\mathbf{W}$ is any $r \times r$ submatrix of $\mathbf{W}$ obtained by crossing out $s - r$ rows and $s - r$ columns; a *principal minor* is a minor whose diagonal elements lie on the diagonal of $\mathbf{W}$. A *cofactor* W_{ij} of the (i, j) element of $\mathbf{W}$ is $(-1)^{i+j}$ times the determinant of the minor of order $s - 1$ obtained by deleting from $\mathbf{W}$ the ith row and the jth column. The determinant $|W|$ of $\mathbf{W}$ can be expanded in terms of the cofactors of any row:

$$|W| = \sum_{j=1}^{s} w_{ij} W_{ij}, \qquad i = 1, \ldots, s; \tag{3.2}$$

or of any column:

$$|W| = \sum_{i=1}^{s} w_{ij} W_{ij}, \qquad j = 1, \ldots, s. \tag{3.3}$$

On the other hand, if the expansion is made in terms of the cofactors of a different row or column, the sum vanishes:

$$\sum_{j=1}^{s} w_{ij} W_{kj} = 0 \qquad \text{if} \qquad k \neq i \tag{3.4}$$

$$\sum_{i=1}^{s} w_{ij} W_{ik} = 0 \qquad \text{if} \qquad k \neq j. \tag{3.5}$$

$\mathbf{W}$ is called a *singular matrix* if its determinant $|W| = 0$, *nonsingular* if $|W| \neq 0$. $\mathbf{W}$ is nonsingular if all the columns (and, in fact, all the rows) are linearly independent. The *rank* of a matrix is the size of its largest nonsingular square submatrix; thus a matrix is of rank r if it has at least one $r \times r$ nonsingular submatrix and if all larger submatrices are singular.

Let W_{ij} be the cofactors of $\mathbf{W}$; the matrix

$$\mathbf{M} = \begin{pmatrix} W_{11} & W_{21} & \cdots & W_{s1} \\ W_{12} & W_{22} & \cdots & W_{s2} \\ \cdot & \cdot & & \cdot \\ \cdot & \cdot & & \cdot \\ \cdot & \cdot & & \cdot \\ W_{1s} & W_{2s} & \cdots & W_{ss} \end{pmatrix} \tag{3.6}$$

is called the *adjoint matrix* of **W** (notice that the indices are transposed!) It is clear from (3.2) to (3.5) that

$$\mathbf{MW} = \begin{pmatrix} W_{11} & W_{21} & \cdots & W_{s1} \\ W_{12} & W_{22} & \cdots & W_{s2} \\ \cdot & \cdot & & \cdot \\ \cdot & \cdot & & \cdot \\ \cdot & \cdot & & \cdot \\ W_{1s} & W_{2s} & \cdots & W_{ss} \end{pmatrix} \begin{pmatrix} w_{11} & w_{12} & \cdots & w_{1s} \\ w_{21} & w_{22} & \cdots & w_{2s} \\ \cdot & \cdot & & \cdot \\ \cdot & \cdot & & \cdot \\ \cdot & \cdot & & \cdot \\ w_{s1} & w_{s2} & \cdots & w_{ss} \end{pmatrix}$$

$$= \begin{pmatrix} |W| & 0 & \cdots & 0 \\ 0 & |W| & \cdots & 0 \\ \cdot & \cdot & & \cdot \\ \cdot & \cdot & & \cdot \\ \cdot & \cdot & & \cdot \\ 0 & 0 & \cdots & |W| \end{pmatrix} = \mathbf{WM}. \tag{3.7}$$

Thus we have

$$\mathbf{MW} = \mathbf{WM} = |W|\,\mathbf{I}, \tag{3.8}$$

where **I** is an $s \times s$ identity matrix.

Suppose **W** is nonsingular so that $|W| \neq 0$, let

$$\mathbf{W}^{-1} = \begin{pmatrix} \dfrac{W_{11}}{|W|} & \dfrac{W_{21}}{|W|} & \cdots & \dfrac{W_{s1}}{|W|} \\ \dfrac{W_{12}}{|W|} & \dfrac{W_{22}}{|W|} & \cdots & \dfrac{W_{s2}}{|W|} \\ \cdot & \cdot & & \cdot \\ \cdot & \cdot & & \cdot \\ \dfrac{W_{1s}}{|W|} & \dfrac{W_{2s}}{|W|} & \cdots & \dfrac{W_{ss}}{|W|} \end{pmatrix}. \tag{3.9}$$

From (3.8)

$$\mathbf{WW}^{-1} = \mathbf{I} = \mathbf{W}^{-1}\mathbf{W}. \tag{3.10}$$

If **W** has an inverse $\mathbf{W}^{-1}$, then $|W|\,|W^{-1}| = |WW^{-1}| = |I| = 1$, therefore the determinant $|W| \neq 0$. Now let w_{jk}^{-1} be the (j, k) element of an inverse of **W** so that $\sum_{j=1}^{s} w_{ij}w_{jk}^{-1} = \delta_{ik}$; from (3.2) and (3.4) we see that $w_{jk}^{-1} = W_{kj}/|W|$. *Thus a square matrix* **W** *has an inverse if, and only if,* $|W| \neq 0$*; and the matrix* $\mathbf{W}^{-1}$ *given in* (3.9) *is the unique inverse of* **W**. ▶

3.1. Eigenvalues and Eigenvectors

If there is a nonzero column vector **t** and a scalar ρ satisfying

$$\mathbf{Wt} = \rho\mathbf{t} \tag{3.11}$$

then ρ is called an *eigenvalue* or *characteristic root* of **W**, and **t** is an *eigenvector* of **W** corresponding to the eigenvalue ρ. Clearly an eigenvector multiplied by a scalar is also an eigenvector corresponding to the same eigenvalue.

Rewriting (3.11) as

$$(\rho\mathbf{I} - \mathbf{W})\mathbf{t} = \mathbf{0} \tag{3.12}$$

we see that (3.11) holds true if and only if the *characteristic matrix* of **W**

$$(\rho\mathbf{I} - \mathbf{W}) = \mathbf{A} \tag{3.13}$$

is singular; or if and only if the determinant $|A| = 0$. Hence ρ is an eigenvalue of **W** if and only if

$$|A| = |\rho\mathbf{I} - \mathbf{W}| = 0. \tag{3.14}$$

Equation (3.14) is known as the *characteristic equation* of the matrix **W**, and the determinant $|A|$ is known as the *characteristic polynomial* of **W**. Since $|A|$ is a polynomial of degree s in the unknown ρ, it follows from a theorem in algebra that (3.14) has exactly s roots, which we shall denote by $\rho_1, \ldots, \rho_s$.

The characteristic polynomial may be expanded in powers of ρ:

$$|\rho\mathbf{I} - \mathbf{W}| = \rho^s - b_1\rho^{s-1} + b_2\rho^{s-2} - \cdots + (-1)^{s-1}b_{s-1}\rho + (-1)^s b_s \tag{3.15}$$

where the b_i are constants. It may be shown by straightforward computations that the coefficient b_r is the sum of the determinants of all principal minors of **W** of order r; for example, $b_1 = w_{11} + \cdots + w_{ss}$; $b_{s-1} = W_{11} + \cdots + W_{ss}$; and $b_s = |W|$. The coefficient b_1 is called the *trace* or *spur* of W, and is usually written tr W or sp W. The characteristic polynomial may also be written in terms of the roots $\rho_1, \ldots, \rho_s$ as

$$|\rho\mathbf{I} - \mathbf{W}| = (\rho - \rho_1)\cdots(\rho - \rho_s). \tag{3.16}$$

Expanding (3.16) in powers of ρ and equating coefficients of corresponding powers in (3.15), we find that for each r the coefficient b_r is equal to the sum of all products of the ρ_i taken r at a time. In particular

$$\begin{aligned} b_1 &= \sum_{i=1}^{s} w_{ii} = \sum_{i=1}^{s} \rho_i \\ b_{s-1} &= \sum_{i=1}^{s} W_{ii} = \rho_1\rho_2\cdots\rho_{s-1} + \cdots + \rho_2\rho_3\cdots\rho_s \\ b_s &= |W| = \rho_1\rho_2\cdots\rho_s. \end{aligned} \tag{3.17}$$

▶

The eigenvalues ρ_j may be real or complex numbers and may or may not be distinct. In any event, the *eigenvectors corresponding to distinct eigenvalues are linearly independent.* This may be easily proved by contradiction as follows: consider r distinct eigenvalues $\rho_1, \ldots, \rho_r$ with the corresponding (column) eigenvectors $\mathbf{T}_1, \ldots, \mathbf{T}_r$, that is,

$$\mathbf{W}\mathbf{T}_j = \rho_j \mathbf{T}_j, \qquad j = 1, \ldots, r. \tag{3.18}$$

Now suppose that the eigenvectors are linearly dependent, so that there exist constants c_j not all of which are zero such that

$$\sum_{j=1}^{r} c_j \mathbf{T}_j = \mathbf{0}. \tag{3.19}$$

Without loss of generality, we may assume that $c_1 \neq 0$. Consider the matrix

$$\mathbf{U} = (\rho_2\mathbf{I} - \mathbf{W})(\rho_3\mathbf{I} - \mathbf{W}) \cdots (\rho_r\mathbf{I} - \mathbf{W}) \tag{3.20}$$

and observe that

$$\mathbf{U}\mathbf{T}_r = (\rho_2\mathbf{I} - \mathbf{W}) \cdots (\rho_{r-1}\mathbf{I} - \mathbf{W})(\rho_r\mathbf{T}_r - \mathbf{W}\mathbf{T}_r) = \mathbf{0} \tag{3.21}$$

since the last factor vanished because of (3.18). Similarly we have

$$\mathbf{U}\mathbf{T}_2 = \mathbf{U}\mathbf{T}_3 = \cdots = \mathbf{U}\mathbf{T}_r = \mathbf{0}. \tag{3.22}$$

On the other hand,

$$\begin{aligned} \mathbf{U}\mathbf{T}_1 &= (\rho_2\mathbf{I} - \mathbf{W}) \cdots (\rho_{r-1}\mathbf{I} - \mathbf{W})(\rho_r\mathbf{T}_1 - \mathbf{W}\mathbf{T}_1) \\ &= (\rho_2\mathbf{I} - \mathbf{W}) \cdots (\rho_{r-1}\mathbf{I} - \mathbf{W})(\rho_r - \rho_1)\mathbf{T}_1 \\ &= (\rho_2\mathbf{I} - \mathbf{W}) \cdots (\rho_{r-1}\mathbf{T}_1 - \mathbf{W}\mathbf{T}_1)(\rho_r - \rho_1) \\ &= (\rho_2\mathbf{I} - \mathbf{W}) \cdots (\rho_{r-1} - \rho_1)(\rho_r - \rho_1)\mathbf{T}_1 \\ &= \cdots \\ &= (\rho_2 - \rho_1) \cdots (\rho_{r-1} - \rho_1)(\rho_r - \rho_1)\mathbf{T}_1. \end{aligned} \tag{3.23}$$

From (3.19), (3.22), and (3.23), we see that

$$\begin{aligned} \mathbf{0} = \mathbf{U}\sum_{j=1}^{r} c_j\mathbf{T}_j = \sum_{j=1}^{r} c_j\mathbf{U}\mathbf{T}_j &= c_1\mathbf{U}\mathbf{T}_1 \\ &= c_1(\rho_2 - \rho_1) \cdots (\rho_r - \rho_1)\mathbf{T}_1. \end{aligned} \tag{3.24}$$

Since the ρ_j are all distinct, (3.24) implies that c_1 must be zero, which contradicts the assumption that $c_1 \neq 0$, and proves that eigenvectors are linearly independent. ▶

3.2. Diagonalization of a Matrix

If all the eigenvalues $\rho_1, \ldots, \rho_s$ of an $s \times s$ matrix **W** are distinct, then the matrix of the corresponding eigenvectors

$$\mathbf{T} = (\mathbf{T}_1, \ldots, \mathbf{T}_s) \tag{3.25}$$

has linearly independent columns and is therefore nonsingular. In this case, we may form the matrix

$$\mathbf{T}^{-1}\mathbf{W}\mathbf{T}. \tag{3.26}$$

To examine the structure of the matrix (3.26), we use (3.25) and (3.18) to write

$$\mathbf{W}\mathbf{T} = (\mathbf{W}\mathbf{T}_1, \ldots, \mathbf{W}\mathbf{T}_s) = (\rho_1\mathbf{T}_1, \ldots, \rho_s\mathbf{T}_s) \tag{3.27}$$

and hence,

$$\mathbf{T}^{-1}\mathbf{W}\mathbf{T} = (\rho_1\mathbf{T}^{-1}\mathbf{T}_1, \ldots, \rho_s\mathbf{T}^{-1}\mathbf{T}_s). \tag{3.28}$$

Now we write the identity $\mathbf{T}^{-1}\mathbf{T} = \mathbf{I}$ as

$$(\mathbf{T}^{-1}\mathbf{T}_1, \ldots, \mathbf{T}^{-1}\mathbf{T}_s) = \begin{pmatrix} 1 & 0 & \cdots & 0 \\ 0 & 1 & \cdots & 0 \\ \cdot & \cdot & & \cdot \\ \cdot & \cdot & & \cdot \\ \cdot & \cdot & & \cdot \\ 0 & 0 & \cdots & 1 \end{pmatrix} \tag{3.29}$$

and apply (3.29) to (3.28) to give

$$\mathbf{T}^{-1}\mathbf{W}\mathbf{T} = \begin{pmatrix} \rho_1 & 0 & \cdots & 0 \\ 0 & \rho_2 & \cdots & 0 \\ \cdot & \cdot & & \cdot \\ \cdot & \cdot & & \cdot \\ \cdot & \cdot & & \cdot \\ 0 & 0 & \cdots & \rho_s \end{pmatrix} = \boldsymbol{\rho}. \tag{3.30}$$

(3.30) is known as the *diagonalized* form of the matrix **W**. *A necessary and sufficient condition for an $s \times s$ matrix to be diagonalizable by a transformation as in* (3.30) *is that it possess s linearly independent eigenvectors; in particular*, *as we have seen*, *it is sufficient that the matrix have s distinct eigenvalues.* ▶

Since an eigenvector $\mathbf{T}_j$ multiplied by a nonzero constant is also an eigenvector corresponding to the same eigenvalue ρ_j, the matrix **T** which performs the diagonalization (3.30) is not unique. Let $c_j \neq 0$ be nonzero constants, and

$$\mathbf{R} = (c_1\mathbf{T}_1, \ldots, c_s\mathbf{T}_s), \tag{3.31}$$

we also have

$$\mathbf{R}^{-1}\mathbf{W}\mathbf{R} = \begin{pmatrix} \rho_1 & 0 & \cdots & 0 \\ 0 & \rho_2 & \cdots & 0 \\ \cdot & \cdot & & \cdot \\ \cdot & \cdot & & \cdot \\ \cdot & \cdot & & \cdot \\ 0 & 0 & \cdots & \rho_s \end{pmatrix}. \tag{3.32}$$

3.3. A Useful Lemma

In further discussion of diagonalized matrices and in deriving solutions for Kolmogorov differential equations, we use the following lemma.

Lemma. *For any distinct numbers,* $\rho_1, \ldots, \rho_s$, *we have*

$$\sum_{i=1}^{s} \frac{\rho_i^{\,r}}{\prod\limits_{\substack{j=1\\ j\neq i}}^{s} (\rho_i - \rho_j)} = 0 \qquad for \qquad 0 \le r < s-1$$

$$= 1 \qquad for \qquad r = s - 1. \tag{3.33}$$

The lemma can be proved by induction (Chiang [1964a]), by partial fractions, and by other methods. In the following proof, we use Lagrange's interpolation formula.[2] A slightly different form of the lemma may be found in Polya and Szegö [1964].

Proof. Consider a polynomial of degree $s - 1$

$$f(\rho) = a_1\rho^{s-1} + a_2\rho^{s-2} + \cdots + a_s \tag{3.34}$$

with $a_i \neq 0$ and let $\rho_1, \ldots, \rho_s$ be distinct numbers. Then

$$f(\rho) = \sum_{i=1}^{s} \frac{(\rho - \rho_1)\cdots(\rho - \rho_{i-1})(\rho - \rho_{i+1})\cdots(\rho - \rho_s)}{(\rho_i - \rho_1)\cdots(\rho_i - \rho_{i-1})(\rho_i - \rho_{i+1})\cdots(\rho_i - \rho_s)} f(\rho_i) \tag{3.35}$$

is an identity in ρ. The right side of (3.35), which is known as Lagrange's interpolation formula, is a polynomial of degree $s - 1$ which reduces to $f(\rho_1), \ldots, f(\rho_s)$ at the s distinct points $\rho_1, \ldots, \rho_s$. It is often used to approximate an arbitrary function $f(\rho)$. In our case, with $f(\rho)$ of the form (3.34), the difference between the two sides of (3.35) is a polynomial of degree $s - 1$ with s distinct roots and hence must be identically zero;, therefore (3.35) holds true for all ρ.

---

[2] I benefited from a useful discussion with Grace Lo Yang on this line of proof of the lemma.

Taking the $(s - 1)$th derivative of (3.35) with respect to ρ, we find on the left side

$$\frac{d^{s-1}}{d\rho^{s-1}} f(\rho) = (s-1)!\, a_1 \tag{3.36}$$

and on the right side

$$(s-1)! \sum_{i=1}^{s} \frac{1}{\prod_{\substack{j=1 \\ j\neq i}}^{s} (\rho_i - \rho_j)} f(\rho_i). \tag{3.37}$$

Substituting

$$f(\rho_i) = a_1 \rho_i^{s-1} + a_2 \rho_i^{s-2} + \cdots + a_s, \tag{3.38}$$

(3.37) can be rewritten as

$$(s-1)! \left[a_1 \sum_{i=1}^{s} \frac{\rho_i^{s-1}}{\prod_{\substack{j=1 \\ j\neq i}}^{s} (\rho_i - \rho_j)} + a_2 \sum_{i=1}^{s} \frac{\rho_i^{s-2}}{\prod_{\substack{j=1 \\ j\neq i}}^{s} (\rho_i - \rho_j)} + \cdots + a_s \sum_{i=1}^{s} \frac{\rho_i^{0}}{\prod_{\substack{j=1 \\ j\neq i}}^{s} (\rho_i - \rho_j)} \right]. \tag{3.39}$$

Since (3.39) must be equal to the right side of (3.36) whatever may be $a_i \neq 0$, the first sum inside the brackets must be one and the other sums must be zero, and the lemma is proved. ►

3.4. Matrix of Eigenvectors

In Section 3.2 we were concerned with some basic properties of eigenvectors and the role they play in diagonalizing a matrix. The essential remaining problem is, for a given eigenvalue ρ_j, how an eigenvector $\mathbf{T}_j$ can be found. The present section provides eigenvectors and the matrix $\mathbf{T}$ which diagonalizes the original matrix $\mathbf{W}$. Consider the characteristic matrix

$$\mathbf{A}(j) = (\rho_j \mathbf{I} - \mathbf{W}) \tag{3.40}$$

corresponding to ρ_j, and let $A_{ki}(j)$ be the cofactors of $\mathbf{A}(j)$. Then the nonzero vector

$$\mathbf{T}_j(k) = \begin{pmatrix} A_{k1}(j) \\ A_{k2}(j) \\ \cdot \\ \cdot \\ \cdot \\ A_{ks}(j) \end{pmatrix} \tag{3.41}$$

is an eigenvector of $\mathbf{W}$ for $\rho = \rho_j$ whatever may be $k = 1, 2, \ldots, s$. In other words, *any nonzero column of the adjoint matrix of* $\mathbf{A}(j)$ *is an eigenvector corresponding to eigenvalue* ρ_j. The proof is straightforward:

$$(\rho_j\mathbf{I} - \mathbf{W})\mathbf{T}_j(k) = \mathbf{A}(j)\begin{pmatrix} A_{k1}(j) \\ A_{k2}(j) \\ \cdot \\ \cdot \\ \cdot \\ A_{ks}(j) \end{pmatrix} = \begin{pmatrix} 0 \\ 0 \\ \cdot \\ \cdot \\ \cdot \\ 0 \end{pmatrix}. \tag{3.42}$$

The last equation is easily justified. For $i \neq k$, the ith component represents an expansion of $|A(j)|$ in terms of cofactors of a different row and therefore must vanish, whereas the kth element is equal to the determinant $|A(j)|$ which also vanishes, since ρ_j is a root of the equation $|\rho\mathbf{I} - \mathbf{W}| = 0$. ▶

When the eigenvalues $\rho_1, \ldots, \rho_s$ are distinct, we may select any nonzero column from each adjoint matrix of $\mathbf{A}(j), j = 1, \ldots, s$, to formulate a matrix $\mathbf{T}$ which will diagonalize $\mathbf{W}$ as in (3.30). However, there is great advantage in selecting the *same* column from each of the adjoint matrices. A typical example is

$$\mathbf{T}(k) = \begin{pmatrix} A_{k1}(1) & A_{k1}(2) & \cdots & A_{k1}(s) \\ A_{k2}(1) & A_{k2}(2) & \cdots & A_{k2}(s) \\ \cdot & \cdot & & \cdot \\ \cdot & \cdot & & \cdot \\ \cdot & \cdot & & \cdot \\ A_{ks}(1) & A_{ks}(2) & \cdots & A_{ks}(s) \end{pmatrix}. \tag{3.43}$$

The jth column in (3.43) is the kth column of the adjoint matrix of $\mathbf{A}(j) = (\rho_j\mathbf{I} - \mathbf{W})$ and k is the same for all j. Since the eigenvalues are assumed to be distinct, the matrix $\mathbf{T}(k)$ consists of s linearly independent nonzero vectors and satisfies the relation

$$\mathbf{T}^{-1}(k)\mathbf{W}\mathbf{T}(k) = \boldsymbol{\rho} \tag{3.44}$$

where the inverse is

$$\mathbf{T}^{-1}(k) = \begin{pmatrix} \dfrac{T_{11}(k)}{|T(k)|} & \cdots & \dfrac{T_{k1}(k)}{|T(k)|} & \cdots & \dfrac{T_{s1}(k)}{|T(k)|} \\ \dfrac{T_{12}(k)}{|T(k)|} & \cdots & \dfrac{T_{k2}(k)}{|T(k)|} & \cdots & \dfrac{T_{s2}(k)}{|T(k)|} \\ \cdot & \cdot & \cdot & \cdot & \cdot \\ \dfrac{T_{1s}(k)}{|T(k)|} & \cdots & \dfrac{T_{ks}(k)}{|T(k)|} & \cdots & \dfrac{T_{ss}(k)}{|T(k)|} \end{pmatrix} \tag{3.45}$$

with $T_{ij}(k)$ being the cofactors of $\mathbf{T}(k)$.

Clearly the matrix $\mathbf{T}(k)$ varies with the choice of k. However, the inverse matrices $\mathbf{T}^{-1}(1), \ldots, \mathbf{T}^{-1}(s)$ have an important feature in common: *the* kth *column of the inverse matrix* $\mathbf{T}^{-1}(k)$ *is independent of* k *and is given by*

$$\begin{pmatrix} \dfrac{T_{11}(1)}{|T(1)|} \\ \dfrac{T_{12}(1)}{|T(1)|} \\ \cdot \\ \cdot \\ \cdot \\ \dfrac{T_{1s}(1)}{|T(1)|} \end{pmatrix} = \cdots = \begin{pmatrix} \dfrac{T_{s1}(s)}{|T(s)|} \\ \dfrac{T_{s2}(s)}{|T(s)|} \\ \cdot \\ \cdot \\ \cdot \\ \dfrac{T_{ss}(s)}{|T(s)|} \end{pmatrix} = \begin{pmatrix} \dfrac{1}{\prod_{l=2}^{s} (\rho_1 - \rho_l)} \\ \dfrac{1}{\prod_{\substack{l=1 \\ l \neq 2}}^{s} (\rho_2 - \rho_l)} \\ \cdot \\ \cdot \\ \cdot \\ \dfrac{1}{\prod_{l=1}^{s-1} (\rho_s - \rho_l)} \end{pmatrix}. \tag{3.46}$$

Before presenting the proof of (3.46), let us illustrate this with an example.

Example. In the illness-death process in Chapter 4, the matrix $\mathbf{V}$ and its characteristic matrix are

$$\mathbf{V} = \begin{pmatrix} \nu_{11} & \nu_{12} \\ \nu_{21} & \nu_{22} \end{pmatrix}$$

and

$$\mathbf{A}(j) = \begin{pmatrix} \rho_j - \nu_{11} & -\nu_{12} \\ -\nu_{21} & \rho_j - \nu_{22} \end{pmatrix}$$

and the eigenvalues are

$$\rho_1 = \tfrac{1}{2}[(\nu_{11} + \nu_{22}) + \sqrt{(\nu_{11} - \nu_{22})^2 + 4\nu_{12}\nu_{21}}]$$

$$\rho_2 = \tfrac{1}{2}[(\nu_{11} + \nu_{22}) - \sqrt{(\nu_{11} - \nu_{22})^2 + 4\nu_{12}\nu_{21}}].$$

Case 1. $k = 1$. If the *first* column of each of the adjoint matrices of $\mathbf{A}(j)$ is taken, we have

$$\mathbf{T}(1) = \begin{pmatrix} A_{11}(1) & A_{11}(2) \\ A_{12}(1) & A_{12}(2) \end{pmatrix} = \begin{pmatrix} \rho_1 - \nu_{22} & \rho_2 - \nu_{22} \\ \nu_{21} & \nu_{21} \end{pmatrix}$$

and the determinant $|T(1)| = \nu_{21}(\rho_1 - \rho_2)$. Direct computations give the inverse

$$\mathbf{T}^{-1}(1) = \begin{pmatrix} \dfrac{T_{11}(1)}{|T(1)|} & \dfrac{T_{21}(1)}{|T(1)|} \\ \dfrac{T_{12}(1)}{|T(1)|} & \dfrac{T_{22}(1)}{|T(1)|} \end{pmatrix} = \begin{pmatrix} \dfrac{1}{\rho_1 - \rho_2} & -\dfrac{\rho_2 - \nu_{22}}{\nu_{21}(\rho_1 - \rho_2)} \\ \dfrac{1}{\rho_2 - \rho_1} & -\dfrac{\rho_1 - \nu_{22}}{\nu_{21}(\rho_2 - \rho_1)} \end{pmatrix}$$

where the *first* column is given in (3.46).

Case 2. $k = 2$. If the *second* column of each of the adjoint matrices of $\mathbf{A}(j)$ is taken, we have

$$\mathbf{T}(2) = \begin{pmatrix} A_{21}(1) & A_{21}(2) \\ A_{22}(1) & A_{22}(2) \end{pmatrix} = \begin{pmatrix} \nu_{12} & \nu_{12} \\ \rho_1 - \nu_{11} & \rho_2 - \nu_{11} \end{pmatrix}$$

and the determinant $|T(2)| = \nu_{12}(\rho_2 - \rho_1)$. The inverse is

$$\mathbf{T}^{-1}(2) = \begin{pmatrix} \dfrac{T_{11}(2)}{|T(2)|} & \dfrac{T_{21}(2)}{|T(2)|} \\ \dfrac{T_{12}(2)}{|T(2)|} & \dfrac{T_{22}(2)}{|T(2)|} \end{pmatrix} = \begin{pmatrix} -\dfrac{\rho_2 - \nu_{11}}{\nu_{12}(\rho_1 - \rho_2)} & \dfrac{1}{\rho_1 - \rho_2} \\ -\dfrac{\rho_1 - \nu_{11}}{\nu_{12}(\rho_2 - \rho_1)} & \dfrac{1}{\rho_2 - \rho_1} \end{pmatrix}$$

where the *second* column is given in (3.46), and is also equal to the first column of $\mathbf{T}^{-1}(1)$. ▶

Proof of (3.46). Use $\mathbf{T}(k)$ in (3.43) to formulate the following system of simultaneous equations

$$\mathbf{T}(k) \begin{pmatrix} z_1 \\ z_2 \\ \cdot \\ \cdot \\ \cdot \\ z_s \end{pmatrix} = \begin{pmatrix} \delta_{1k} \\ \delta_{2k} \\ \cdot \\ \cdot \\ \cdot \\ \delta_{sk} \end{pmatrix} \tag{3.47}$$

where the components z_j of the column vector $\mathbf{z}$ are unknown constants and δ_{jk} is the Kronecker delta. Obviously the solution of (3.47) is

$$z_j = \frac{T_{kj}(k)}{|T(k)|}, \qquad j = 1, \ldots, s. \tag{3.48}$$

In scalar notation, equation (3.47) can be written

$$\left.\begin{array}{l} A_{k1}(1)z_1 + \cdots + A_{k1}(s)z_s = 0 \\ \cdots\cdots\cdots\cdots\cdots\cdots\cdots\cdots \\ A_{kk}(1)z_1 + \cdots + A_{kk}(s)z_s = 1 \\ \cdots\cdots\cdots\cdots\cdots\cdots\cdots\cdots \\ A_{ks}(1)z_1 + \cdots + A_{ks}(s)z_s = 0. \end{array}\right\} \tag{3.49}$$

The cofactors $A_{ki}(j)$ in (3.49) may be expanded in polynomials of ρ_j as in (3.15),

$$\begin{aligned} A_{kk}(j) &= \rho_j^{s-1} + b_{kk2}\rho_j^{s-2} + \cdots + b_{kks}\rho_j^{\,0} \\ A_{ki}(j) &= \qquad\quad b_{ki2}\rho_j^{s-2} + \cdots + b_{kis}\rho_j^{\,0}, \qquad i \neq k \end{aligned} \tag{3.50}$$

where the coefficients b are functions of w_{ij} but are independent of ρ_j and j. When (3.50) is introduced in (3.49) and the terms rearranged according to the coefficients b, the resulting equations are

$$\left.\begin{array}{r} b_{k12}\sum_{j=1}^{s} z_j\rho_j^{s-2} + \cdots + b_{k1s}\sum_{j=1}^{s} z_j\rho_j^{\,0} = 0 \\ \cdots\cdots\cdots\cdots\cdots\cdots\cdots\cdots\cdots\cdots \\ \sum_{j=1}^{s} z_j\rho_j^{s-1} + b_{kk2}\sum_{j=1}^{s} z_j\rho_j^{s-2} + \cdots + b_{kks}\sum_{j=1}^{s} z_j\rho_j^{\,0} = 1 \\ \cdots\cdots\cdots\cdots\cdots\cdots\cdots\cdots\cdots\cdots \\ b_{ks2}\sum_{j=1}^{s} z_j\rho_j^{s-2} + \cdots + b_{kss}\sum_{j=1}^{s} z_j\rho_j^{\,0} = 0. \end{array}\right\} \tag{3.51}$$

Our problem is to solve (3.51) for z_j. Recalling from the lemma in Section 3.3 that whatever the s distinct numbers $\rho_1, \ldots, \rho_s$,

$$\begin{aligned} \sum_{j=1}^{s} \frac{\rho_j^{\,r}}{\prod_{\substack{l=1 \\ l\neq j}}^{s} (\rho_j - \rho_l)} &= 1, \qquad r = s-1 \\ &= 0, \qquad 0 \le r < s-1 \end{aligned} \tag{3.33}$$

we see that the solution of (3.51) is

$$z_j = \frac{1}{\prod_{\substack{l=1 \\ l\neq j}}^{s} (\rho_j - \rho_l)}, \qquad j = 1, \ldots, s. \tag{3.52}$$

Since the determinant of $\mathbf{T}(k)$ is not zero, system (3.47) has a unique solution, hence solutions (3.48) and (3.52) are equal, proving Equation (3.46) (Chiang [1968]). ▶

4. EXPLICIT SOLUTIONS FOR KOLMOGOROV DIFFERENTIAL EQUATIONS

In this section we give two explicit solutions for the Kolmogorov differential equations for finite Markov processes homogeneous in time, and we show that the two solutions are identical. The argument used in deriving these solutions applies to any finite system of linear, first-order, homogeneous differential equations with constant coefficients; the general solutions in Sections 4.2 and 4.3 hold true also for functions other than transition probabilities.

Let a Markov process $\{X(t)\}$ consist of a finite number s of states denoted by $1, \ldots, s$, with intensity functions ν_{ij} independent of t. We assume that the system is closed so that for every i and whatever may be $t > 0$

$$\sum_{j=1}^{s} P_{ij}(t) = 1, \qquad i, j = 1, \ldots, s \tag{4.1}$$

or

$$\nu_{ii} = -\sum_{\substack{j=1 \\ j \neq i}}^{s} \nu_{ij}, \qquad i = 1, \ldots, s \tag{4.2}$$

so that the matrix

$$\mathbf{V} = \begin{pmatrix} \nu_{11} & \cdots & \nu_{1s} \\ \cdot & & \cdot \\ \cdot & & \cdot \\ \cdot & & \cdot \\ \nu_{s1} & \cdots & \nu_{ss} \end{pmatrix} \tag{4.3}$$

is singular.

In this chapter we also assume that in the system there are no absorbing states i for which $\nu_{ii} = 0$. The case in which there are absorbing states will be treated in Chapter 7.

In finite Markov processes the Kolmogorov differential equations are

$$\frac{d}{dt} P_{ik}(t) = \sum_{j=1}^{s} P_{ij}(t)\nu_{jk} \tag{4.4}$$

$$\frac{d}{dt} P_{ik}(t) = \sum_{j=1}^{s} \nu_{ij} P_{jk}(t) \tag{4.5}$$

with the initial conditions at $t = 0$

$$P_{ik}(0) = \delta_{ik}. \tag{4.6}$$

The corresponding matrix equations are

$$D\mathbf{P}(t) = \mathbf{P}(t)\mathbf{V} \qquad \text{or} \qquad (D - \mathbf{V}')\mathbf{P}'(t) = 0 \tag{4.7}$$

$$D\mathbf{P}(t) = \mathbf{V}\mathbf{P}(t) \qquad \text{or} \qquad (D - \mathbf{V})\mathbf{P}(t) = 0 \tag{4.8}$$

and

$$\mathbf{P}(0) = \mathbf{I} \tag{4.9}$$

where D is a diagonal matrix with the differentiation operator d/dt on the diagonal line, and $\mathbf{P}(t)$ is the transition probability matrix

$$\mathbf{P}(t) = \begin{pmatrix} P_{11}(t) & \cdots & P_{1s}(t) \\ \cdot & & \cdot \\ \cdot & & \cdot \\ \cdot & & \cdot \\ P_{s1}(t) & \cdots & P_{ss}(t) \end{pmatrix}. \tag{4.10}$$

Our problem is to derive explicit functions $P_{ik}(t)$ satisfying (4.4) and (4.5) and the Chapman-Kolmogorov equations

$$P_{ik}(\tau + t) = \sum_{j=1}^{s} P_{ij}(\tau)P_{jk}(t). \tag{1.7}$$

4.1. Intensity Matrix V and Its Eigenvalues

Let

$$\mathbf{A} = (\rho\mathbf{I} - \mathbf{V}) \tag{4.11}$$

be the characteristic matrix of $\mathbf{V}$ and

$$|A| = |\rho\mathbf{I} - \mathbf{V}| = 0 \tag{4.12}$$

be the corresponding characteristic equation. The roots of (4.12) are the eigenvalues of $\mathbf{V}$, denoted by $\rho_1, \ldots, \rho_s$. Because of (4.2), it is apparent that one eigenvalue, say ρ_1, is zero. In deriving an explicit solution, we assume that all the eigenvalues $\rho_1, \ldots, \rho_s$ are real and distinct; in particular, none of $\rho_2, \ldots, \rho_s$ is zero. We now show that $\rho_2, \ldots, \rho_s$ are negative.[3] Generally, *if a matrix* $\mathbf{V}$ *satisfies the conditions that* $v_{ij} > 0$ *for* $i \neq j$ *and* $\sum_j v_{ij} \leq 0$ *for each* i, *then any real eigenvalue of* $\mathbf{V}$ *is nonpositive.* Suppose ρ is any real eigenvalue of $\mathbf{V}$, then there is at least one real nonzero vector

$$\mathbf{C} = \begin{pmatrix} c_1 \\ \cdot \\ \cdot \\ \cdot \\ c_s \end{pmatrix} \tag{4.13}$$

[3] I am grateful to Myra Jordan Samuels for useful discussions on the sign of eigenvalue of $\mathbf{V}$. An alternative proof due to Mrs. Samuels is given in Problem 27.

such that

$$(\rho \mathbf{I} - \mathbf{V})\mathbf{C} = 0 \tag{4.14}$$

or, equivalently,

$$\rho c_j = v_{j1}c_1 + \cdots + v_{js}c_s, \qquad j = 1, \ldots, s. \tag{4.15}$$

Let c_α be the largest component in absolute value of the vector $\mathbf{C}$ in (4.13) so that

$$c_\alpha^{\ 2} \geq c_\alpha c_j, \qquad j = 1, \ldots, s. \tag{4.16}$$

For $j = \alpha$, we have from (4.15)

$$\rho c_\alpha^{\ 2} = v_{\alpha 1}c_1 c_\alpha + \cdots + v_{\alpha\alpha}c_\alpha^{\ 2} + \cdots + v_{\alpha s}c_s c_\alpha. \tag{4.17}$$

Using (4.16) and the relation $\sum_j v_{\alpha j} \leq 0$, we may rewrite (4.17) to obtain

$$\rho c_\alpha^{\ 2} \leq \left(\sum_{j=1}^{s} v_{\alpha j}\right) c_\alpha^{\ 2} \leq 0. \tag{4.18}$$

Therefore $\rho \leq 0$, proving the assertion. ▶

It is interesting to note that assumption (4.2) implies not only that the determinant $|V| = 0$ but also that *all elements in each row of* $\mathbf{V}$ *have the same cofactor*; that is, if we denote the cofactor of v_{ij} by V_{ij}, then we have for each i

$$V_{ij} = V_{ik}, \qquad j, k = 1, \ldots, s. \tag{4.19}$$

When $i = 1, j = 1$, and $k = 2$, for example, we have

$$V_{11} = \begin{vmatrix} v_{22} & v_{23} & \cdots & v_{2s} \\ v_{32} & v_{33} & \cdots & v_{3s} \\ \cdot & \cdot & & \cdot \\ \cdot & \cdot & & \cdot \\ \cdot & \cdot & & \cdot \\ v_{s2} & v_{s3} & \cdots & v_{ss} \end{vmatrix} \tag{4.20}$$

$$V_{12} = (-1) \begin{vmatrix} v_{21} & v_{23} & \cdots & v_{2s} \\ v_{31} & v_{33} & \cdots & v_{3s} \\ \cdot & \cdot & & \cdot \\ \cdot & \cdot & & \cdot \\ \cdot & \cdot & & \cdot \\ v_{s1} & v_{s3} & \cdots & v_{ss} \end{vmatrix} \tag{4.21}$$

where the only difference is in the first column. Replacing the first column in (4.21) by the sum of all the columns[4] and making the substitution

$$v_{i1} + v_{i3} + \cdots + v_{is} = -v_{i2} \tag{4.22}$$

for each row, we recover (4.20) and hence $V_{11} = V_{12}$. ▶

The eigenvalues and the cofactors V_{ij} of $\mathbf{V}$ have the relations in (3.17). In the present case, $\rho_1 = 0$, we have

$$\rho_2\rho_3 \cdots \rho_s = V_{11} + \cdots + V_{ss} \tag{4.23}$$

or, in view of (4.19),

$$\rho_2\rho_3 \cdots \rho_s = V_{1j} + \cdots + V_{sj}, \tag{4.24}$$

whatever may be $j = 1, \ldots, s$.

4.2. First Solution for Individual Transition Probabilities $P_{ij}(t)$

This solution was given by Chiang [1964a] in a study of competing risks; it is adapted here to a more general case. As in (2.13) of Chapter 4, we look for solutions of (4.4) of the form

$$P_{ij}(t) = c_{ij}e^{\rho t}, \qquad i, j = 1, \ldots, s, \tag{4.25}$$

or, in matrix notation,

$$\mathbf{P}(t) = \mathbf{C}e^{\rho t} \tag{4.26}$$

where ρ and the constant matrix

$$\mathbf{C} = \begin{pmatrix} c_{11} & \cdots & c_{1s} \\ \cdot & & \cdot \\ \cdot & & \cdot \\ \cdot & & \cdot \\ c_{s1} & \cdots & c_{ss} \end{pmatrix} \tag{4.27}$$

are to be determined. Substituting (4.26) in the second equation in (4.7) and canceling out the nonvanishing scalar factor $e^{\rho t}$, we obtain

$$(\rho\mathbf{I} - \mathbf{V}')\mathbf{C}' = \mathbf{0}. \tag{4.28}$$

In order for $\mathbf{C}'$ to have a nontrivial solution, it is necessary that the matrix $\mathbf{A}' = (\rho\mathbf{I} - \mathbf{V}')$ be singular, that is, the determinant $|A'| = 0$. This means

[4] Here we use the well-known result in algebra that replacement of any column (or any row) of a determinant by the sum of all the columns (or rows) does not change the value of the determinant. See Problem 4 for proof.

that the roots of the characteristic equation

$$|\rho\mathbf{I} - \mathbf{V}'| = 0 \tag{4.29}$$

are the only values of ρ for which (4.25) is a valid solution of (4.7). Since the matrix $\mathbf{V}$ and its transpose $\mathbf{V}'$ have the same eigenvalues, the roots of (4.29) are also the eigenvalues of $\mathbf{V}$ and these are denoted by $\rho_1, \ldots, \rho_s$.

For each ρ_l, the corresponding constant matrix $\mathbf{C}(l)$ is determined from

$$(\rho_l\mathbf{I} - \mathbf{V}')\mathbf{C}'(l) = \mathbf{0}. \tag{4.30}$$

Since ρ_l is a solution of (4.29), the matrix

$$\mathbf{A}'(l) = \begin{pmatrix} \rho_l - \nu_{11} & -\nu_{21} & \cdots & -\nu_{s1} \\ -\nu_{12} & \rho_l - \nu_{22} & \cdots & -\nu_{s2} \\ \cdot & \cdot & & \cdot \\ \cdot & \cdot & & \cdot \\ \cdot & \cdot & & \cdot \\ -\nu_{1s} & -\nu_{2s} & \cdots & \rho_l - \nu_{ss} \end{pmatrix} \tag{4.31}$$

is singular. Using Cramer's rule, we see that (4.30) has nontrivial solutions in which the constants c_{ijl} are proportional to the corresponding cofactors $A'_{ij}(l)$ of $\mathbf{A}'(l)$, or[5]

$$c_{ijl} = k_{il}A'_{ij}(l). \tag{4.32}$$

If all the s eigenvalues of $\mathbf{V}$ are real and distinct, the general solution of (4.4) is

$$P_{ij}(t) = \sum_{l=1}^{s} k_{il}A'_{ij}(l)e^{\rho_l t}, \qquad i, j = 1, \ldots, s. \tag{4.33}$$

Equation (4.33) is the general solution for any finite system of differential equations (4.4) in which $P_{ij}(t)$ need not be transition probabilities, as pointed out earlier.

To derive a particular solution corresponding to the initial condition (4.6), we set $t = 0$ in (4.33) to obtain

$$\begin{aligned} \sum_{l=1}^{s} k_{il}A'_{ij}(l) &= 1 \qquad j = i \\ &= 0 \qquad j \neq i. \end{aligned} \tag{4.34}$$

[5] The cofactors $A'_{\alpha j}(l)$ of any row α would be solutions for the c_{ijl}; but the choice of $\alpha = i$ is convenient for evaluation of the proportionality constants k_{il}.

From this we can determine k_{il}. The cofactors $A'_{ij}(l)$, when expanded in polynomials of ρ_l, are

$$\begin{aligned} A'_{ii}(l) &= \rho_l^{s-1} + b_{ii2}\rho_l^{s-2} + \cdots + b_{iis} \\ A'_{ij}(l) &= \qquad\quad b_{ij2}\rho_l^{s-2} + \cdots + b_{ijs}, \qquad j \neq i \end{aligned} \tag{4.35}$$

where the coefficients b are functions of ν_{ij} but are independent of ρ_l and l. Substituting (4.35) in (4.34) and rearranging terms according to the coefficients b, the resulting equations are

$$\begin{aligned} b_{i12}\sum_{l=1}^{s} k_{il}\rho_l^{s-2} + \cdots + b_{i1s}\sum_{l=1}^{s} k_{il} &= 0 \\ \cdots\cdots\cdots\cdots\cdots\cdots\cdots\cdots\cdots& \\ \sum_{l=1}^{s} k_{il}\rho_l^{s-1} + b_{ii2}\sum_{l=1}^{s} k_{il}\rho_l^{s-2} + \cdots + b_{iis}\sum_{l=1}^{s} k_{il} &= 1 \\ \cdots\cdots\cdots\cdots\cdots\cdots\cdots\cdots\cdots& \\ b_{is2}\sum_{l=1}^{s} k_{il}\rho_l^{s-2} + \cdots + b_{iss}\sum_{l=1}^{s} k_{il} &= 0. \end{aligned} \tag{4.36}$$

According to the lemma in Section 3.3,

$$\begin{aligned} \sum_{l=1}^{s} \frac{\rho_l^r}{\prod_{\substack{m=1 \\ m\neq l}}^{s} (\rho_l - \rho_m)} &= 1 \qquad r = s - 1 \\ &= 0 \qquad 0 \leq r < s - 1. \end{aligned} \tag{4.37}$$

Therefore the values of k_{il} satisfying (4.36) are

$$k_{il} = \frac{1}{\prod_{\substack{m=1 \\ m\neq l}}^{s} (\rho_l - \rho_m)}, \qquad i, l = 1, \ldots, s. \tag{4.38}$$

Substituting (4.38) in (4.33) gives the explicit solution

$$P_{ij}(t) = \sum_{l=1}^{s} \frac{A'_{ij}(l)e^{\rho_l t}}{\prod_{\substack{m=1 \\ m\neq l}}^{s} (\rho_l - \rho_m)}, \qquad i, j = 1, \ldots, s, \tag{4.39}$$

for the Kolmogorov forward differential equations (4.4). ▶

The backward differential equations (4.5) can be solved in the same way, giving the same solution as in (4.39).

4.3. Second Solution for Individual Transition Probabilities $P_{ij}(t)$

If we ignore the fact that $\mathbf{P}(t)$ and $\mathbf{V}$ are matrices, the differential equation

$$D\mathbf{P}(t) = \mathbf{P}(t)\mathbf{V} \tag{4.7}$$

has the appearance of an ordinary first-order differential equation with a constant coefficient. Formally, this suggests the solution

$$\mathbf{P}(t) = e^{\mathbf{V}t}\mathbf{P}(0) \tag{4.40}$$

where $\mathbf{P}(0) = \mathbf{I}$. If the matrix exponential is defined by

$$e^{\mathbf{V}t} = \sum_{n=0}^{\infty} \frac{\mathbf{V}^n t^n}{n!} \tag{4.41}$$

then (4.40) is indeed a solution of (4.7). It can be shown that the matrix series in (4.41) converges uniformly in t (see Problem 20, also Bellman [1960] and Doob [1953]). Therefore, we can take the derivative of the infinite sum in (4.40) with respect to t term by term

$$\begin{aligned} D\mathbf{P}(t) &= \frac{d}{dt} e^{\mathbf{V}t}\mathbf{P}(0) = \frac{d}{dt} \sum_{n=0}^{\infty} \frac{\mathbf{V}^n t^n}{n!} \mathbf{P}(0) \\ &= \sum_{n=1}^{\infty} \frac{\mathbf{V}^{n-1} t^{n-1}}{(n-1)!} \mathbf{P}(0)\mathbf{V} = e^{\mathbf{V}t}\mathbf{P}(0)\mathbf{V} = \mathbf{P}(t)\mathbf{V}, \end{aligned} \tag{4.42}$$

and recover the differential equation (4.7). Furthermore, substituting $t = 0$ in (4.40) produces an identity. ▶

The formal solution (4.40), however, is not very useful from a practical point of view. For the purpose of application, we need explicit functions for the individual transition probabilities $P_{ij}(t)$ that will satisfy differential equations (4.4) and (4.5). Such functions depend on the eigenvalues of $\mathbf{V}$. When $\mathbf{V}$ has multiple eigenvalues the problem is quite complicated; but if it has real and single eigenvalues, simple explicit functions can be found. Some material relevant to the following computation may be found in Hochstadt [1964].

Following the discussion on diagonalization of a matrix in Sections 3.2 and 3.4, we let $A_{ij}(l)$ be the cofactor of the matrix $\mathbf{A}(l) = (\rho_l\mathbf{I} - \mathbf{V})$ and

$$\mathbf{T}_l(k) = \begin{pmatrix} A_{k1}(l) \\ \cdot \\ \cdot \\ \cdot \\ A_{ks}(l) \end{pmatrix} \tag{4.43}$$

be an eigenvector of $\mathbf{V}$ for $\rho = \rho_l$. The matrix

$$\mathbf{T}(k) = [\mathbf{T}_1(k), \ldots, \mathbf{T}_s(k)] \tag{4.44}$$

diagonalizes $\mathbf{V}$

$$\mathbf{T}^{-1}(k)\mathbf{V}\mathbf{T}(k) = \boldsymbol{\rho} \tag{4.45}$$

where $\boldsymbol{\rho}$ is the diagonal matrix in (3.30).

Formula (4.45) may be rewritten in the form

$$\mathbf{V} = \mathbf{T}(k)\boldsymbol{\rho}\mathbf{T}^{-1}(k) \tag{4.46}$$

from which we compute

$$\mathbf{V}^2 = [\mathbf{T}(k)\boldsymbol{\rho}\mathbf{T}^{-1}(k)][\mathbf{T}(k)\boldsymbol{\rho}\mathbf{T}^{-1}(k)] = \mathbf{T}(k)\boldsymbol{\rho}^2\mathbf{T}^{-1}(k). \tag{4.47}$$

Induction shows that in general

$$\mathbf{V}^n = \mathbf{T}(k)\boldsymbol{\rho}^n\mathbf{T}^{-1}(k), \qquad n = 1, 2, \ldots . \tag{4.48}$$

Using (4.48) we have

$$e^{\mathbf{V}t} = \sum_{n=0}^{\infty} \frac{\mathbf{V}^n t^n}{n!} = \sum_{n=0}^{\infty} \frac{\mathbf{T}(k)\boldsymbol{\rho}^n\mathbf{T}^{-1}(k)t^n}{n!} = \mathbf{T}(k)\sum_{n=0}^{\infty} \frac{\boldsymbol{\rho}^n t^n}{n!}\,\mathbf{T}^{-1}(k). \tag{4.49}$$

Now $\boldsymbol{\rho}$ is a diagonal matrix,

$$\boldsymbol{\rho}^n = \begin{pmatrix} \rho_1{}^n & 0 & \cdots & 0 \\ 0 & \rho_2{}^n & \cdots & 0 \\ \cdot & \cdot & & \cdot \\ \cdot & \cdot & & \cdot \\ \cdot & \cdot & & \cdot \\ 0 & 0 & \cdots & \rho_s{}^n \end{pmatrix} \tag{4.50}$$

and hence,

$$\sum_{n=0}^{\infty} \frac{\boldsymbol{\rho}^n t^n}{n!} = \begin{pmatrix} \sum_{n=0}^{\infty} \frac{\rho_1{}^n t^n}{n!} & 0 & \cdots & 0 \\ 0 & \sum_{n=0}^{\infty} \frac{\rho_2{}^n t^n}{n!} & \cdots & 0 \\ \cdot & \cdot & & \cdot \\ \cdot & \cdot & & \cdot \\ 0 & 0 & \cdots & \sum_{n=0}^{\infty} \frac{\rho_s{}^n t^n}{n!} \end{pmatrix}$$

$$= \begin{pmatrix} e^{\rho_1 t} & 0 & \cdots & 0 \\ 0 & e^{\rho_2 t} & \cdots & 0 \\ \cdot & \cdot & & \cdot \\ \cdot & \cdot & & \cdot \\ \cdot & \cdot & & \cdot \\ 0 & 0 & \cdots & e^{\rho_s t} \end{pmatrix}. \tag{4.51}$$

Letting

$$\mathbf{E}(t) = \begin{pmatrix} e^{\rho_1 t} & 0 & \cdots & 0 \\ 0 & e^{\rho_2 t} & \cdots & 0 \\ \cdot & \cdot & & \cdot \\ \cdot & \cdot & & \cdot \\ \cdot & \cdot & & \cdot \\ 0 & 0 & \cdots & e^{\rho_s t} \end{pmatrix} \tag{4.52}$$

and using (4.49) and (4.51), solution (4.40) can be written

$$e^{\mathbf{V}t}\mathbf{P}(0) = \mathbf{T}(k)\mathbf{E}(t)\mathbf{T}^{-1}(k)\mathbf{P}(0). \tag{4.53}$$

In our case

$$\mathbf{P}(0) = \mathbf{I} \tag{4.54}$$

so that the particular solution for the differential equation (4.7) corresponding to initial condition (4.54) is

$$\mathbf{P}(t) = \mathbf{T}(k)\mathbf{E}(t)\mathbf{T}^{-1}(k). \tag{4.55}$$

Expanding (4.55) we obtain the second explicit solution for the transition probabilities $P_{ij}(t)$,

$$P_{ij}(t) = \sum_{l=1}^{s} A_{ki}(l)\frac{T_{jl}(k)}{|T(k)|}e^{\rho_l t}, \qquad i, j = 1, \ldots, s. \tag{4.56}$$

Equation (4.56) holds true whatever may be $k = 1, \ldots, s$, as noted in (3.41). ▶

The differential equations (4.5) can also be solved in the same way; we leave this to the reader as an exercise.

4.4. Identity of the Two Solutions

We must now show that the two explicit solutions in (4.39) and (4.56) are identical, or that

$$\sum_{l=1}^{s}\frac{A'_{ij}(l)}{\prod\limits_{\substack{m=1\\m\neq l}}^{s}(\rho_l - \rho_m)}e^{\rho_l t} = \sum_{l=1}^{s} A_{ki}(l)\frac{T_{jl}(k)}{|T(k)|}e^{\rho_l t}. \tag{4.57}$$

Since (4.56) holds true for any k, we may let $k = j$ to rewrite (4.57) as

$$\sum_{l=1}^{s}\frac{A'_{ij}(l)}{\prod\limits_{\substack{m=1\\m\neq l}}^{s}(\rho_l - \rho_m)}e^{\rho_l t} = \sum_{l=1}^{s} A_{ji}(l)\frac{T_{jl}(j)}{|T(j)|}e^{\rho_l t}. \tag{4.58}$$

Clearly (4.58) holds true if, for each l,

$$\frac{A'_{ij}(l)}{\prod_{\substack{m=1\\m\neq l}}^{s}(\rho_l - \rho_m)} = A_{ji}(l)\,\frac{T_{jl}(j)}{|T(j)|}, \tag{4.59}$$

or, since $\mathbf{A}'(l)$ is the transpose of $\mathbf{A}(l)$ so that $A'_{ij}(l) = A_{ji}(l)$,

$$\frac{1}{\prod_{\substack{m=1\\m\neq l}}^{s}(\rho_l - \rho_m)} = \frac{T_{jl}(j)}{|T(j)|}. \tag{4.60}$$

Therefore, in order to prove that the two solutions are identical, it is sufficient to prove (4.60) for each l. Equation (4.60), however, has been proved in Section 3.4 [see Equation (3.46)]; hence the proof of the identity is complete. ▶

4.5. Chapman-Kolmogorov Equations

In deriving Kolmogorov differential equations in Section 2, we used the Chapman-Kolmogorov equation (1.6) as the point of departure. In the present case of homogeneous processes, we have

$$P_{ik}(\tau + t) = \sum_{j=1}^{s} P_{ij}(\tau)P_{jk}(t). \tag{1.7}$$

The corresponding matrix representation is

$$\mathbf{P}(\tau + t) = \mathbf{P}(\tau)\mathbf{P}(t). \tag{4.61}$$

We leave it for the reader to verify that solutions (4.39) and (4.56) satisfy (1.7) and we proceed here to verify (4.61). Substituting solution (4.55) in the right side of (4.61), we compute

$$\begin{aligned}\mathbf{P}(\tau)\mathbf{P}(t) &= [\mathbf{T}(k)\mathbf{E}(\tau)\mathbf{T}^{-1}(k)][\mathbf{T}(k)\mathbf{E}(t)\mathbf{T}^{-1}(k)]\\ &= \mathbf{T}(k)\mathbf{E}(\tau)\mathbf{E}(t)\mathbf{T}^{-1}(k) = \mathbf{T}(k)\mathbf{E}(\tau + t)\mathbf{T}^{-1}(k) = \mathbf{P}(\tau + t).\end{aligned} \tag{4.62}$$

Equation (4.61) is verified. ▶

The results obtained so far in this section may be summarized in

Theorem 2. *If the eigenvalues of the intensity matrix* $\mathbf{V}$ *are real and distinct, then the solution of Kolmogorov differential equations* (4.4) *and* (4.5) *corresponding to the initial condition* (4.6) *is given in* (4.39) *and* (4.56); *the two solutions are identical and satisfy the Chapman-Kolmogorov equation* (1.7). ▶

Corollary. *The limiting transition probability $P_{ij}(t)$ as t approaches infinity is independent of the initial state i and is given by*

$$\lim_{t\to\infty} P_{ij}(t) = \frac{V_{j\alpha}}{\sum_{l=1}^{s} V_{l\alpha}}, \qquad i, j, \alpha = 1, \ldots, s. \tag{4.63}$$

Therefore the process is ergodic.

Proof of corollary. Since $\rho_k < 0$ for $k = 2, \ldots, s$, $e^{\rho_k t}$ tends to zero as t tends to infinity; from solution (4.39) we have

$$\lim_{t\to\infty} P_{ij}(t) = \frac{A'_{ij}(1)}{\prod_{m=2}^{s} (\rho_1 - \rho_m)}. \tag{4.64}$$

Here $\mathbf{A}'(1) = (\rho_1\mathbf{I} - \mathbf{V}')$, $\rho_1 = 0$; hence

$$A'_{ij}(1) = (-1)^{s-1} V'_{ij} = (-1)^{s-1} V_{ji}. \tag{4.65}$$

Using (4.24) we write the denominator in (4.64) as

$$(-1)^{s-1}\rho_2 \cdots \rho_s = (-1)^{s-1}[V_{1i} + \cdots + V_{si}]. \tag{4.66}$$

According to (4.19), $V_{j1} = \cdots = V_{js}$; therefore (4.65) and (4.66) may be rewritten as

$$A'_{ij}(1) = (-1)^{s-1} V_{j\alpha} \tag{4.65a}$$

and

$$(-1)^{s-1}\rho_2 \cdots \rho_s = (-1)^{s-1}[V_{1\alpha} + \cdots + V_{s\alpha}], \tag{4.66a}$$

whatever may be $\alpha = 1, \ldots, s$. Substituting (4.65*a*) and (4.66*a*) in (4.64) yields (4.63), proving the corollary. ▶

PROBLEMS

1. Use Equations (2.1) and (2.2) to show that

$$\frac{\partial}{\partial t} P_{ij}(\tau, t)\bigg|_{t=\tau} = \nu_{ij}(\tau)$$

2. Show that if

$$\sum_j P_{ij}(\tau, t) = 1 \tag{2.8}$$

whatever may be $t > \tau$, then

$$\nu_{ii}(\tau) = -\sum_{j\neq i} \nu_{ij}(\tau) \tag{2.9}$$

3. Justify that the linear-growth model in Section 6.1, Chapter 3, is homogeneous with respect to time, whereas the time-dependent birth-death process in Section 6.2 is nonhomogeneous.

4. Show that (a) a determinant with two rows (or columns) proportional to one another is equal to zero, and (b) the value of a determinant is unchanged if a row (or column) is replaced by the sum of all rows (columns).

5. *Linearly independent vectors*. Show that all the column vectors of a square matrix **W** are linearly independent if and only if all the row vectors are linearly independent.

6. Show that if matrix **W** is nonsingular, then the transpose of the inverse is the inverse of the transpose, $(\mathbf{W}^{-1})' = (\mathbf{W}')^{-1}$.

7. The sum of the diagonal elements of a square matrix **W** is called the *trace* of **W**, tr **W**. In general the sum of the determinants of all principal minors of **W** of order r is called the *spur* of **W**, sp_r **W** (see, for example, Aitken [1962]). Verify the relations in (3.17) among the coefficients b, the spurs, and the eigenvalues for the following matrix.

$$\mathbf{W} = \begin{pmatrix} 0 & 1 & 0 \\ 0 & 0 & 2 \\ -3 & -\frac{1}{2} & 4 \end{pmatrix}$$

8. *Cramer's rule*. Given a system of s simultaneous equations

$$\begin{gathered} a_{11}Z_1 + \cdots + a_{1s}Z_s = c_1 \\ \cdots\cdots\cdots \\ a_{s1}Z_1 + \cdots + a_{ss}Z_s = c_s. \end{gathered}$$

If the c's are not all zero and if the determinant

$$|A| = \begin{vmatrix} a_{11} & \cdots & a_{1s} \\ \cdot & & \cdot \\ \cdot & & \cdot \\ a_{s1} & \cdots & a_{ss} \end{vmatrix} \neq 0$$

then the system has a unique solution. Find the solution and show that the solution is unique.

9. *Continuation*. Show that if in Problem 8 the c's are all zero, then a nontrivial solution exists if and only if the determinant $|A| = 0$ and that the solution is proportional to the corresponding cofactors of A; namely

$$Z_j = kA_{ij}, \qquad j = 1, \ldots, s$$

whatever may be $i = 1, \ldots, s$.

10. *Diagonalization of matrix.* Let a, b, and c be nonzero real numbers with $a \neq c$ and let

$$\mathbf{W} = \begin{pmatrix} a & 0 \\ b & c \end{pmatrix}.$$

(a) Find the eigenvalues ρ_1 and ρ_2 of $\mathbf{W}$.

(b) Let $\mathbf{A}(l) = (\rho_l \mathbf{I} - \mathbf{W})$ and take the first column of each adjoint matrix of $\mathbf{A}(l)$

$$\mathbf{T}_l(1) = \begin{pmatrix} A_{11}(l) \\ A_{12}(l) \end{pmatrix}, \qquad l = 1, 2$$

to form the matrix $\mathbf{T}(1) = (\mathbf{T}_1(1), \mathbf{T}_2(1))$. Show that

$$\mathbf{T}^{-1}(1)\mathbf{W}\mathbf{T}(1) = \begin{pmatrix} \rho_1 & 0 \\ 0 & \rho_2 \end{pmatrix}.$$

(c) Show that the first column of the inverse $\mathbf{T}^{-1}(1)$ of $\mathbf{T}(1)$ is given by

$$\begin{pmatrix} \dfrac{1}{\rho_1 - \rho_2} \\ \dfrac{1}{\rho_2 - \rho_1} \end{pmatrix}.$$

(d) Can a statement similar to (c) be made about the matrix $\mathbf{T}(2) = (\mathbf{T}_2(1), \mathbf{T}_2(2))$? Explain.

11. *Continuation.* Given a matrix

$$\mathbf{W} = \begin{pmatrix} -1 & 1 \\ 2 & -2 \end{pmatrix}$$

(a) Determine the eigenvalues.

(b) For each eigenvalue ρ_l, $l = 1, 2$, formulate the matrix $\mathbf{A}(l) = (\rho_l \mathbf{I} - \mathbf{W})$, and show that

$$(\rho_l \mathbf{I} - \mathbf{W})\begin{pmatrix} A_{k1}(l) \\ A_{k2}(l) \end{pmatrix} = \begin{pmatrix} 0 \\ 0 \end{pmatrix}, \qquad k = 1, 2.$$

(c) For any k of your choice formulate the matrix

$$\mathbf{T}(k) = \begin{pmatrix} A_{k1}(1) & A_{k1}(2) \\ A_{k2}(1) & A_{k2}(2) \end{pmatrix}$$

and show that

$$\mathbf{T}^{-1}(k)\mathbf{W}\mathbf{T}(k) = \begin{pmatrix} \rho_1 & 0 \\ 0 & \rho_2 \end{pmatrix}.$$

(d) Verify that the kth column of the inverse matrix $\mathbf{T}^{-1}(k)$ of $\mathbf{T}(k)$ is equal to

$$\begin{pmatrix} \dfrac{1}{\rho_1 - \rho_2} \\ \dfrac{1}{\rho_2 - \rho_1} \end{pmatrix}.$$

12. For $\rho_1 = 1$, $\rho_2 = 2$, $\rho_3 = 3$, show that

$$\frac{\rho_1^{\,r}}{(\rho_1 - \rho_2)(\rho_1 - \rho_3)} + \frac{\rho_2^{\,r}}{(\rho_2 - \rho_1)(\rho_2 - \rho_3)} + \frac{\rho_3^{\,r}}{(\rho_3 - \rho_1)(\rho_3 - \rho_2)} = 0 \qquad r = 0, 1$$

$$= 1 \qquad r = 2.$$

13. *Vandermonde Determinant.* (a) Deduce the following expansion of the Vandermonde determinant

$$\begin{vmatrix} 1 & 1 & 1 & 1 \\ \rho_1 & \rho_2 & \rho_3 & \rho_4 \\ \rho_1^{\,2} & \rho_2^{\,2} & \rho_3^{\,2} & \rho_4^{\,2} \\ \rho_1^{\,3} & \rho_2^{\,3} & \rho_3^{\,3} & \rho_4^{\,3} \end{vmatrix} = (\rho_1 - \rho_2)(\rho_1 - \rho_3)(\rho_1 - \rho_4)(\rho_2 - \rho_3)(\rho_2 - \rho_4)(\rho_3 - \rho_4)$$

(b) Show that in general

$$\begin{vmatrix} 1 & 1 & \cdots & 1 \\ \rho_1 & \rho_2 & \cdots & \rho_s \\ \cdot & \cdot & & \cdot \\ \cdot & \cdot & & \cdot \\ \rho_1^{s-1} & \rho_2^{s-1} & \cdots & \rho_s^{s-1} \end{vmatrix} = \prod_{i=1}^{s-1} \prod_{j=i+1}^{s} (\rho_i - \rho_j).$$

(c) Let V_{ij} be the cofactor of a Vandermonde determinant of order s. Show that $\sum_{j=1}^{s} V_{ij} = 0$ for $1 < i \le s$.

14. Use part (c) in Problem 13 to show that, for distinct numbers $\rho_1, \rho_2, \ldots, \rho_s$,

$$\sum_{i=1}^{s} \frac{1}{\prod_{\substack{j=1 \\ j \neq i}}^{s} (\rho_i - \rho_j)} = 0.$$

15. *The lemma in Section* 3.3. Let $\rho_1, \ldots, \rho_s$ be distinct numbers. Prove the following by double induction on s and r (Chiang [1964a]).

$$\sum_{i=1}^{s} \frac{\rho_i^{\,r}}{\prod_{\substack{j=1 \\ j \neq i}}^{s} (\rho_i - \rho_j)} = 0 \qquad \text{for} \qquad 0 \le r < s - 1$$

$$= 1 \qquad \text{for} \qquad r = s - 1$$

$$\left[\textit{Hint.}\ \frac{\rho_i^{r+1}}{\rho_i - \rho_{s+1}} = \rho_i^{\,r} + \rho_i^{r-1}\rho_{s+1} + \rho_i^{r-2}\rho_{s+1}^2 + \cdots + \rho_{s+1}^{r} + \frac{\rho_{s+1}^{r+1}}{\rho_i - \rho_{s+1}}. \right]$$

16. *Continuation.* Prove by induction on s (John McGuire[6]) that

$$\sum_{i=1}^{s} \frac{\rho_i^{s-1}}{\prod_{\substack{j=1 \\ j \neq i}}^{s} (\rho_i - \rho_j)} = 1$$

[*Hint.* Let x be any real number different from $\rho_1, \ldots, \rho_s$ and substitute $(\rho_i - x)$ for ρ_i.]

17. *Continuation.* Show that if

$$\sum_{i=1}^{s} \frac{\rho_i^{s-1}}{\prod_{\substack{j=1 \\ j \neq i}}^{s} (\rho_i - \rho_j)} = 1$$

then

$$\sum_{i=1}^{s} \frac{\rho_i^{r}}{\prod_{\substack{j=1 \\ j \neq i}}^{s} (\rho_i - \rho_j)} = 0, \qquad 0 \leq r < s - 1.$$

18. *Continuation.* Prove the lemma in Problem 15 by the method of partial fractions (M. Hills[6]) with

$$G(\rho) = \frac{\rho^r}{(\rho - \rho_1) \cdots (\rho - \rho_s)}.$$

[See Equation (5.1), Chapter 2.]

19. Consider the matrix

$$\mathbf{V} = \begin{pmatrix} \nu_{11} & \cdots & \nu_{1s} \\ \cdot & & \cdot \\ \cdot & & \cdot \\ \cdot & & \cdot \\ \nu_{s1} & \cdots & \nu_{ss} \end{pmatrix} \tag{4.3}$$

with

$$\nu_{ii} = -\sum_{\substack{j=1 \\ j \neq i}}^{s} \nu_{ij}, \qquad i = 1, \ldots, s. \tag{4.2}$$

Show that the cofactors of $\mathbf{V}$ satisfy the relation

$$V_{ij} = V_{ik}$$

whatever may be $i, j, k = 1, \ldots, s$.

20. *The matrix exponential.* (See Bellman [1960] pp. 162–4) For any $s \times s$ matrix $\mathbf{W}$ of elements w_{ij}, define

$$\|\mathbf{W}\| = \sum_{i=1}^{s} \sum_{j=1}^{s} |w_{ij}|.$$

[6] Personal communication.

(a) Show that if $\mathbf{W}$ and $\mathbf{V}$ are two $s \times s$ matrices, then

$$\|\mathbf{W} + \mathbf{V}\| \leq \|\mathbf{W}\| + \|\mathbf{V}\|$$

and

$$\|\mathbf{W}\mathbf{V}\| \leq \|\mathbf{W}\| \; \|\mathbf{V}\|.$$

(b) Show that a sufficient condition for the convergence of the matrix series

$\sum_{n=1}^{\infty} \mathbf{W}_n$ is that the numerical series $\sum_{n=1}^{\infty} \|\mathbf{W}_n\|$ converges.

(c) Using (a) and (b), show that the matrix exponential defined in (4.41) converges uniformly in any bounded interval $-T \leq t \leq T$.

(d) Using this result, show that (4.40) is a solution of the backward differential equation (4.8).

21. *Solution of backward differential equations.* Derive a solution for the backward differential equations

$$\frac{d}{dt} P_{ik}(t) = \sum_{j=1}^{s} v_{ij} P_{jk}(t) \tag{4.5}$$

satisfying the initial conditions at $t = 0$, $P_{ik}(0) = \delta_{ik}$, using the approach in Section 4.2.

22. *Continuation.* Derive a solution for the backward differential equations (4.5) in Problem 21 using the approach in Section 4.3.

23. Show that if the eigenvalues ρ_l of $\mathbf{V}$ are real and distinct, then whatever may be $i, j, l = 1, \ldots, s$, the functions

$$P_{ij}(t) = \sum_{l=1}^{s} k_{il} A'_{ij}(l) e^{\rho_l t} \tag{4.33}$$

satisfy the system of differential equations

$$\frac{d}{dt} P_{ih}(t) = \sum_{j=1}^{s} P_{ij}(t) v_{jh}. \tag{4.4}$$

24. *Chapman-Kolmogorov equation.* Show that both

$$P_{ij}(t) = \sum_{l=1}^{s} \frac{A_{ij}'(l)}{\prod_{\substack{m=1 \\ m \neq l}}^{s} (\rho_l - \rho_m)} e^{\rho_l t} \tag{4.39}$$

and

$$P_{ij}(t) = \sum_{l=1}^{s} A_{ki}(l) \frac{T_{jl}(k) \, e^{\rho_l t}}{|T(k)|} \tag{4.56}$$

satisfy the Chapman-Kolmogorov equation

$$P_{ik}(\tau + t) = \sum_{j=1}^{s} P_{ij}(\tau)P_{jk}(t), \qquad i, k = 1, \ldots, s.$$

25. To illustrate that the solution in (4.56) for the transition probabilities is independent of k, consider the simple illness-death process in Chapter 4 with

$$\mathbf{V} = \begin{pmatrix} \nu_{11} & \nu_{12} \\ \nu_{21} & \nu_{22} \end{pmatrix}, \qquad \mathbf{A}(l) = (\rho_l \mathbf{I} - \mathbf{V}),$$

where $\nu_{11} + \nu_{12} < 0$ and $\nu_{21} + \nu_{22} < 0$. Find $P_{\alpha\beta}(t)$ from

$$P_{\alpha\beta}(t) = \sum_{l=1}^{2} A_{1\alpha}(l) \frac{T_{\beta l}(1)}{|T(1)|} e^{\rho_l t} \tag{4.56a}$$

and

$$P_{\alpha\beta}(t) = \sum_{l=1}^{2} A_{2\alpha}(l) \frac{T_{\beta l}(2)}{|T(2)|} e^{\rho_l t}, \tag{4.56b}$$

and show that they are equal for given α and β.

26. Consider a singular matrix

$$\mathbf{V} = \begin{vmatrix} -0.2 & 0.1 & 0.1 \\ 0.1 & -0.4 & 0.3 \\ 0.2 & 0.1 & -0.3 \end{vmatrix}.$$

(a) Show that the cofactors

$$V_{i1} = V_{i2} = V_{i3}, \qquad i = 1, 2, 3.$$

(b) Matrix **V** has one zero eigenvalue and two negative eigenvalues. Determine these eigenvalues.

(c) For each eigenvalue ρ_l, formulate the characteristic matrix

$$\mathbf{A}(l) = (\rho_l \mathbf{I} - \mathbf{V})$$

and compute the cofactors $A_{ij}(l)$.

(d) Show that whatever may be $k = 1, 2, 3$, the nonzero vector

$$\mathbf{T}_l(k) = \begin{pmatrix} A_{k1}(l) \\ A_{k2}(l) \\ A_{k3}(l) \end{pmatrix}$$

is an eigenvector of **V** corresponding to ρ_l.

(e) For a particular k of your choice formulate a matrix

$$\mathbf{T}(k) = [\mathbf{T}_1(k), \mathbf{T}_2(k), \mathbf{T}_3(k)].$$

Verify that

$$\mathbf{T}^{-1}(k)\mathbf{V}\mathbf{T}(k) = \begin{pmatrix} \rho_1 & 0 & 0 \\ 0 & \rho_2 & 0 \\ 0 & 0 & \rho_3 \end{pmatrix}$$

and

$$\mathbf{T}(k)\begin{pmatrix} \rho_1 & 0 & 0 \\ 0 & \rho_2 & 0 \\ 0 & 0 & \rho_3 \end{pmatrix}\mathbf{T}^{-1}(k) = \mathbf{V}.$$

(f) Compute $\mathbf{V}^2$ and

$$\mathbf{T}(k)\begin{pmatrix} \rho_1^2 & 0 & 0 \\ 0 & \rho_2^2 & 0 \\ 0 & 0 & \rho_3^2 \end{pmatrix}\mathbf{T}^{-1}(k)$$

and show that the two matrices are equal.

(g) Compute $\mathbf{T}^{-1}(k)\mathbf{V}^2\mathbf{T}(k)$ and show that it is equal to the diagonal matrix

$$\begin{pmatrix} \rho_1^2 & 0 & 0 \\ 0 & \rho_2^2 & 0 \\ 0 & 0 & \rho_3^2 \end{pmatrix}.$$

(h) Substitute the numerical values of v_{ij} to formulate the system of forward differential equations

$$\frac{d}{dt}P_{ik}(t) = \sum_{j=1}^{3} P_{ij}(t)v_{jk}$$

and the backward differential equations

$$\frac{d}{dt}P_{ik}(t) = \sum_{j=1}^{3} v_{ij}P_{jk}(t)$$

assuming $P_{ij}(0) = \delta_{ij}$.

(i) Compute

$$P_{ij}(t) = \sum_{l=1}^{3} \frac{A'_{ij}(l)}{\prod_{\substack{m=1 \\ m\neq l}}^{3} (\rho_l - \rho_m)} e^{\rho_l t}, \qquad i, j = 1, 2, 3$$

and show that they satisfy the differential equations and the initial conditions in (h).

(j) Show that $P_{ij}(t)$ in (i) satisfy

$$\sum_{j=1}^{3} P_{ij}(\tau)P_{jk}(t) = P_{ik}(\tau + t)$$

for some particular i and k of your choice.

(k) For two values of k of your choice compute

$$\sum_{l=1}^{3} A_{ki}(l) \frac{T_{jl}(k)}{|T(k)|} e^{\rho_l t}, \qquad j = 1, 2, 3.$$

Do they depend on the value of k?

(l) Compare your results in (k) with (i).

(m) Show that

$$\lim_{t\to\infty} P_{ij}(t) = \frac{V_{j\alpha}}{\prod_{l=1}^{s} V_{l\alpha}}$$

27. In Section 4.1 we proved that any real eigenvalue of $\mathbf{V}$ is nonpositive. We now prove the assertion with a method devised by Myra Jordan Samuels. Let

$$\mathbf{C} = \begin{pmatrix} c_1 \\ \cdot \\ \cdot \\ \cdot \\ c_s \end{pmatrix} \tag{4.13}$$

be a nonzero vector such that

$$(\rho\mathbf{I} - \mathbf{V}')\mathbf{C} = 0, \tag{4.14}$$

or equivalently

$$c_1 v_{1j} + \cdots + c_s v_{sj} = \rho c_j, \qquad j = 1, \ldots, s. \tag{4.15}$$

Adding (4.15) over those j for which $c_j > 0$, denoting this partial sum by Σ^*, we have

$$c_1 \Sigma^* v_{1j} + \cdots + c_s \Sigma^* v_{sj} = \rho \Sigma^* c_j.$$

Show that each of the terms on the left side of the last equation is nonpositive and hence $\rho \le 0$.

CHAPTER 7

A General Model of Illness-Death Process

1. INTRODUCTION

We can now use the results of the preceding chapter to extend the simple illness-death process in Chapter 4 to include the variability of illness and causes of death. In this general model there are s states of illness, which we shall denote by S_α, $\alpha = 1, \ldots, s$, and r states of death denoted by R_δ, $\delta = 1, \ldots, r$, where s and r are positive integers. An illness state may be broadly defined to include the absence of illness (a health state), a physical impairment, a single specific disease or stage of disease, or any combination of diseases. A death state, on the other hand, is defined by the cause of death, whether single or multiple; emigration may also be treated as a cause of death. Transition from an illness state to a death state specifies not only the cause of death but also any disease present. For example, an individual in illness state S_α of tuberculosis-pneumonia who passes to the death state R_δ of pneumonia dies of pneumonia in the presence of tuberculosis. A slightly different classification of illness and death states may be found in Chiang [1964a].

Transition from one state to another, and indeed the model itself, is determined by the intensities of risks of illness $\nu_{\alpha\beta}$ and the intensities of risks of death $\mu_{\alpha\delta}$, as defined in (2.1) and (2.2) in Chapter 4. For easy reference, they are restated below. For $0 < \tau < t$, let

$$\nu_{\alpha\beta}\Delta + 0(\Delta) = \Pr\{\text{an individual in state } S_\alpha \text{ at time } \tau \text{ will be in state } S_\beta \text{ at time } \tau + \Delta\}, \qquad \alpha \neq \beta;\ \alpha, \beta = 1, \ldots, s;$$

$$\mu_{\alpha\delta}\Delta + 0(\Delta) = \Pr\{\text{an individual in state } S_\alpha \text{ at time } \tau \text{ will be in state } R_\delta \text{ at time } \tau + \Delta\}, \qquad \delta = 1, \ldots, R.$$

These intensities are assumed to be independent of time. Again, we let

$$\nu_{\alpha\alpha} = -\left(\sum_{\substack{\beta=1\\ \beta\neq\alpha}}^{s} \nu_{\alpha\beta} + \sum_{\delta=1}^{r} \mu_{\alpha\delta}\right). \tag{1.1}$$

For convenience we introduce the *illness intensity matrix*

$$\mathbf{V} = \begin{pmatrix} \nu_{11} & \cdot & \cdot & \cdot & \nu_{1s} \\ \cdot & & & & \cdot \\ \cdot & & & & \cdot \\ \cdot & & & & \cdot \\ \nu_{s1} & \cdot & \cdot & \cdot & \nu_{ss} \end{pmatrix} \tag{1.2}$$

and the *death intensity matrix*

$$\mathbf{U} = \begin{pmatrix} \mu_{11} & \cdot & \cdot & \cdot & \mu_{1r} \\ \cdot & & & & \cdot \\ \cdot & & & & \cdot \\ \cdot & & & & \cdot \\ \mu_{s1} & \cdot & \cdot & \cdot & \mu_{sr} \end{pmatrix}. \tag{1.3}$$

For a time interval $(0, t)$ for $0 \le t < \infty$, let us define the transition probabilities:

$$P_{\alpha\beta}(t) = \Pr\{\text{an individual in state } S_\alpha \text{ at time 0 will be in illness state } S_\beta \text{ at time } t\}, \quad \alpha, \beta = 1, \ldots, s \tag{1.4}$$

$$Q_{\alpha\delta}(t) = \Pr\{\text{an individual in state } S_\alpha \text{ at time 0 will be in death state } R_\delta \text{ at time } t\}, \quad \alpha = 1, \ldots, s; \delta = 1, \ldots, r \tag{1.5}$$

with

$$P_{\alpha\alpha}(0) = 1, \quad P_{\alpha\beta}(0) = 0, \quad Q_{\alpha\delta}(0) = 0, \quad \beta \ne \alpha; \quad \alpha, \beta = 1, \ldots, s; \delta = 1, \ldots, r, \tag{1.6}$$

and the corresponding matrices

$$\mathbf{P}(t) = \begin{pmatrix} P_{11}(t) & \cdot & \cdot & \cdot & P_{1s}(t) \\ \cdot & & & & \cdot \\ \cdot & & & & \cdot \\ \cdot & & & & \cdot \\ P_{s1}(t) & \cdot & \cdot & \cdot & P_{ss}(t) \end{pmatrix} \tag{1.7}$$

$$\mathbf{Q}(t) = \begin{pmatrix} Q_{11}(t) & \cdot & \cdot & \cdot & Q_{1r}(t) \\ \cdot & & & & \cdot \\ \cdot & & & & \cdot \\ \cdot & & & & \cdot \\ Q_{s1}(t) & \cdot & \cdot & \cdot & Q_{sr}(t) \end{pmatrix} \tag{1.8}$$

with

$$\mathbf{P}(0) = \mathbf{I} \qquad \mathbf{Q}(0) = \mathbf{0}. \tag{1.9}$$

The strict formulation of a real illness-death process requires the assumption that no death takes place without a cause, and that at the time of death the individual is affected with the disease from which he dies; consequently, in practice many of the intensities $\mu_{\alpha\delta}$ will be zero. Furthermore, if an individual is unlikely [with probability $o(\Delta)$] to contract or recover from more than one disease within an infinitesimal interval Δ, many of the terms $\nu_{\alpha\beta}$ will also be zero. However, our model will be presented in general form without deleting any term $\nu_{\alpha\beta}$ or $\mu_{\alpha\delta}$, but it will satisfy the following assumptions.

Assumption 1. The system is closed: whatever may be $t \geq 0$ and for every α, we have

$$\sum_{\beta=1}^{s} P_{\alpha\beta}(t) + \sum_{\delta=1}^{r} Q_{\alpha\delta}(t) = 1 \tag{1.10}$$

so that the intensities and the transition probabilities have the relations:

$$\nu_{\alpha\beta} = \frac{d}{dt} P_{\alpha\beta}(t)\bigg|_{t=0}, \qquad \alpha, \beta = 1, \ldots, s \tag{1.11}$$

and

$$\mu_{\alpha\delta} = \frac{d}{dt} Q_{\alpha\delta}(t)\bigg|_{t=0}, \qquad \delta = 1, \ldots, r. \tag{1.12}$$

Assumption 2. The illness intensity matrix $\mathbf{V}$ is of rank s and the death intensity matrix $\mathbf{U}$ is not a zero matrix. Therefore none of the illness states S_α is an absorbing state, and there is at least one death state R_δ in the system.

2. TRANSITION PROBABILITIES

In this section we present explicit formulas for the transition probabilities and related problems. Because of the similarity of the illness transition probabilities $P_{\alpha\beta}(t)$ and the transition probabilities in Chapter 6, we shall avoid repetition by not giving minute details.

2.1. Illness Transition Probabilities, $P_{\alpha\beta}(t)$

Using the same reasoning that led to Kolmogorov differential equations in Chapter 6, we find that the transition probabilities $P_{\alpha\beta}(t)$ satisfy the forward differential equations

$$\frac{d}{dt} P_{\alpha\gamma}(t) = \sum_{\beta=1}^{s} P_{\alpha\beta}(t)\nu_{\beta\gamma} \tag{2.1}$$

and the backward differential equations

$$\frac{d}{dt}P_{\alpha\gamma}(t) = \sum_{\beta=1}^{s} \nu_{\alpha\beta}P_{\beta\gamma}(t), \qquad \alpha, \gamma = 1, \ldots, s, \tag{2.2}$$

with initial conditions $P_{\alpha\gamma}(0) = \delta_{\alpha\gamma}$. As in Chapter 6, we let $\rho_1, \ldots, \rho_s$ be the roots of the characteristic equation

$$|\rho\mathbf{I} - \mathbf{V}'| = 0 \tag{2.3}$$

and $\mathbf{A}'(l) = (\rho_l\mathbf{I} - \mathbf{V}')$ and let $A'_{\alpha\beta}(l)$ be the cofactors of $\mathbf{A}'(l)$. If the roots (or eigenvalues) $\rho_1, \ldots, \rho_s$ are real and distinct, then equations (2.1) and (2.2) have the common solution

$$P_{\alpha\gamma}(t) = \sum_{l=1}^{s} \frac{A'_{\alpha\gamma}(l)}{\prod_{\substack{j=1\\ j\neq l}}^{s} (\rho_l - \rho_j)} e^{\rho_l t}, \qquad \alpha, \beta = 1, \ldots, s. \tag{2.4}$$

An alternative form of (2.4) is

$$P_{\alpha\gamma}(t) = \sum_{l=1}^{s} A_{k\alpha}(l)\frac{T_{\gamma l}(k)}{|T(k)|} e^{\rho_l t} \tag{2.5}$$

where $A_{\alpha\beta}(l)$ is the transpose of $A'_{\alpha\beta}(l)$ and $T_{\gamma l}(k)$ is the cofactor of the matrix

$$\mathbf{T}(k) = \begin{pmatrix} A_{k1}(1) & \cdots & A_{k1}(s) \\ \cdot & & \cdot \\ \cdot & & \cdot \\ A_{ks}(1) & \cdots & A_{ks}(s) \end{pmatrix}. \tag{2.6}$$

Equation (2.5) may be compactly presented in matrix notation as

$$\mathbf{P}(t) = \mathbf{T}(k)\mathbf{E}(t)\mathbf{T}^{-1}(k) \tag{2.7}$$

where $\mathbf{E}(t)$ is a diagonal matrix

$$\mathbf{E}(t) = \begin{pmatrix} e^{\rho_1 t} & 0 & \cdots & 0 \\ 0 & e^{\rho_2 t} & \cdots & 0 \\ \cdot & \cdot & & \cdot \\ \cdot & \cdot & & \cdot \\ 0 & 0 & \cdots & e^{\rho_s t} \end{pmatrix}. \tag{2.8}$$

Equations (2.4), (2.5), and (2.7) are similar to those in (4.39), (4.56), and (4.55) in Chapter 6; the reader is referred to Chapter 6 for the derivation of these equations and for the proof of the identity of (2.4) and (2.5) [see Equation (4.57) in Chapter 6]. For completeness, we shall show that (2.4) satisfies equation (2.1). Substituting (2.4) in (2.1) yields

$$\sum_{l=1}^{s} \frac{A'_{\alpha\gamma}(l)\rho_l e^{\rho_l t}}{\prod_{\substack{j=1 \\ j\neq l}}^{s}(\rho_l - \rho_j)} = \sum_{\beta=1}^{s}\sum_{l=1}^{s} \frac{A'_{\alpha\beta}(l)e^{\rho_l t}}{\prod_{\substack{j=1 \\ j\neq l}}^{s}(\rho_l - \rho_j)} \nu_{\beta\gamma}. \tag{2.9}$$

Therefore it is sufficient to show that, for each l,

$$A'_{\alpha\gamma}(l)\rho_l = \sum_{\beta=1}^{s} A'_{\alpha\beta}(l)\nu_{\beta\gamma} \tag{2.10}$$

or

$$A'_{\alpha\gamma}(l)[\rho_l - \nu_{\gamma\gamma}] - \sum_{\substack{\beta=1 \\ \beta\neq\gamma}}^{s} A'_{\alpha\beta}(l)\nu_{\beta\gamma} = 0. \tag{2.11}$$

Relation (2.11) is easy to prove. For $\gamma = \alpha$, we have

$$A'_{\alpha\alpha}(l)[\rho_l - \nu_{\alpha\alpha}] - \sum_{\substack{\beta=1 \\ \beta\neq\alpha}}^{s} A'_{\alpha\beta}(l)\nu_{\beta\alpha} = |A'(l)| \tag{2.12}$$

since the left side of (2.12) is an expansion of the determinant $|A'(l)|$ using the αth row. Obviously $|A'(l)| = 0$, because ρ_l is a root of the characteristic equation (2.3). For $\nu \neq \alpha$, the left side of (2.11) is an expansion of $|A'(l)|$ in terms of cofactors of a different row, and hence must be zero [see (3.5) in Chapter 6]. The verification is complete.

In a similar way, the reader may show that (2.4) satisfies also the backward equations (2.2).

The solutions in (2.4) and (2.5) also satisfy the Chapman-Kolmogorov equation

$$P_{\alpha\gamma}(\tau + t) = \sum_{\beta=1}^{s} P_{\alpha\beta}(\tau)P_{\beta\gamma}(t), \qquad \alpha, \gamma = 1, \ldots, s. \tag{2.13}$$

We have proved the corresponding matrix representation

$$\mathbf{P}(\tau + t) = \mathbf{P}(\tau)\mathbf{P}(t) \tag{2.14}$$

in (4.62) in Chapter 6. For the scalar-oriented reader, we shall verify (2.13) directly, although the reasoning is exactly the same as in (4.62). By substituting solution (2.5) we can rewrite (2.13) as

$$\sum_{l} A_{k\alpha}(l)\frac{T_{\gamma l}(k)}{|T(k)|}e^{\rho_l(\tau+t)} = \sum_{\beta}\left[\sum_{l} A_{k\alpha}(l)\frac{T_{\beta l}(k)}{|T(k)|}e^{\rho_l\tau}\right]\left[\sum_{l} A_{k\beta}(l)\frac{T_{\gamma l}(k)}{|T(k)|}e^{\rho_l t}\right]. \tag{2.15}$$

All the summations are taken from 1 to s. The right side of (2.15) after multiplication becomes

$$\sum_{l}\sum_{\beta} A_{k\alpha}(l)\frac{T_{\beta l}(k)}{|T(k)|}A_{k\beta}(l)\frac{T_{\gamma l}(k)}{|T(k)|}e^{\rho_l(\tau+t)}$$
$$+\sum_{l}\sum_{\substack{j\\ l\neq j}}\sum_{\beta} A_{k\alpha}(l)\frac{T_{\beta l}(k)}{|T(k)|}A_{k\beta}(j)\frac{T_{\gamma j}(k)}{|T(k)|}e^{\rho_l\tau+\rho_j t}. \tag{2.16}$$

For the first term of (2.16), for each l, we write

$$\sum_{\beta} A_{k\alpha}(l)\frac{T_{\beta l}(k)}{|T(k)|}A_{k\beta}(l)\frac{T_{\gamma l}(k)}{|T(k)|} = A_{k\alpha}(l)\frac{T_{\gamma l}(k)}{|T(k)|}\sum_{\beta}\frac{T_{\beta l}(k)}{|T(k)|}A_{k\beta}(l)$$
$$= A_{k\alpha}(l)\frac{T_{\gamma l}(k)}{|T(k)|} \tag{2.17}$$

since the second sum in (2.17) is equal to unity; for the second term, for $l \neq j$, we have

$$\sum_{\beta} A_{k\alpha}(l)\frac{T_{\beta l}(k)}{|T(k)|}A_{k\beta}(j)\frac{T_{\gamma j}(k)}{|T(k)|} = A_{k\alpha}(l)\frac{T_{\gamma j}(k)}{|T(k)|}\frac{\sum_{\beta} T_{\beta l}(k)A_{k\beta}(j)}{|T(k)|} = 0 \tag{2.18}$$

since in the last sum the elements $A_{k\beta}(j)$ and the cofactors $T_{\beta l}(k)$ belong to two different columns of $|T(k)|$. Substituting (2.17) and (2.18) in (2.16) yields the left side of (2.15), proving (2.13).

For the initial condition (1.6), we need to show that

$$\sum_{l=1}^{s}\frac{A'_{\alpha\gamma}(l)}{\prod\limits_{\substack{j=1\\ j\neq l}}^{s}(\rho_l-\rho_j)} = 1 \qquad \text{for} \qquad \gamma=\alpha$$
$$= 0 \qquad \text{for} \qquad \gamma\neq\alpha. \tag{2.19}$$

Proof of (2.19) is left to the reader.

2.2. Transition Probabilities Leading to Death, $Q_{\alpha\delta}(t)$

The probability $Q_{\alpha\delta}(t)$ that an individual in an illness state S_α at time 0 will be in a death state R_δ at time t and the corresponding transition probability matrix $\mathbf{Q}(t)$ can be derived from probability $P_{\alpha\beta}(t)$ and the matrix $\mathbf{P}(t)$. An individual in an illness state S_α at time 0 may enter the death state R_δ directly from S_α, or by way of some other state S_β, $\beta \neq \alpha$, $\beta = 1, \ldots, s$, at any time τ prior to t. As in Equation (2.29) of Chapter 4, we have

$$Q_{\alpha\delta}(t) = \int_0^t \sum_{\beta=1}^{s} P_{\alpha\beta}(\tau)\mu_{\beta\delta}\, d\tau \tag{2.20}$$

and the corresponding matrix equation

$$\mathbf{Q}(t) = \int_0^t \mathbf{P}(\tau)\mathbf{U}\, d\tau, \tag{2.21}$$

where the intensity matrix $\mathbf{U}$ is $s \times r$ as defined in (1.3) and the integral in (2.21) means the integration of each element of the matrix product $\mathbf{P}(\tau)\mathbf{U}$. To derive $\mathbf{Q}(t)$ from (2.21), we substitute formula (2.7) in (2.21) and integrate to obtain

$$\begin{aligned}\mathbf{Q}(t) &= \int_0^t \mathbf{T}(k)\mathbf{E}(\tau)\mathbf{T}^{-1}(k)\mathbf{U}\, d\tau \\ &= \mathbf{T}(k)[\mathbf{E}(t) - \mathbf{I}]\boldsymbol{\rho}^{-1}\mathbf{T}^{-1}(k)\mathbf{U}.\end{aligned} \tag{2.22}$$

Here the unit matrix $\mathbf{I}$ is $s \times s$ and $\boldsymbol{\rho}$ is an $s \times s$ diagonal matrix defined by

$$\boldsymbol{\rho} = \begin{pmatrix} \rho_1 & 0 & \cdot & \cdot & \cdot & 0 \\ 0 & \rho_2 & \cdot & \cdot & \cdot & 0 \\ \cdot & \cdot & & & & \cdot \\ \cdot & \cdot & & & & \cdot \\ \cdot & \cdot & & & & \cdot \\ 0 & 0 & \cdot & \cdot & \cdot & \rho_s \end{pmatrix}. \tag{2.23}$$

The individual transition probability $Q_{\alpha\delta}(t)$ can be obtained either from (2.20) or by expanding (2.22); in either way it is given by

$$Q_{\alpha\delta}(t) = \sum_{l=1}^{s}\sum_{\beta=1}^{s} A_{k\alpha}(l)\frac{T_{\beta l}(k)}{|T(k)|}\rho_l^{-1}(e^{\rho_l t} - 1)\mu_{\beta\delta} \tag{2.24}$$

or its equivalent form

$$Q_{\alpha\delta}(t) = \sum_{l=1}^{s}\sum_{\beta=1}^{s} \frac{A'_{\alpha\beta}(l)}{\prod_{\substack{j=1\\ j\neq l}}^{s}(\rho_l - \rho_j)\rho_l}(e^{\rho_l t} - 1)\mu_{\beta\delta}$$

$$\alpha = 1, \ldots, s;\ \delta = 1, \ldots, r. \tag{2.25}$$

A Chapman-Kolmogorov type equation can be established also for the transition probabilities $Q_{\alpha\delta}(t)$. For an individual in S_α at time 0 to be in

R_δ at time $\tau + t$, he must be either in R_δ prior to τ or in some illness state S_β at τ and enter R_δ at some time between τ and $\tau + t$. Therefore, the corresponding transition probabilities satisfy the equation [see Equation (3.6) in Chapter 4]:

$$Q_{\alpha\delta}(\tau + t) = Q_{\alpha\delta}(\tau) + \sum_{\beta=1}^{s} P_{\alpha\beta}(\tau)Q_{\beta\delta}(t). \tag{2.26}$$

Equation (2.26) holds true for $\alpha = 1, \ldots, s$ and for $\delta = 1, \ldots, r$ if and only if the corresponding matrices satisfy

$$\mathbf{Q}(\tau + t) = \mathbf{Q}(\tau) + \mathbf{P}(\tau)\mathbf{Q}(t). \tag{2.27}$$

To verify (2.27), we recall (2.7) and (2.22) and write

$$\begin{aligned}\mathbf{Q}(\tau) + \mathbf{P}(\tau)\mathbf{Q}(t) = {} & \mathbf{T}(k)[\mathbf{E}(\tau) - \mathbf{I}]\boldsymbol{\rho}^{-1}\mathbf{T}^{-1}(k)\mathbf{U} \\ & + [\mathbf{T}(k)\mathbf{E}(\tau)\mathbf{T}^{-1}(k)][\mathbf{T}(k)\{\mathbf{E}(t) - \mathbf{I}\}\boldsymbol{\rho}^{-1}\mathbf{T}^{-1}(k)\mathbf{U}].\end{aligned} \tag{2.28}$$

The right side, after multiplication, becomes

$$\begin{aligned}\mathbf{T}(k)[\mathbf{E}(\tau) - \mathbf{I}]\boldsymbol{\rho}^{-1}\mathbf{T}^{-1}(k)\mathbf{U} + \mathbf{T}(k)[\mathbf{E}(\tau + t) - \mathbf{E}(\tau)]\boldsymbol{\rho}^{-1}\mathbf{T}^{-1}(k)\mathbf{U} \\ = \mathbf{T}(k)[\mathbf{E}(\tau + t) - \mathbf{I}]\boldsymbol{\rho}^{-1}\mathbf{T}^{-1}(k)\mathbf{U} \\ = \mathbf{Q}(\tau + t)\end{aligned} \tag{2.29}$$

proving equation (2.27) and hence also (2.26).

2.3. An Equality Concerning Transition Probabilities

An individual in an illness state S_α at time 0 must at time t be either in one of the illness states S_β or in one of the death states R_δ; therefore the corresponding transition probabilities must satisfy the equation

$$\sum_{\beta=1}^{s} P_{\alpha\beta}(t) + \sum_{\delta=1}^{r} Q_{\alpha\delta}(t) = 1, \qquad \alpha = 1, \ldots, s. \tag{2.30}$$

Having derived explicit formulas for the transition probabilities, we may verify Equation (2.30) directly. Substituting (2.4) and (2.25) in (2.30) we obtain

$$\sum_{\beta=1}^{s}\sum_{l=1}^{s}\frac{A'_{\alpha\beta}(l)}{\prod\limits_{\substack{j=1\\ j\neq l}}^{s}(\rho_l - \rho_j)}e^{\rho_l t} + \sum_{\delta=1}^{r}\sum_{\beta=1}^{s}\sum_{l=1}^{s}\frac{A'_{\alpha\beta}(l)}{\prod\limits_{\substack{j=1\\ j\neq l}}^{s}(\rho_l - \rho_j)\rho_l}(e^{\rho_l t} - 1)\mu_{\beta\delta} = 1, \tag{2.31}$$

which can be shown by direct computation and is left to the reader. The corresponding matrix equation is verified below.

Equation (2.30) may be summarized in the following matrix notation

$$\mathbf{P}(t)\begin{pmatrix}1\\ \cdot\\ \cdot\\ \cdot\\ 1\end{pmatrix}_s + \mathbf{Q}(t)\begin{pmatrix}1\\ \cdot\\ \cdot\\ \cdot\\ 1\end{pmatrix}_r = \begin{pmatrix}1\\ \cdot\\ \cdot\\ \cdot\\ 1\end{pmatrix}_s . \tag{2.32}$$

Here the first and the third column vectors are $s \times 1$, the second one is $r \times 1$, and all the components are unity. We substitute (2.7) for $\mathbf{P}(t)$ and (2.22) for $\mathbf{Q}(t)$ in the left side of (2.32), which becomes

$$\mathbf{T}(k)\mathbf{E}(t)\mathbf{T}^{-1}(k)\begin{pmatrix}1\\ \cdot\\ \cdot\\ \cdot\\ 1\end{pmatrix}_s + \mathbf{T}(k)[\mathbf{E}(t) - \mathbf{I}]\boldsymbol{\rho}^{-1}\mathbf{T}^{-1}(k)\mathbf{U}\begin{pmatrix}1\\ \cdot\\ \cdot\\ \cdot\\ 1\end{pmatrix}_r \tag{2.33}$$

where the unit matrix $\mathbf{I}$ is $s \times s$. From (1.1) and (1.2) and (1.3) we have the relation

$$\mathbf{U}\begin{pmatrix}1\\ \cdot\\ \cdot\\ \cdot\\ 1\end{pmatrix}_r = -\mathbf{V}\begin{pmatrix}1\\ \cdot\\ \cdot\\ \cdot\\ 1\end{pmatrix}_s \tag{2.34}$$

and hence we can rewrite (2.33) as

$$\mathbf{T}(k)\mathbf{E}(t)\mathbf{T}^{-1}(k)\begin{pmatrix}1\\ \cdot\\ \cdot\\ \cdot\\ 1\end{pmatrix}_s - \mathbf{T}(k)[\mathbf{E}(t) - \mathbf{I}]\boldsymbol{\rho}^{-1}\mathbf{T}^{-1}(k)\mathbf{V}\begin{pmatrix}1\\ \cdot\\ \cdot\\ \cdot\\ 1\end{pmatrix}_s . \tag{2.35}$$

From Equation (3.30) in Chapter 6 we see that $\mathbf{T}(k)$, as defined in (2.6), diagonalizes matrix $\mathbf{V}$ in the form

$$\mathbf{T}^{-1}(k)\mathbf{V}\mathbf{T}(k) = \boldsymbol{\rho} \tag{2.36}$$

where $\boldsymbol{\rho}$ is the diagonal matrix (2.23), so that

$$\boldsymbol{\rho}^{-1}\mathbf{T}^{-1}(k)\mathbf{V} = \mathbf{T}^{-1}(k). \tag{2.37}$$

Substituting (2.37) in (2.35) and canceling out the first two terms give us

$$\mathbf{T}(k)\,\mathbf{I}\,\mathbf{T}^{-1}(k)\begin{pmatrix}1\\ \cdot\\ \cdot\\ \cdot\\ 1\end{pmatrix}_s = \begin{pmatrix}1\\ \cdot\\ \cdot\\ \cdot\\ 1\end{pmatrix}_s$$

which proves equations (2.32) and (2.31).

2.4. Limiting Transition Probabilities

We have proved in Section 4.1 of Chapter 6 that if $\nu_{\alpha\beta} \geq 0$ for $\alpha \neq \beta$ and $\sum_{\beta} \nu_{\alpha\beta} \leq 0$ for each α, then any real eigenvalue of $\mathbf{V}$ is nonpositive. In the present case $\mathbf{V}$ is of rank s and all the eigenvalues are assumed to be real; they are therefore negative. Consequently, as t approaches infinity, the transition probabilities become

$$P_{\alpha\beta}(\infty) = \lim_{t\to\infty} \sum_{l=1}^{s} \frac{A'_{\alpha\beta}(l)}{\prod_{\substack{j=1\\ j\neq l}}^{s} (\rho_l - \rho_j)} e^{\rho_l t} = 0, \qquad \alpha, \beta = 1, \ldots, s \tag{2.38}$$

and

$$Q_{\alpha\delta}(\infty) = \lim_{t\to\infty} \sum_{\beta=1}^{s}\sum_{l=1}^{s} \frac{A'_{\alpha\beta}(l)\mu_{\beta\delta}}{\prod_{\substack{j=1\\ j\neq l}}^{s} (\rho_l - \rho_j)\rho_l} (e^{\rho_l t} - 1) = -\sum_{\beta=1}^{s}\sum_{l=1}^{s} \frac{A'_{\alpha\beta}(l)\mu_{\beta\delta}}{\prod_{\substack{j=1\\ j\neq l}}^{s} (\rho_l - \rho_j)\rho_l}$$

$$\delta = 1, \ldots, r. \tag{2.39}$$

It follows from (2.31) that

$$\sum_{\delta=1}^{r} Q_{\alpha\delta}(\infty) = 1. \tag{2.40}$$

Thus, starting from any illness state S_α at any given time, an individual will eventually enter a death state with a probability of one.

2.5. Expected Durations of Stay in Illness and Death States

Given an individual in an illness state S_α at time 0, how long is he expected to stay in each of the states within a time interval $(0, t)$? This important question led to the concept of expected duration of stay introduced in the simple illness-death process in Chapter 4. For convenience, let us reintroduce the symbols. For an individual in S_α at time 0, let

$$e_{\alpha\beta}(t) = \text{the expected duration of stay in } S_\beta \text{ in the interval } (0, t),$$
$$\beta = 1, \ldots, s \tag{2.41}$$

$$\varepsilon_{\alpha\delta}(t) = \text{the expected duration of stay in } R_\delta \text{ in the interval } (0, t),$$
$$\delta = 1, \ldots, r. \tag{2.42}$$

By the same reasoning used in Chapter 4, we find that

$$e_{\alpha\beta}(t) = \int_0^t P_{\alpha\beta}(\tau)\, d\tau \tag{2.43}$$

$$\varepsilon_{\alpha\delta}(t) = \int_0^t Q_{\alpha\delta}(\tau)\, d\tau. \tag{2.44}$$

Using formulas (2.4) and (2.25) for the transition probabilities, we compute

$$e_{\alpha\beta}(t) = \int_0^t \sum_{l=1}^{s} \frac{A'_{\alpha\beta}(l)}{\prod_{\substack{j=1\\ j\neq l}}^{s} (\rho_l - \rho_j)} e^{\rho_l \tau}\, d\tau$$

$$= \sum_{l=1}^{s} \frac{A'_{\alpha\beta}(l)}{\prod_{\substack{j=1\\ j\neq l}}^{s} (\rho_l - \rho_j)\rho_l} [e^{\rho_l t} - 1], \qquad \alpha, \beta = 1, \ldots, s, \tag{2.45}$$

and

$$\varepsilon_{\alpha\delta}(t) = \int_0^t \sum_{l=1}^{s} \sum_{\beta=1}^{s} \frac{A'_{\alpha\beta}(l)\mu_{\beta\delta}}{\prod_{\substack{j=1\\ j\neq l}}^{s} (\rho_l - \rho_j)\rho_l} (e^{\rho_l \tau} - 1)\, d\tau$$

$$= \sum_{l=1}^{s} \sum_{\beta=1}^{s} \frac{A'_{\alpha\beta}(l)\mu_{\beta\delta}}{\prod_{\substack{j=1\\ j\neq l}}^{s} (\rho_l - \rho_j)\rho_l} \left[\frac{1}{\rho_l}(e^{\rho_l t} - 1) - t\right].$$

$$\alpha = 1, \ldots, s;\ \delta = 1, \ldots, r. \tag{2.46}$$

Since an individual in S_α at time 0 must be in one and only one of the $s + r$ states at any given instant during the interval $(0, t)$, it follows that

$$\sum_{\beta=1}^{s} e_{\alpha\beta}(t) + \sum_{\delta=1}^{r} \varepsilon_{\alpha\delta}(t) = t. \tag{2.47}$$

Equation (2.47) may be shown directly with the explicit formulas in (2.45) and (2.46), but a simpler approach is to use (2.43) and (2.44). Thus

$$\sum_{\beta=1}^{s} e_{\alpha\beta}(t) + \sum_{\delta=1}^{r} \varepsilon_{\alpha\delta}(t) = \int_0^t \left[\sum_{\beta=1}^{s} P_{\alpha\delta}(\tau) + \sum_{\delta=1}^{r} Q_{\alpha\delta}(\tau) \right] d\tau \tag{2.48}$$

where the integrand, according to (2.30), is unity, and therefore the integral is equal to t.

2.6. Population Sizes in Illness States and Death States

At time 0, let there be $x_\alpha(0)$ individuals in state S_α, $\alpha = 1, \ldots, s$, so that

$$x(0) = x_1(0) + \cdots + x_s(0) \tag{2.49}$$

is the initial size of the total population. These individuals will travel independently from one illness state to another and some will enter a death state during the time interval $(0, t)$. At the end of the interval, there will be $X_\beta(t)$ individuals in S_β and $Y_\delta(t)$ in R_δ, $\beta = 1, \ldots, s$; $\delta = 1, \ldots, r$. We are interested in the joint probability distribution of all the random variables $X_\beta(t)$ and $Y_\delta(t)$. As in Section 5 of Chapter 4, random variables $X_\beta(t)$ and $Y_\delta(t)$ can be characterized according to their initial state at time 0. This is expressed by the formula

$$X_\beta(t) = X_{1\beta}(t) + \cdots + X_{s\beta}(t) \tag{2.50}$$

where $X_{\alpha\beta}(t)$ is the number of people in state S_β at time t who were in state S_α at time 0, and

$$Y_\delta(t) = Y_{1\delta}(t) + \cdots + Y_{s\delta}(t) \tag{2.51}$$

where $Y_{\alpha\delta}(t)$ is the number of people in state R_δ at time t who were in state S_α at time 0. On the other hand, each of the $x_\alpha(0)$ people in state S_α at time 0 must be in one of the illness states or in one of the death states at time t; this gives us

$$x_\alpha(0) = \sum_{\beta=1}^{s} X_{\alpha\beta}(t) + \sum_{\delta=1}^{r} Y_{\alpha\delta}(t). \tag{2.52}$$

According to (2.30)

$$1 = \sum_{\beta=1}^{s} P_{\alpha\beta}(t) + \sum_{\delta=1}^{r} Q_{\alpha\delta}(t) \tag{2.30}$$

therefore, given $x_\alpha(0)$, the random variables on the right side of (2.52) have joint multinomial distribution. Letting $X_{\alpha\beta}(t)$ assume values $x_{\alpha\beta}$ and $Y_{\alpha\delta}(t)$ assume values $y_{\alpha\delta}$, we have the joint probability distribution

$$\frac{x_\alpha(0)!}{\prod_{\beta=1}^{s} x_{\alpha\beta}! \prod_{\delta=1}^{r} y_{\alpha\delta}!} \prod_{\beta=1}^{s} [P_{\alpha\beta}(t)]^{x_{\alpha\beta}} \prod_{\delta=1}^{r} [Q_{\alpha\delta}(t)]^{y_{\alpha\delta}} \tag{2.53}$$

where $x_{\alpha\beta}$ and $y_{\alpha\delta}$ are nonnegative integers satisfying the condition

$$\sum_{\beta} x_{\alpha\beta} + \sum_{\delta} y_{\alpha\delta} = x_{\alpha}(0). \tag{2.54}$$

The joint probability distribution of $X_\beta(t)$ and $Y_\delta(t)$ is the convolution of the distribution in (2.53) for $\alpha = 1, \ldots, s$, and is given by

$$\Pr\{X_1(t) = x_1, \ldots, X_s(t) = x_s; Y_1(t) = y_1, \ldots, Y_r(t) = y_r \mid x_1(0), \ldots, x_s(0)\}$$

$$= \sum \prod_{\alpha=1}^{s} \frac{x_\alpha(0)!}{\prod_{\beta=1}^{s} x_{\alpha\beta}! \prod_{\delta=1}^{r} y_{\alpha\delta}!} \prod_{\beta=1}^{s} [P_{\alpha\beta}(t)]^{x_{\alpha\beta}} \prod_{\delta=1}^{r} [Q_{\alpha\beta}(t)]^{y_{\alpha\delta}}, \tag{2.55}$$

where $x_{\alpha\beta}$ and $y_{\alpha\delta}$ satisfy (2.54); the summation is taken over $x_{\alpha\beta}$ and $y_{\alpha\delta}$ so that for all β and δ

$$\sum_{\alpha=1}^{s} x_{\alpha\beta} = x_\beta \qquad \text{and} \qquad \sum_{\alpha=1}^{s} y_{\alpha\delta} = y_\delta. \tag{2.56}$$

The expected population sizes of the states and the corresponding variances and covariances are easily obtained from (2.55) by using familiar theorems in the multinomial distribution given in Chapter 1. The expectations, for example, are

$$E[X_\beta(t)] = \sum_{\alpha=1}^{s} x_\alpha(0) P_{\alpha\beta}(t) \tag{2.57}$$

and

$$E[Y_\delta(t)] = \sum_{\alpha=1}^{s} x_\alpha(0) Q_{\alpha\delta}(t). \tag{2.58}$$

As t approaches infinity, $P_{\alpha\beta}(t)$ tends to zero, and the probability distribution (2.55) becomes

$$\sum \prod_{\alpha=1}^{s} \frac{x_\alpha(0)!}{\prod_{\delta=1}^{r} y_{\alpha\delta}!} \prod_{\delta=1}^{r} [Q_{\alpha\delta}(\infty)]^{y_{\alpha\delta}} \tag{2.59}$$

where $Q_{\alpha\delta}(\infty)$ is given in (2.39).

3. MULTIPLE TRANSITION PROBABILITIES

The transition probabilities $P_{\alpha\beta}(t)$ discussed in the preceding section relate only to the end results of an individual's movement in a given time interval, and they are inadequate for describing the pattern of his movement during the interval. This problem was resolved for the simple illness-death process by studying the multiple transition probabilities, the details of which were given in Chapter 5. We now consider the same problem for the general model.

Let us first review the symbols and definitions. $P_{\alpha\beta}(t)$ was defined as the probability that an individual in state S_α at time 0 will be in state S_β at time t. During the interval $(0, t)$ the individual will travel continuously from one state to another; he may, for example, return to and *exit from* S_α a number of times and he may *enter* S_β a number of times. We define $M_{\alpha\beta}(t)$ as the *number of exit transitions from* S_α, and $N_{\alpha\beta}(t)$ as the *number of entrance transitions* to S_β. Both $M_{\alpha\beta}(t)$ and $N_{\alpha\beta}(t)$ are random variables with the corresponding probabilities denoted by $P_{\alpha\beta}^{(m)}(t)$ and $p_{\alpha\beta}^{(n)}(t)$, respectively, so that

$$\begin{aligned} P_{\alpha\beta}^{(m)}(t) &= \Pr\{M_{\alpha\beta}(t) = m\} \\ &= \Pr\{\text{an individual in } S_\alpha \text{ at time 0 will } \textit{leave } S_\alpha\ m \text{ times during} \\ &\qquad (0, t) \text{ and will be in } S_\beta \text{ at time } t\} \qquad m = 0, 1, \ldots; \end{aligned} \tag{3.1}$$

$$\begin{aligned} p_{\alpha\beta}^{(n)}(t) &= \Pr\{N_{\alpha\beta}(t) = n\} \\ &= \Pr\{\text{an individual in } S_\alpha \text{ at time 0 will } \textit{enter } S_\beta\ n \text{ times during} \\ &\qquad (0, t) \text{ and will be in } S_\beta \text{ at time } t\} \qquad n = 0, 1, \ldots; \\ &\qquad\qquad \alpha,\beta = 1, \ldots, s. \end{aligned} \tag{3.2}$$

Because an individual may travel between S_α (or S_β) and other states in the system without ever entering S_β (or returning to S_α), the random variables $M_{\alpha\beta}(t)$ and $N_{\alpha\beta}(t)$ have different probability distributions. When $\alpha = \beta$, however, we have

$$\Pr\{M_{\alpha\alpha}(t) = m\} = \Pr\{N_{\alpha\alpha}(t) = m\} \tag{3.3}$$

whatever may be $m = 0, 1, \ldots$. In any case,

$$\sum_{m=0}^{\infty} P_{\alpha\beta}^{(m)}(t) = \sum_{n=0}^{\infty} p_{\alpha\beta}^{(n)}(t) = P_{\alpha\beta}(t). \tag{3.4}$$

Since $P_{\alpha\beta}(t) < 1$, $M_{\alpha\beta}(t)$ and $N_{\alpha\beta}(t)$ are improper random variables. Explicit formulas for the multiple transition probabilities are difficult to obtain for an arbitrary number of illness states, but formulas for the corresponding p.g.f.'s can be derived.

3.1. Multiple Exit Transition Probabilities, $P_{\alpha\beta}^{(m)}(t)$

Let the p.g.f. of $P_{\alpha\beta}^{(m)}(t)$ be defined by

$$g_{\alpha\beta}(u; t) = \sum_{m=0}^{\infty} u^m P_{\alpha\beta}^{(m)}(t). \tag{3.5}$$

Obviously, at $u = 1$,

$$g_{\alpha\beta}(1, t) = P_{\alpha\beta}(t). \tag{3.6}$$

Using an approach similar to that in Section 2 of Chapter 5, we establish the following differential equations for the multiple exit transition probabilities [see Equation (2.2) in Chapter 5],

$$\frac{d}{dt} P_{\alpha\alpha}^{(m)}(t) = P_{\alpha\alpha}^{(m)}(t)\nu_{\alpha\alpha} + \sum_{\substack{\beta=1\\ \beta\neq\alpha}}^{s} P_{\alpha\beta}^{(m)}(t)\nu_{\beta\alpha}$$

$$\frac{d}{dt} P_{\alpha\gamma}^{(m)}(t) = P_{\alpha\alpha}^{(m-1)}(t)\nu_{\alpha\gamma} + \sum_{\substack{\beta=1\\ \beta\neq\alpha}}^{s} P_{\alpha\beta}^{(m)}(t)\nu_{\beta\gamma}, \tag{3.7}$$

$$m = 0, 1, \ldots; \gamma \neq \alpha; \alpha, \gamma = 1, \ldots, s.$$

From (3.7) direct computations show that the p.g.f.'s satisfy the differential equations

$$\frac{d}{dt} g_{\alpha\alpha}(u; t) = \sum_{\beta=1}^{s} g_{\alpha\beta}(u; t)\nu_{\beta\alpha}$$

$$\frac{d}{dt} g_{\alpha\gamma}(u; t) = u g_{\alpha\alpha}(u; t)\nu_{\alpha\gamma} + \sum_{\substack{\beta=1\\ \beta\neq\alpha}}^{s} g_{\alpha\beta}(u; t)\nu_{\beta\gamma}. \tag{3.8}$$

Since $P_{\alpha\alpha}^{(0)}(0) = 1$, $P_{\alpha\gamma}^{(0)}(0) = 0$, $\gamma \neq \alpha$, and $P_{\alpha\gamma}^{(m)}(0) = 0$, $m = 1, \ldots,$ the initial conditions are

$$g_{\alpha\alpha}(u; 0) = 1, \qquad g_{\alpha\gamma}(u; 0) = 0. \tag{3.9}$$

System (3.8) represents s^2 differential equations and can be solved in the same way as system (2.1), although the presence of u as the coefficient of $g_{\alpha\gamma}(u; t)$ in some but not all of the equations in (3.8) makes it difficult to use matrix notation and to present the solution in a compact form.

For each α let us rewrite (3.8) as follows:

$$\begin{array}{lllllll}
(D-\nu_{11})g_{\alpha1} & - \nu_{21}g_{\alpha2} & - \cdots - & u\nu_{\alpha1}g_{\alpha\alpha} & - \cdots - & \nu_{s1}g_{\alpha s} & = 0\\
-\nu_{12}g_{\alpha1} & + (D-\nu_{22})g_{\alpha2} & - \cdots - & u\nu_{\alpha2}g_{\alpha\alpha} & - \cdots - & \nu_{s2}g_{\alpha s} & = 0\\
\cdots & \cdots & \cdots & \cdots & \cdots & \cdots & \cdots\\
-\nu_{1\alpha}g_{\alpha1} & - \nu_{2\alpha}g_{\alpha2} & - \cdots + & (D-\nu_{\alpha\alpha})g_{\alpha\alpha} & - \cdots - & \nu_{s\alpha}g_{\alpha s} & = 0\\
\cdots & \cdots & \cdots & \cdots & \cdots & \cdots & \cdots\\
-\nu_{1s}g_{\alpha1} & - \nu_{2s}g_{\alpha2} & - \cdots - & u\nu_{\alpha s}g_{\alpha\alpha} & - \cdots + & (D-\nu_{ss})g_{\alpha s} & = 0
\end{array} \tag{3.10}$$

where, for simplicity, the symbol D is written for the differentiation d/dt and $g_{\alpha\beta}$ for $g_{\alpha\beta}(u; t)$. The corresponding characteristic equation is

$$|A'(\alpha)| = \begin{vmatrix} (\rho - \nu_{11}) & -\nu_{21} & \cdots & -u\nu_{\alpha 1} & \cdots & -\nu_{s1} \\ -\nu_{12} & (\rho - \nu_{22}) & \cdots & -u\nu_{\alpha 2} & \cdots & -\nu_{s2} \\ \cdots & \cdots & \cdots & \cdots & \cdots & \cdots \\ -\nu_{1\alpha} & -\nu_{2\alpha} & \cdots & (\rho - \nu_{\alpha\alpha}) & \cdots & -\nu_{s\alpha} \\ \cdots & \cdots & \cdots & \cdots & \cdots & \cdots \\ -\nu_{1s} & -\nu_{2s} & \cdots & -u\nu_{\alpha s} & \cdots & (\rho - \nu_{ss}) \end{vmatrix} = 0, \tag{3.11}$$

with the argument u appearing in the αth column except in the diagonal element. Suppose that the characteristic equation has distinct real roots $\rho_i(\alpha)$ for $i = 1, \ldots, s$. Following the steps in the solution of the system (4.4) of the Kolmogorov differential equations in Section 4.2 of Chapter 6, we obtain the general solution of (3.10),

$$g_{\alpha\gamma}(u; t) = \sum_{i=1}^{s} k_{\alpha i} A'_{\alpha\gamma}(\alpha, i) e^{\rho_i(\alpha)t}, \tag{3.12}$$

where $A'_{\alpha\gamma}(\alpha, i)$ are cofactors of $|A'(\alpha, i)|$ in (3.11) with $\rho = \rho_i(\alpha)$, $\alpha, \gamma, i = 1, \ldots, s$. Substituting (3.12) in (3.9) and solving the resulting simultaneous equations for $k_{\alpha i}$ give

$$k_{\alpha i} = \frac{1}{\prod_{\substack{j=1 \\ j \neq i}}^{s} [\rho_i(\alpha) - \rho_j(\alpha)]}. \tag{3.13}$$

From (3.12) and (3.13) follows the solution for the p.g.f.

$$g_{\alpha\gamma}(u; t) = \sum_{i=1}^{s} \frac{A'_{\alpha\gamma}(\alpha, i)}{\prod_{\substack{j=1 \\ j \neq i}}^{s} [\rho_i(\alpha) - \rho_j(\alpha)]} e^{\rho_i(\alpha)t}. \tag{3.14}$$

For given α and γ, the p.g.f. in (3.14) is an implicit function of the argument u which is involved in the eigenvalues $\rho_i(\alpha)$ and the cofactors $A'_{\alpha\gamma}(\alpha, i)$. When $u = 1$, $|A'(\alpha)|$ in (3.11) becomes identical to $|A'|$ in (2.3), and the p.g.f. $g_{\alpha\gamma}(1, t)$ is equal to the transition probability $P_{\alpha\gamma}(t)$ in (2.4).

Individual multiple transition probabilities may be obtained from the corresponding p.g.f.'s in (3.14) by taking appropriate derivatives,

$$P_{\alpha\gamma}^{(m)}(t) = \frac{1}{m!} \frac{\partial^m}{\partial u^m} g_{\alpha\gamma}(u; t) \big|_{u=0}. \tag{3.15}$$

The computations, however, may be overwhelming when the number of illness states and the number of transitions are large.

3.2. Multiple Transition Probabilities Leading to Death, $Q_{\alpha\delta}^{(m)}(t)$

Consider again an individual in S_α at time 0. Let $D_{\alpha\delta}(t)$ be the number of times that he leaves S_α before reaching a death state R_δ prior to t. Let

$$Q_{\alpha\delta}^{(m)}(t) = \Pr\{D_{\alpha\delta}(t) = m\}, \qquad m = 1, \ldots ; \alpha = 1, \ldots, s; \delta = 1, \ldots, r, \tag{3.16}$$

be the probability and

$$h_{\alpha\delta}(u; t) = \sum_{m=1}^{\infty} u^m Q_{\alpha\delta}^{(m)}(t) \tag{3.17}$$

be the corresponding p.g.f. We shall derive explicit formulas for $h_{\alpha\delta}(u; t)$. Obviously, $Q_{\alpha\delta}^{(m)}(t)$ and $P_{\alpha\beta}^{(m)}(t)$ have the relation

$$Q_{\alpha\delta}^{(m)}(t) = \int_0^t P_{\alpha\alpha}^{(m-1)}(\tau)\mu_{\alpha\delta}\, d\tau + \int_0^t \sum_{\substack{\beta=1 \\ \beta\neq\alpha}}^{s} P_{\alpha\beta}^{(m)}(\tau)\mu_{\beta\delta}\, d\tau \tag{3.18}$$

[see Equation (3.3) in Chapter 5]. Substituting (3.18) in (3.17) yields

$$h_{\alpha\delta}(u; t) = \sum_{m=1}^{\infty} u^m \left(\int_0^t P_{\alpha\alpha}^{(m-1)}(\tau)\mu_{\alpha\delta}\, d\tau + \int_0^t \sum_{\substack{\beta=1 \\ \beta\neq\alpha}}^{s} P_{\alpha\beta}^{(m)}(\tau)\mu_{\beta\delta}\, d\tau \right). \tag{3.19}$$

Here the infinite series converges absolutely. We interchange the summation and integral signs and write

$$h_{\alpha\delta}(u; t) = \int_0^t \left\{ \sum_{m=1}^{\infty} u^{m-1} P_{\alpha\alpha}^{(m-1)}(\tau) \right\} u\, \mu_{\alpha\delta}\, d\tau + \int_0^t \sum_{\substack{\beta=1 \\ \beta\neq\alpha}}^{s} \left\{ \sum_{m=1}^{\infty} u^m P_{\alpha\beta}^{(m)}(\tau) \right\} \mu_{\beta\delta}\, d\tau. \tag{3.20}$$

Each infinite sum in (3.20) is the p.g.f. in (3.14). Substitution of (3.14) in (3.20) and integration of the resulting expression give the formulas

$$h_{\alpha\delta}(u; t) = \sum_{i=1}^{s} \frac{A'_{\alpha\alpha}(\alpha; i)\, u\, \mu_{\alpha\delta}}{\prod_{\substack{j=1 \\ j\neq i}}^{s} [\rho_i(\alpha) - \rho_j(\alpha)]\rho_i(\alpha)} [e^{\rho_i(\alpha)t} - 1] + \sum_{\substack{\beta=1 \\ \beta\neq\alpha}}^{s} \sum_{i=1}^{s} \frac{A'_{\alpha\beta}(\alpha; i)\mu_{\beta\delta}}{\prod_{\substack{j=1 \\ j\neq i}}^{s} [\rho_i(\alpha) - \rho_j(\alpha)]\rho_i(\alpha)} [e^{\rho_i(\alpha)t} - 1]. \tag{3.21}$$

When the argument $u = 1$, $h_{\alpha\delta}(1; t) = Q_{\alpha\delta}(t)$, as given in (2.25).

3.3. Multiple Entrance Transition Probabilities, $p_{\alpha\beta}^{(n)}(t)$

In the general illness-death model of more than two illness states the number of exit transitions from a state S_α and the number of entrance transitions into another state S_β not only are conceptually different but also have different probability distributions. This can be shown from the corresponding p.g.f.'s. Let

$$\mathbf{g}_{\alpha\beta}(u;t)=\sum_{n=0}^{\infty}u^n p_{\alpha\beta}^{(n)}(t) \tag{3.22}$$

be the p.g.f. of the entrance transition probability $p_{\alpha\beta}^{(n)}(t)$ defined in (3.2). Following the procedure in Section 6 of Chapter 5, we see that the probabilities defined in (3.2) satisfy the backward differential equations

$$\begin{aligned}\frac{d}{dt}p_{\alpha\gamma}^{(n)}(t)&=\sum_{\substack{\beta=1\\ \beta\neq\gamma}}^{s}\nu_{\alpha\beta}p_{\beta\gamma}^{(n)}(t)+\nu_{\alpha\gamma}p_{\gamma\gamma}^{(n-1)}(t)\\ \frac{d}{dt}p_{\gamma\gamma}^{(n)}(t)&=\sum_{\beta=1}^{s}\nu_{\gamma\beta}p_{\beta\gamma}^{(n)}(t).\end{aligned} \tag{3.23}$$

It follows that the p.g.f.'s satisfy the differential equations

$$\begin{aligned}\frac{d}{dt}\mathbf{g}_{\alpha\gamma}(u;t)&=\sum_{\substack{\beta=1\\ \beta\neq\gamma}}^{s}\nu_{\alpha\beta}\mathbf{g}_{\beta\gamma}(u;t)+\nu_{\alpha\gamma}u\mathbf{g}_{\gamma\gamma}(u;t)\\ \frac{d}{dt}\mathbf{g}_{\gamma\gamma}(u;t)&=\sum_{\beta=1}^{s}\nu_{\gamma\beta}\mathbf{g}_{\beta\gamma}(u;t)\end{aligned} \tag{3.24}$$

with the conditions

$$\mathbf{g}_{\alpha\gamma}(u;0)=0,\qquad \mathbf{g}_{\gamma\gamma}(u;0)=1. \tag{3.25}$$

The corresponding characteristic equation is

$$|A(\gamma)|=\begin{vmatrix}(\rho-\nu_{11}) & -\nu_{12} & \cdots & -\nu_{1\gamma}u & \cdots & -\nu_{1s}\\ -\nu_{21} & (\rho-\nu_{22}) & \cdots & -\nu_{2\gamma}u & \cdots & -\nu_{2s}\\ \cdots & \cdots & \cdots & \cdots & \cdots & \cdots\\ -\nu_{\gamma1} & -\nu_{\gamma2} & \cdots & (\rho-\nu_{\gamma\gamma}) & \cdots & -\nu_{\gamma s}\\ \cdots & \cdots & \cdots & \cdots & \cdots & \cdots\\ -\nu_{s1} & -\nu_{s2} & \cdots & -\nu_{s\gamma}u & \cdots & (\rho-\nu_{ss})\end{vmatrix}=0, \tag{3.26}$$

which plays the same role as the characteristic equation (3.11) for the p.g.f.'s of the exit transition probabilities. Here the argument u appears in the γth column of the determinant $|A(\gamma)|$, whereas in (3.11) u was in the αth column of the transpose $|A'(\alpha)|$. Therefore the roots $\rho_1(\gamma), \ldots, \rho_s(\gamma)$ of (3.26) differ from $\rho_1(\alpha), \ldots, \rho_s(\alpha)$ of (3.11). They are equal only when $\gamma = \alpha$. If the roots of (3.26) are real and distinct, we have the solution

$$\mathbf{g}_{\alpha\gamma}(u;t) = \sum_{i=1}^{s} \frac{A'_{\alpha\gamma}(\gamma;i)}{\prod_{\substack{j=1\\j\neq i}}^{s} [\rho_i(\gamma) - \rho_j(\gamma)]} e^{\rho_i(\gamma)t}, \qquad \alpha, \gamma = 1, \ldots, s. \quad (3.27)$$

The p.g.f.'s in (3.27) are different from the p.g.f.'s $g_{\alpha\gamma}(u;t)$ of the multiple exit transition probabilities in (3.14). Thus the probability distribution of the number of exit transitions from a state is different from that of the number of entrance transitions into another state.

PROBLEMS

1. Show that

$$P_{\alpha\gamma}(t) = \sum_{l=1}^{s} A_{k\alpha}(l) \frac{T_{\gamma l}(k)}{|T(k)|} e^{\rho_l t} \quad (2.5)$$

satisfy the backward differential equations

$$\frac{d}{dt} P_{\alpha\gamma}(t) = \sum_{l=1}^{s} \nu_{\alpha\beta} P_{\beta\gamma}(t). \quad (2.2)$$

2. Show that

$$\sum_{l=1}^{s} \frac{A'_{\alpha\gamma}(l)}{\prod_{\substack{j=1\\j\neq l}}^{s} (\rho_l - \rho_j)} = 1 \quad \text{for} \quad \gamma = \alpha$$

$$= 0 \quad \text{for} \quad \gamma \neq \alpha. \quad (2.19)$$

3. Verify the equation

$$\int_0^t \mathbf{T}(k)\mathbf{E}(\tau)\mathbf{T}^{-1}(k)\mathbf{U}\, d\tau = \mathbf{T}(k)[\mathbf{E}(t) - \mathbf{I}]\boldsymbol{\rho}^{-1}\mathbf{T}^{-1}(k)\mathbf{U}. \quad (2.22)$$

4. Derive the individual transition probabilities $Q_{\alpha\delta}(t)$ in (2.24) from the matrix product

$$\mathbf{Q}(t) = \mathbf{T}(k)[\mathbf{E}(t) - \mathbf{I}]\boldsymbol{\rho}^{-1}\mathbf{T}^{-1}(k)\mathbf{U}. \quad (2.22)$$

5. Verify that the transition probabilities $P_{\alpha\gamma}(t)$ in (2.4) and $Q_{\alpha\delta}(t)$ in (2.25) satisfy the equation

$$Q_{\alpha\delta}(\tau + t) = Q_{\alpha\delta}(\tau) + \sum_{\beta=1}^{s} P_{\alpha\beta}(\tau) Q_{\beta\delta}(t). \quad (2.26)$$

6. Verify equation (2.31).

7. The matrix

$$\mathbf{V} = \begin{pmatrix} -4 & 1 & 2 \\ 1 & -3 & 1 \\ 1 & 0 & -2 \end{pmatrix}$$

is of rank 3 and satisfies the conditions

$$\sum_{\beta=1}^{3} \nu_{\alpha\beta} < 0 \qquad \text{and} \qquad \nu_{\alpha\beta} > 0 \text{ for } \beta \neq \alpha.$$

Therefore, all its eigenvalues are negative. Find these eigenvalues.

8. Given the intensity matrix $\mathbf{V}$ in Problem 7, compute the transition probabilities $P_{\alpha\beta}(t)$, $\alpha, \beta = 1, 2, 3$.

9. Show that the limiting probabilities in (2.39) satisfy the equation

$$-\sum_{\delta=1}^{r} \sum_{\beta=1}^{s} \sum_{l=1}^{s} \frac{A'_{\alpha\beta}(l)\mu_{\beta\delta}}{\prod_{\substack{j=1 \\ j \neq l}}^{s} (\rho_l - \rho_j)\rho_l} = 1.$$

10. Substitute (2.45) and (2.46) in (2.47) and prove the resulting equation.

11. Verify the system of differential equations in (3.23) for the multiple entrance transition probabilities.

CHAPTER 8

Migration Processes and Birth-Illness-Death Process

1. INTRODUCTION

In Chapter 7 emphasis was placed on transition probabilities rather than on the distribution of population sizes, which was obtained only as a by-product. This chapter is concerned exclusively with population sizes in illness states, and we shall derive directly the corresponding probability distributions. Furthermore, the populations of various states, which were previously treated as being entirely based on the initial size, are now allowed to change because of birth and migration. Movement between states (internal migration) is retained through illness transition as before.

The illness states are again denoted by S_α, $\alpha = 1, \ldots, s$. For a time interval (τ, t), let the populations of the states at the initial time τ be represented by a constant column vector

$$\mathbf{x}_\tau = \begin{pmatrix} x_{1\tau} \\ \cdot \\ \cdot \\ \cdot \\ x_{s\tau} \end{pmatrix} \tag{1.1}$$

and at time t by a random vector

$$\mathbf{X}_t = \begin{pmatrix} X_{1t} \\ \cdot \\ \cdot \\ \cdot \\ X_{st} \end{pmatrix}. \tag{1.2}$$

The value that $\mathbf{X}_t$ assumes is represented by a constant vector

$$\mathbf{x}_t = \begin{pmatrix} x_{1t} \\ \cdot \\ \cdot \\ \cdot \\ x_{st} \end{pmatrix} \tag{1.3}$$

where the components $x_{\alpha t}$ are nonnegative integers. We are interested in the conditional probability

$$P(\mathbf{x}_\tau, \mathbf{x}_t; \tau, t) = \Pr\{\mathbf{X}_t = \mathbf{x}_t \text{ at } t \quad \text{given } \mathbf{x}_\tau \text{ at } \tau\}. \tag{1.4}$$

When there is no danger of ambiguity, formulation is simplified by writing

$$P(\mathbf{x}_\tau, \mathbf{x}_t) = P(\mathbf{x}_\tau, \mathbf{x}_t; \tau, t). \tag{1.5}$$

As it stands, the probability distribution in (1.4) differs from that given in formula (2.55) in Chapter 7 only in that it does not take into account the sizes of death states R_δ.

In the models considered in this chapter the probability distribution (1.4) depends not only on the initial size $\mathbf{x}_\tau$ and the illness-death transitions, but also on birth and migration occurring within the interval (τ, t). In Section 2 migration processes are discussed in detail and their relation with the general model of illness-death process is noted. In Section 3 a birth-illness-death process is briefly presented. For the moment let us consider a general case.

For each ξ, $\tau < \xi < t$, a change in the population size of each state S_α in the time interval $(\xi, \xi + \Delta)$ is assumed to take place according to the following instantaneous probabilities:

$$\begin{aligned}
\lambda_\alpha^*(\xi)\Delta + o(\Delta) &= \Pr\{\text{the size of state } S_\alpha \text{ will increase by one in the interval } (\xi, \xi + \Delta)\} \\
\nu_{\alpha\beta}^*(\xi)\Delta + o(\Delta) &= \Pr\{\text{one individual will move from illness state } S_\alpha \text{ to } S_\beta \text{ in interval } (\xi, \xi + \Delta)\}, \quad \beta \neq \alpha \\
\mu_\alpha^*(\xi)\Delta + o(\Delta) &= \Pr\{\text{the size of state } S_\alpha \text{ will decrease by one in interval } (\xi, \xi + \Delta)\}.
\end{aligned} \tag{1.6}$$

We let

$$\nu_{\alpha\alpha}^*(\xi) = -\left[\sum_{\beta \neq \alpha} \nu_{\alpha\beta}^*(\xi) + \mu_\alpha^*(\xi)\right] \tag{1.7}$$

and introduce the column vector

$$\boldsymbol{\delta}_\alpha = \begin{pmatrix} \delta_{1\alpha} \\ \cdot \\ \cdot \\ \cdot \\ \delta_{s\alpha} \end{pmatrix} \tag{1.8}$$

where the components are Kronecker deltas. It is easily seen that the probabilities in (1.4) satisfy the following system of differential equations

$$\frac{\partial}{\partial t} P(\mathbf{x}_\tau, \mathbf{x}_t) = -P(\mathbf{x}_\tau, \mathbf{x}_t) \sum_{\alpha=1}^{s} [\lambda_\alpha^*(t) - \nu_{\alpha\alpha}^*(t)] + \sum_{\alpha=1}^{s} P(\mathbf{x}_\tau, \mathbf{x}_t - \boldsymbol{\delta}_\alpha)\lambda_\alpha^*(t)$$

$$+ \sum_{\alpha=1}^{s} \sum_{\substack{\beta=1 \\ \alpha\neq\beta}}^{s} P(\mathbf{x}_\tau, \mathbf{x}_t + \boldsymbol{\delta}_\alpha - \boldsymbol{\delta}_\beta)\nu_{\alpha\beta}^*(t) + \sum_{\alpha=1}^{s} P(\mathbf{x}_\tau, \mathbf{x}_t + \boldsymbol{\delta}_\alpha)\mu_\alpha^*(t) \tag{1.9}$$

with the initial conditions at $t = \tau$

$$P(\mathbf{x}_\tau, \mathbf{x}_\tau) = 1, \qquad P(\mathbf{x}_\tau, \mathbf{x}_t) = 0, \qquad \text{for} \qquad \mathbf{x}_t \neq \mathbf{x}_\tau. \tag{1.10}$$

System (1.9) describes in general the growth of a population. Various models can be derived from (1.9) by making appropriate assumptions regarding $\lambda_\alpha^*(t)$, $\nu_{\alpha\beta}^*(t)$, and $\mu_\alpha^*(t)$. Table 1 lists the stochastic processes discussed in this book in terms of the general system of differential equations in (1.9).

2. EMIGRATION-IMMIGRATION PROCESSES—POISSON-MARKOV PROCESSES

It was pointed out early in Chapter 7 that emigration may be treated the same as death. But the term "immigration" has a different meaning from "birth" in stochastic processes, although the distinction is arbitrary. An increase in a population within a small time interval $(t, t + \Delta)$ is regarded as the result of birth if the corresponding instantaneous probability is a function of the size of the population at time t; otherwise it is treated as immigration. Migration processes have applications to practical problems. An early study of a migration process was made by Bartlett [1949]. Patil considered a model in the study of spermatozoa counts [1957]. The following discussion is based on a paper by Chiang [1964b].

Table 1. Stochastic processes concerned with birth, illness, migration, and death

Type of Process	Number of States s	$\lambda_\alpha^*(t)$	$\nu_{\alpha\beta}^*(t)$	$\mu_\alpha^*(t)$
Poisson process	1	λ	—	0
	1	$\lambda(t)$	—	0
Pure birth process	1	λx_t	—	0
	1	$x_t\lambda(t)$	—	0
Pure death process	1	0	—	$x_t\mu(t)$
Polya process	1	$\dfrac{\lambda + \lambda a x_t}{1 + \lambda a t}$	—	0
Linear growth	1	$x_t\lambda$	—	$x_t\mu$
General birth-death process	1	$x_t\lambda(t)$	—	$x_t\mu(t)$
Simple illness-death process	2	0	$x_{\alpha t}\nu_{\alpha\beta}$	$x_{\alpha t}(\mu_{\alpha 1} + \cdots + \mu_{\alpha r})$
Finite Markov process	s	—	$x_{it}\nu_{ij}$	—
General illness-death process	s	0	$x_{\alpha t}\nu_{\alpha\beta}$	$x_{\alpha t}(\mu_{\alpha 1} + \cdots + \mu_{\alpha r})$
Migration process	s	$\lambda_\alpha(t)$	$x_{\alpha t}\nu_{\alpha\beta}$	$x_{\alpha t}\mu_\alpha$
Birth-illness-death process	s	$x_{\alpha t}\lambda_\alpha$	$x_{\alpha t}\nu_{\alpha\beta}$	$x_{\alpha t}\mu_\alpha$

x_t ($x_{\alpha t}$) is the population size (in state S_α) at time t.

2.1. The Differential Equations

Let us consider the case in which an increase in the population during a time interval $(t, t + \Delta)$ is independent of the population size and for state S_α

$$\lambda_\alpha^*(t) = \lambda_\alpha(t). \tag{2.1}$$

The transition of an individual from one state to another is assumed, as before, to be independent of the transitions of other individuals so that

$$\nu^*_{\alpha\beta}(t) = x_{\alpha t}\nu_{\alpha\beta}, \qquad \mu^*_\alpha(t) = x_{\alpha t}\mu_\alpha \quad \text{and} \quad \nu_{\alpha\alpha} = -\left(\sum_{\substack{\beta=1\\ \beta\neq\alpha}}^{s} \nu_{\alpha\beta} + \mu_\alpha\right) \tag{2.2}$$

where $x_{\alpha t}$ is the number of individuals in state S_α at time t. Here $\lambda_\alpha(t)$ is a function of t, whereas $\nu_{\alpha\beta}$ and μ_α are independent of time. Under these assumptions the differential equations corresponding to (1.9) are

$$\begin{aligned}\frac{\partial}{\partial t} P(\mathbf{x}_\tau, \mathbf{x}_t) = {} & -P(\mathbf{x}_\tau, \mathbf{x}_t)\sum_{\alpha=1}^{s} [\lambda_\alpha(t) - x_{\alpha t}\nu_{\alpha\alpha}] + \sum_{\alpha=1}^{s} P(\mathbf{x}_\tau, \mathbf{x}_t - \boldsymbol{\delta}_\alpha)\lambda_\alpha(t)\\ & + \sum_{\substack{\alpha=1\\ \alpha\neq\beta}}^{s}\sum_{\beta=1}^{s} P(\mathbf{x}_\tau, \mathbf{x}_t + \boldsymbol{\delta}_\alpha - \boldsymbol{\delta}_\beta)(x_{\alpha t} + 1)\nu_{\alpha\beta}\\ & + \sum_{\alpha=1}^{s} P(\mathbf{x}_\tau, \mathbf{x}_t + \boldsymbol{\delta}_\alpha)(x_{\alpha t} + 1)\mu_\alpha. \end{aligned} \tag{2.3}$$

Our purpose is to derive an explicit solution for the probabilities $P(\mathbf{x}_\tau, \mathbf{x}_t)$.
Let the p.g.f. of $P(\mathbf{x}_\tau, \mathbf{x}_t)$ be defined by

$$G_{\mathbf{x}_t}(\mathbf{s}; \tau, t) = \sum_{x_{1t}} \cdots \sum_{x_{st}} u_1^{x_{1t}} \cdots u_s^{x_{st}} P(\mathbf{x}_\tau, \mathbf{x}_t). \tag{2.4}$$

Direct calculations show that the p.g.f. satisfies the following partial differential equation

$$\begin{aligned}\frac{\partial}{\partial t} G_{\mathbf{x}_t}(\mathbf{u}; \tau, t) = {} & -\sum_{\alpha=1}^{s} \lambda_\alpha(t) G_{\mathbf{x}_t}(\mathbf{u}; \tau, t) + \sum_{\alpha=1}^{s} \nu_{\alpha\alpha} u_\alpha \frac{\partial}{\partial u_\alpha} G_{\mathbf{x}_t}(\mathbf{u}; \tau, t)\\ & + \sum_{\alpha=1}^{s} \lambda_\alpha(t) u_\alpha G_{\mathbf{x}_t}(\mathbf{u}; \tau, t) + \sum_{\substack{\alpha=1\\ \alpha\neq\beta}}^{s}\sum_{\beta=1}^{s} \nu_{\alpha\beta} u_\beta \frac{\partial}{\partial u_\alpha} G_{\mathbf{x}_t}(\mathbf{u}; \tau, t)\\ & + \sum_{\alpha=1}^{s} \mu_\alpha \frac{\partial}{\partial u_\alpha} G_{\mathbf{x}_t}(\mathbf{u}; \tau, t). \end{aligned} \tag{2.5}$$

Equation (2.5) may be simplified. After collecting terms, we find that

$$\begin{aligned}\frac{\partial}{\partial t} G_{\mathbf{x}_t}(\mathbf{u}; \tau, t) = {} & -\sum_{\alpha=1}^{s} \lambda_\alpha(t)(1 - u_\alpha) G_{\mathbf{x}_t}(\mathbf{u}; \tau, t)\\ & + \sum_{\alpha=1}^{s}\left(\sum_{\beta=1}^{s} \nu_{\alpha\beta} u_\beta + \mu_\alpha\right)\frac{\partial}{\partial u_\alpha} G_{\mathbf{x}_t}(\mathbf{u}; \tau, t). \end{aligned} \tag{2.6}$$

Since (2.2) implies that $\mu_\alpha = -(\nu_{\alpha 1} + \cdots + \nu_{\alpha s})$, Equation (2.6) may be rewritten as

$$\frac{\partial}{\partial t} G_{\mathbf{X}_t}(\mathbf{u}; \tau, t) = -\sum_{\alpha=1}^{s} \lambda_\alpha(t)(1 - u_\alpha) G_{\mathbf{X}_t}(\mathbf{u}; \tau, t) - \sum_{\alpha=1}^{s} \sum_{\beta=1}^{s} \nu_{\alpha\beta}(1 - u_\beta) \frac{\partial}{\partial u_\alpha} G_{\mathbf{X}_t}(\mathbf{u}; \tau, t) \quad (2.7)$$

which, upon substitution of $z_\alpha = 1 - u_\alpha$, becomes

$$\frac{\partial}{\partial t} G_{\mathbf{X}_t}(\mathbf{z}; \tau, t) = -\sum_{\alpha=1}^{s} \lambda_\alpha(t) z_\alpha G_{\mathbf{X}_t}(\mathbf{z}; \tau, t) + \sum_{\alpha=1}^{s} \sum_{\beta=1}^{s} \nu_{\alpha\beta} z_\beta \frac{\partial}{\partial z_\alpha} G_{\mathbf{X}_t}(\mathbf{z}; \tau, t) \quad (2.8)$$

with the initial condition

$$G_{\mathbf{X}_t}(\mathbf{z}; \tau, t) = \prod_{\alpha=1}^{s} (1 - z_\alpha)^{x_{\alpha\tau}}. \quad (2.9)$$

2.2. Solution for the Probability Generating Function

Differential Equation (2.8) can be solved in the usual way. The auxiliary equations are

$$-\frac{dt}{1} = \frac{dz_1}{\sum_\beta \nu_{1\beta} z_\beta} = \cdots = \frac{dz_s}{\sum_\beta \nu_{s\beta} z_\beta} = \frac{dG_{\mathbf{X}_t}(\mathbf{z}; \tau, t)}{\sum_\beta \lambda_\beta(t) z_\beta G_{\mathbf{X}_t}(\mathbf{z}; \tau, t)} \quad (2.10)$$

which may be rewritten as

$$\frac{dz_\alpha}{dt} + \sum_{\beta=1}^{s} \nu_{\alpha\beta} z_\beta = 0, \qquad \alpha = 1, \ldots, s \quad (2.10a)$$

$$\frac{d}{dt} \log G_{\mathbf{X}_t}(\mathbf{z}; \tau, t) = -\sum_{\beta=1}^{s} \lambda_\beta(t) z_\beta. \quad (2.10b)$$

We shall use matrix notation to simplify the presentation. Let

$$\boldsymbol{\lambda}(t) = \begin{pmatrix} \lambda_1(t) \\ \cdot \\ \cdot \\ \cdot \\ \lambda_s(t) \end{pmatrix} \qquad \mathbf{z} = \begin{pmatrix} z_1 \\ \cdot \\ \cdot \\ \cdot \\ z_s \end{pmatrix} \quad (2.11)$$

and

$$\mathbf{V} = \begin{pmatrix} \nu_{11} & \cdot & \cdot & \cdot & \nu_{1s} \\ \cdot & & & & \cdot \\ \cdot & & & & \cdot \\ \cdot & & & & \cdot \\ \nu_{s1} & \cdot & \cdot & \cdot & \nu_{ss} \end{pmatrix} \quad (2.12)$$

and let D be a diagonal matrix with the differentiation operator d/dt on the diagonal line. Equations (2.10a) and (2.10b) become, respectively,

$$(D + \mathbf{V})\mathbf{z} = 0 \tag{2.13a}$$

and

$$d \log G_{\mathbf{X}_t}(\mathbf{z}; \tau, t) = -\boldsymbol{\lambda}'(t)\mathbf{z}\, dt \tag{2.13b}$$

where $\boldsymbol{\lambda}'(t)$ is the transpose of $\boldsymbol{\lambda}(t)$.

Equation (2.13a) is a system of linear, first-order, homogeneous differential equations with constant coefficients, and may be solved for $\mathbf{z}$ with the procedure described in Chapter 6. For clarity, although at the expense of some repetition, Equation (2.13a) is solved directly. Consider the characteristic equation

$$|\rho\mathbf{I} + \mathbf{V}| = 0 \tag{2.14}$$

in which $\mathbf{I}$ is an $s \times s$ unit matrix, and let the characteristic roots of (2.14) be denoted by $-\rho_1, \ldots, -\rho_s$. For each ρ_j we consider the characteristic matrix

$$\mathbf{A}(j) = (\rho_j\mathbf{I} - \mathbf{V}) \tag{2.15}$$

with its cofactors denoted by $A_{k\alpha}(j)$; then the nonzero vector

$$\mathbf{T}_j(k) = \begin{pmatrix} A_{k1}(j) \\ \cdot \\ \cdot \\ \cdot \\ A_{ks}(j) \end{pmatrix} \tag{2.16}$$

is an eigenvector of $\mathbf{V}$ corresponding to ρ_j with

$$(\rho_j\mathbf{I} - \mathbf{V})\mathbf{T}_j(k) = 0. \tag{2.17}$$

Finally we formulate the matrix $\mathbf{T}(k)$

$$\mathbf{T}(k) = \begin{pmatrix} A_{k1}(1) & A_{k1}(2) & \cdots & A_{k1}(s) \\ A_{k2}(1) & A_{k2}(2) & \cdots & A_{k2}(s) \\ \cdot & \cdot & & \cdot \\ \cdot & \cdot & & \cdot \\ \cdot & \cdot & & \cdot \\ A_{ks}(1) & A_{ks}(2) & \cdots & A_{ks}(s) \end{pmatrix} \tag{2.18}$$

where the jth column is the kth column of the adjoint matrix of $\mathbf{A}(j)$, and k is arbitrary but the same for all columns [see Equation (3.43) in Chapter 6].

Now for each ρ_j there is a corresponding solution for (2.13a),

$$\mathbf{z} = e^{-\rho_j t}\mathbf{c}(j) \tag{2.19}$$

where

$$\mathbf{c}(j) = \begin{pmatrix} c_{1j} \\ \cdot \\ \cdot \\ \cdot \\ c_{sj} \end{pmatrix} \tag{2.20}$$

is a column vector with constant components. When (2.19) is substituted in (2.13a) we obtain

$$(\rho_j \mathbf{I} - \mathbf{V})\mathbf{c}(j) = \mathbf{0} \tag{2.21}$$

or, by (2.15),

$$\mathbf{A}(j)\mathbf{c}(j) = \mathbf{0}. \tag{2.22}$$

Since $-\rho_j$ is a solution of (2.14), the determinant $|A(j)|$ of the matrix $\mathbf{A}(j)$ is equal to zero. Cramer's rule implies that

$$\mathbf{c}(j) = b_j \begin{pmatrix} A_{k1}(j) \\ \cdot \\ \cdot \\ \cdot \\ A_{ks}(j) \end{pmatrix} \tag{2.23}$$

or, from (2.16),

$$\mathbf{c}(j) = b_j \mathbf{T}_j(k) \tag{2.24}$$

where b_j is an arbitrary constant. Substituting (2.24) in (2.19) gives

$$\mathbf{z} = \mathbf{T}_j(k) e^{-\rho_j t} b_j. \tag{2.25}$$

If the eigenvalues $-\rho_j$ are real and distinct, then we have the general solution for the first set of auxiliary equations (2.13a)

$$\mathbf{z} = \sum_{j=1}^{s} \mathbf{T}_j(k) e^{-\rho_j t} b_j. \tag{2.26}$$

To express (2.26) in matrix form, we introduce the diagonal matrix

$$\mathbf{E}(t) = \begin{pmatrix} e^{\rho_1 t} & 0 & \cdot & \cdot & \cdot & 0 \\ 0 & e^{\rho_2 t} & \cdot & \cdot & \cdot & 0 \\ \cdot & \cdot & & & & \cdot \\ \cdot & \cdot & & & & \cdot \\ \cdot & \cdot & & & & \cdot \\ 0 & 0 & \cdot & \cdot & \cdot & e^{\rho_s t} \end{pmatrix} \tag{2.27}$$

and the column vector

$$\mathbf{b} = \begin{pmatrix} b_1 \\ \cdot \\ \cdot \\ \cdot \\ b_s \end{pmatrix} \tag{2.28}$$

and write

$$\mathbf{z} = \mathbf{T}(k)\mathbf{E}^{-1}(t)\mathbf{b}. \tag{2.29}$$

Equation (2.29) is now substituted into the second auxiliary equation (2.13*b*) to yield

$$d \log G_{\mathbf{X}_t}(\mathbf{z}; \tau, t) = -\boldsymbol{\lambda}'(t)\mathbf{T}(k)\mathbf{E}^{-1}(t)\mathbf{b}\, dt \tag{2.30}$$

and hence

$$G_{\mathbf{X}_t}(\mathbf{z}; \tau, t) = c \exp\left\{-\int_\tau^t \boldsymbol{\lambda}'(\xi)\mathbf{T}(k)\mathbf{E}^{-1}(\xi)\mathbf{b}\, d\xi\right\} \tag{2.31}$$

where log c is the constant of integration. The integrand in (2.31) may be expanded

$$\boldsymbol{\lambda}'(\xi)\mathbf{T}(k)\mathbf{E}^{-1}(\xi)\mathbf{b} = \left[\sum_{\alpha=1}^{s} \lambda_\alpha(\xi)A_{k\alpha}(1)e^{-\rho_1\xi}, \ldots, \sum_{\alpha=1}^{s} \lambda_\alpha(\xi)A_{k\alpha}(s)e^{-\rho_s\xi}\right]\mathbf{b} \tag{2.32}$$

so that the integral becomes

$$\int_\tau^t \boldsymbol{\lambda}'(\xi)\mathbf{T}(k)\mathbf{E}^{-1}(\xi)\mathbf{b}\, d\xi = \boldsymbol{\eta}'(t)\mathbf{b} \tag{2.33}$$

where

$$\boldsymbol{\eta}'(t) = [\eta_1(t), \ldots, \eta_s(t)] \tag{2.34}$$

and

$$\eta_j(t) = \sum_{\alpha=1}^{s} A_{k\alpha}(j) \int_\tau^t \lambda_\alpha(\xi)e^{-\rho_j\xi}\, d\xi. \tag{2.35}$$

Obviously at $t = \tau$

$$\boldsymbol{\eta}'(\tau) = (0, \ldots, 0). \tag{2.36}$$

Substituting (2.33) in (2.31) gives

$$G_{\mathbf{X}_t}(\mathbf{z}; \tau, t) = c \exp\{-\boldsymbol{\eta}'(t)\mathbf{b}\}. \tag{2.37}$$

Rearranging (2.29) as

$$\mathbf{b} = \mathbf{E}(t)\mathbf{T}^{-1}(k)\mathbf{z} \tag{2.38}$$

and expressing the constant c in (2.37) as a function Φ of the constant vector $\mathbf{b}$, we have

$$c = \Phi\{\mathbf{b}\} = \Phi\{\mathbf{E}(t)\mathbf{T}^{-1}(k)\mathbf{z}\}. \tag{2.39}$$

Substituting (2.38) and (2.39) in (2.37) gives the general solution for the differential equation (2.8),

$$G_{\mathbf{X}_t}(\mathbf{z}; \tau, t) = \Phi\{\mathbf{E}(t)\mathbf{T}^{-1}(k)\mathbf{z}\} \exp\{-\boldsymbol{\eta}'(t)\mathbf{E}(t)\mathbf{T}^{-1}(k)\mathbf{z}\} \tag{2.40}$$

For the particular solution corresponding to the initial condition

$$G_{\mathbf{X}_t}(\mathbf{z}; \tau, \tau) = \prod_{\alpha=1}^{s} [1 - z_\alpha]^{x_{\alpha\tau}} \tag{2.9}$$

we need to determine the function Φ. If we set $t = \tau$, we see that $\boldsymbol{\eta}(\tau) = \mathbf{0}$ and the exponential function in (2.40) vanishes, so that (2.40) becomes

$$G_{\mathbf{X}_t}(\mathbf{z}; \tau, \tau) = \Phi\{\mathbf{E}(\tau)\mathbf{T}^{-1}(k)\mathbf{z}\}. \tag{2.41}$$

Substituting (2.41) in (2.9), we arrive at a relation

$$\Phi\{\mathbf{E}(\tau)\mathbf{T}^{-1}(k)\mathbf{z}\} = \prod_{\alpha=1}^{s} [1 - z_\alpha]^{x_{\alpha\tau}} \tag{2.42}$$

from which Φ can be determined. Let

$$\mathbf{E}(\tau)\mathbf{T}^{-1}(k)\mathbf{z} = \mathbf{b}_\tau \tag{2.43}$$

and solve for $\mathbf{z}$ to get

$$\mathbf{z} = \mathbf{T}(k)\mathbf{E}^{-1}(\tau)\mathbf{b}_\tau. \tag{2.44}$$

Hence for all $\mathbf{b}_\tau$ such that every component of $\mathbf{z}$ in (2.44) is less than one in absolute value,

$$\Phi\{\mathbf{b}_\tau\} = \prod_{\alpha=1}^{s} [1 - z_\alpha]^{x_{\alpha\tau}} \tag{2.45}$$

where z_α are components of $\mathbf{z}$ as given in (2.44). For

$$\mathbf{b}_t = \mathbf{E}(t)\mathbf{T}^{-1}(k)\mathbf{z} \tag{2.46}$$

we let

$$\boldsymbol{\zeta} = \mathbf{T}(k)\mathbf{E}^{-1}(\tau)\mathbf{E}(t)\mathbf{T}^{-1}(k)\mathbf{z} \tag{2.47}$$

and hence

$$\Phi\{\mathbf{E}(t)\mathbf{T}^{-1}(k)\mathbf{z}\} = \prod_{\alpha=1}^{s} [1 - \zeta_\alpha]^{x_{\alpha\tau}} \tag{2.48}$$

where ζ_α are components of $\boldsymbol{\zeta}$ in (2.47). Substituting (2.48) in (2.40) gives the required particular solution,

$$G_{\mathbf{X}_t}(\mathbf{z}; \tau, t) = \prod_{\alpha=1}^{s} [1 - \zeta_\alpha]^{x_{\alpha\tau}} \exp\{-\boldsymbol{\eta}'(t)\mathbf{E}(t)\mathbf{T}^{-1}(k)\mathbf{z}\}. \tag{2.49}$$

2.3. Relation to the Illness–Death Process and Solution for the Probability Distribution

Formula (2.47) has an interesting relation to the transition probability $P_{\alpha\beta}(t-\tau)$ discussed in Chapter 7. In fact, since $\mathbf{E}(\tau)$ and $\mathbf{E}(t)$ are diagonal matrices, we refer to (2.7) in Chapter 7 and write

$$\boldsymbol{\zeta} = \mathbf{T}(k)\mathbf{E}(t-\tau)\mathbf{T}^{-1}(k)\mathbf{z} = \mathbf{P}(t-\tau)\mathbf{z} \tag{2.50}$$

where $\mathbf{P}(t-\tau) = \mathbf{T}(k)\mathbf{E}(t-\tau)\mathbf{T}^{-1}(k)$ is the transition probability matrix given by (2.7) in Chapter 7 defined as

$$\mathbf{P}(t-\tau) = \begin{pmatrix} P_{11}(t-\tau) & \cdot & \cdot & \cdot & P_{1s}(t-\tau) \\ \cdot & & & & \cdot \\ \cdot & & & & \cdot \\ \cdot & & & & \cdot \\ P_{s1}(t-\tau) & \cdot & \cdot & \cdot & P_{ss}(t-\tau) \end{pmatrix} \tag{2.51}$$

with the elements

$$P_{\alpha\beta}(t-\tau) = \sum_{j=1}^{s} A_{k\alpha}(j)\frac{T_{\beta j}(k)}{|T(k)|}\, e^{\rho_j(t-\tau)}.$$

Therefore, since $z_\beta = 1 - u_\beta$,

$$\zeta_\alpha = \sum_{\beta=1}^{s} P_{\alpha\beta}(t-\tau)z_\beta = \sum_{\beta=1}^{s} P_{\alpha\beta}(t-\tau)(1-u_\beta) \tag{2.52}$$

and

$$[1-\zeta_\alpha]^{x_{\alpha\tau}} = \left[1 - \sum_{\beta=1}^{s} P_{\alpha\beta}(t-\tau) + \sum_{\beta=1}^{s} P_{\alpha\beta}(t-\tau)u_\beta\right]^{x_{\alpha\tau}} \tag{2.53}$$

where the difference [see Equation (2.30), Chapter 7]

$$1 - \sum_{\beta=1}^{s} P_{\alpha\beta}(t-\tau) = \sum_{\delta=1}^{r} Q_{\alpha\delta}(t-\tau)$$

is the probability that an individual in the illness state S_α at time τ will be in one of the death states R_δ at or before time t. In (2.53) we have the p.g.f. of a multinomial distribution with the probabilities $P_{\alpha\beta}(t-\tau)$.

The exponential part of the p.g.f. in (2.49) may be written a little differently. Let

$$\boldsymbol{\eta}'(t)\mathbf{E}(t)\mathbf{T}^{-1}(k)\mathbf{z} = \boldsymbol{\theta}'(t)\mathbf{z} = \sum_{\beta=1}^{s} \theta_\beta(t)(1-u_\beta) \tag{2.54}$$

when $\boldsymbol{\theta}'(t) = [\theta_1(t), \ldots, \theta_s(t)]$ with

$$\theta_\beta(t) = \sum_{j=1}^{s} \eta_j(t)\frac{T_{\beta j}(k)}{|T(k)|}\, e^{\rho_j t} \tag{2.55}$$

and $\eta_j(t)$ as given in (2.35). Consequently

$$G_{\mathbf{X}_t}(\mathbf{u};\tau,t)=\prod_{\alpha=1}^{s}\left[1-\sum_{\beta=1}^{s}P_{\alpha\beta}(t-\tau)+\sum_{\beta=1}^{s}u_\beta P_{\alpha\beta}(t-\tau)\right]^{x_{\alpha\tau}}$$
$$\times\exp\left\{-\sum_{\beta=1}^{s}\theta_\beta(t)(1-u_\beta)\right\}. \quad (2.56)$$

Thus the probability distribution of the random vector $\mathbf{X}_t$ at time t, given $\mathbf{x}_\tau$ at τ, consists of two components: one with s multinomial distributions associated with the initial population $\mathbf{x}_\tau$ in the s illness states, and the second with a Poisson-type distribution originating from immigration into the s states. If there is no immigration so that $\lambda_\alpha(t)=0$ and $\theta_\beta(t)=0$, then the probability distribution of $\mathbf{X}_t$ reduces to that of the illness-death process discussed in Chapter 7. If on the other hand the initial population is $\mathbf{x}_\tau=\mathbf{0}$, the distribution of $\mathbf{X}_t$ will be of a pure Poisson type.

As it stands now, the probability distribution, expectation, variance, and covariance derived from the p.g.f. (2.56) are the following:

$$\Pr\{X_{1t}=x_{1t},\ldots,X_{st}=x_{st}\mid\mathbf{x}_\tau\}$$
$$=\sum_{k_{0\beta}}\cdots\sum_{k_{s\beta}}\left\{\prod_{\alpha=1}^{s}\frac{x_{\alpha\tau}!}{\prod_{\beta=1}^{s}k_{\alpha\beta}!\,[x_{\alpha\tau}-\sum_\beta k_{\alpha\beta}]!}\prod_{\beta=1}^{s}[P_{\alpha\beta}(t-\tau)]^{k_{\alpha\beta}}\right.$$
$$\left.\times\left[1-\sum_{\beta=1}^{s}P_{\alpha\beta}(t-\tau)\right]^{x_{\alpha\tau}-\sum_\beta k_{\alpha\beta}}\right\}\left\{\prod_{\beta=1}^{s}\frac{e^{-\theta_\beta(t)}[\theta_\beta(t)]^{k_{0\beta}}}{k_{0\beta}!}\right\} \quad (2.57)$$

where $k_{0\beta}+k_{1\beta}+\cdots+k_{s\beta}=x_{\beta t},\quad \beta=1,\ldots,s,$

$$E[X_{\beta t}\mid\mathbf{x}_\tau]=\sum_{\alpha=1}^{s}x_{\alpha\tau}P_{\alpha\beta}(t-\tau)+\theta_\beta(t),$$

$$\operatorname{Var}(X_{\beta t}\mid\mathbf{x}_\tau)=\sum_{\alpha=1}^{s}x_{\alpha\tau}P_{\alpha\beta}(t-\tau)[1-P_{\alpha\beta}(t-\tau)]+\theta_\beta(t) \quad (2.58)$$

and

$$\operatorname{Cov}(X_{\beta t},X_{\gamma t}\mid\mathbf{x}_\tau)=-\sum_{\alpha=1}^{s}x_{\alpha\tau}P_{\alpha\beta}(t-\tau)P_{\alpha\gamma}(t-\tau).$$

2.4. Constant Immigration

The probability distribution of the random vector $\mathbf{X}_t$ presented thus far in this section depends upon the function $\boldsymbol{\lambda}(t)$. Generally the process is nonhomogeneous in time. When $\lambda_\alpha(t)=\lambda_\alpha$ is constant for each α, however, we have a time-homogeneous process. In this case we may let $\tau=0$, and let t be the length of a time interval; then $\boldsymbol{\eta}'(t)$ in (2.33) becomes

$$\boldsymbol{\eta}'(t) = \boldsymbol{\lambda}'\mathbf{T}(k)\int_0^t \mathbf{E}^{-1}(\xi)\,d\xi = \boldsymbol{\lambda}'\mathbf{T}(k)[\mathbf{I} - \mathbf{E}^{-1}(t)]\boldsymbol{\rho}^{-1} \tag{2.59}$$

and $\boldsymbol{\theta}'(t)\mathbf{z}$ in (2.54) becomes

$$\boldsymbol{\theta}'(t)\mathbf{z} = \boldsymbol{\lambda}'\mathbf{T}(k)[\mathbf{E}(t) - \mathbf{I}]\boldsymbol{\rho}^{-1}\mathbf{T}^{-1}(k)\mathbf{z}. \tag{2.60}$$

Matrix $\mathbf{T}(k)$ diagonalizes $\mathbf{V}$ in the form $\mathbf{T}^{-1}(k)\mathbf{V}\mathbf{T}(k) = \boldsymbol{\rho}$ [cf. Equations (3.30) and (4.45) in Chapter 6]; therefore

$$\mathbf{T}(k)\boldsymbol{\rho}^{-1}\mathbf{T}^{-1}(k) = \mathbf{V}^{-1}. \tag{2.61}$$

Using (2.61) and the relation $\mathbf{T}(k)\mathbf{E}(t)\mathbf{T}^{-1}(k) = \mathbf{P}(t)$, we may rearrange the right side of (2.60) to obtain

$$\begin{aligned}\boldsymbol{\lambda}'\mathbf{T}(k)[\mathbf{E}(t) - \mathbf{I}]\boldsymbol{\rho}^{-1}\mathbf{T}^{-1}(k)\mathbf{z} &= \boldsymbol{\lambda}'\mathbf{T}(k)\boldsymbol{\rho}^{-1}\mathbf{T}^{-1}(k)[\mathbf{T}(k)\mathbf{E}(t)\mathbf{T}^{-1}(k) - \mathbf{I}]\mathbf{z}\\ &= \boldsymbol{\lambda}'\mathbf{V}^{-1}[\mathbf{P}(t) - \mathbf{I}]\mathbf{z}.\end{aligned} \tag{2.62}$$

Equation (2.60) thus becomes

$$\boldsymbol{\theta}'(t)\mathbf{z} = \boldsymbol{\lambda}'\mathbf{V}^{-1}[\mathbf{P}(t) - \mathbf{I}]\mathbf{z} \tag{2.63}$$

with the components

$$\theta_\beta(t) = \sum_{\alpha=1}^{s}\lambda_\alpha\left[\sum_{\gamma=1}^{s}\frac{V_{\gamma\alpha}}{|V|}P_{\gamma\beta}(t) - \frac{V_{\beta\alpha}}{|V|}\right] \tag{2.64}$$

where $V_{\alpha\beta}$ is the cofactor and $|V|$ is the determinant of the intensity matrix $\mathbf{V}$. Corresponding to (2.49), the p.g.f. has an alternative form

$$G_{\mathbf{X}_t}(\mathbf{u}; t) = \prod_{\alpha=1}^{s}\left[1 - \sum_{\beta=1}^{s}P_{\alpha\beta}(t) + \sum_{\beta=1}^{s}P_{\alpha\beta}(t)u_\beta\right]^{x_{\alpha\tau}}\exp\{-\boldsymbol{\lambda}'\mathbf{V}^{-1}[\mathbf{P}(t) - \mathbf{I}]\mathbf{z}\}. \tag{2.65}$$

The probability distribution and moments of $\mathbf{X}_t$ have the same expressions as those given in (2.57) and (2.58) with $\theta_\beta(t)$ defined in (2.64).

3. A BIRTH-ILLNESS-DEATH PROCESS

When $\lambda_\alpha^*(t)$ is a function of the population size $x_{\alpha t}$ of state S_α at time t, we have a birth-illness-death process. A simple example is

$$\lambda_\alpha^*(t) = x_{\alpha t}\lambda_\alpha, \qquad \nu_{\alpha\beta}^* = x_{\alpha t}\nu_{\alpha\beta}, \qquad \mu_\alpha^* = x_{\alpha t}\mu_\alpha. \tag{3.1}$$

In this case, the process is homogeneous with respect to time. Let a random vector $\mathbf{X}_t$ be the population at time t, and write the probability of $\mathbf{X}_t$ as

$$P(\mathbf{x}_\tau, \mathbf{x}_t) = \Pr\{\mathbf{X}_t = \mathbf{x}_t \mid \mathbf{x}_\tau \text{ at } \tau\}. \tag{3.2}$$

Substituting (3.1) in (1.9) gives the following differential equation for the probability in (3.2)

$$\frac{\partial}{\partial t} P(\mathbf{x}_\tau, \mathbf{x}_t) = -P(\mathbf{x}_\tau, \mathbf{x}_t)\sum_{\alpha=1}^{s} x_{\alpha t}[\lambda_\alpha - \nu_{\alpha\alpha}] + \sum_{\alpha=1}^{s} P(\mathbf{x}_\tau, \mathbf{x}_t - \boldsymbol{\delta}_\alpha)(x_{\alpha t} - 1)\lambda_\alpha$$
$$+ \sum_{\substack{\alpha=1 \\ \alpha\neq\beta}}^{s} \sum_{\beta=1}^{s} P(\mathbf{x}_\tau, \mathbf{x}_t + \boldsymbol{\delta}_\alpha - \boldsymbol{\delta}_\beta)(x_{\alpha t} + 1)\nu_{\alpha\beta}$$
$$+ \sum_{\alpha=1}^{s} P(\mathbf{x}_\tau, \mathbf{x}_t + \boldsymbol{\delta}_\alpha)(x_{\alpha t} + 1)\mu_\alpha. \tag{3.3}$$

Direct computation shows that the corresponding p.g.f. satisfies the differential equation

$$\frac{\partial}{\partial t} G_{\mathbf{X}_t}(\mathbf{u}; t) = -\sum_{\alpha=1}^{s} \lambda_\alpha u_\alpha(1 - u_\alpha) \frac{\partial}{\partial u_\alpha} G_{\mathbf{X}_t}(\mathbf{u}; t) + \sum_{\alpha=1}^{s} u_\alpha \nu_{\alpha\alpha} \frac{\partial}{\partial u_\alpha} G_{\mathbf{X}_t}(\mathbf{u}; t)$$
$$+ \sum_{\substack{\alpha=1 \\ \alpha\neq\beta}}^{s} \sum_{\beta=1}^{s} u_\beta \nu_{\alpha\beta} \frac{\partial}{\partial u_\alpha} G_{\mathbf{X}_t}(\mathbf{u}; t) + \sum_{\alpha=1}^{s} \mu_\alpha \frac{\partial}{\partial u_\alpha} G_{\mathbf{X}_t}(\mathbf{u}; t). \tag{3.4}$$

Making the substitutions of $\mu_\alpha = -\sum_\beta \nu_{\alpha\beta}$ and $z_\alpha = 1 - u_\alpha$, (3.4) may be simplified to

$$\frac{\partial}{\partial t} G_{\mathbf{X}_t}(\mathbf{z}; t) = \sum_{\alpha=1}^{s} \left[\lambda_\alpha z_\alpha(1 - z_\alpha) + \sum_{\beta=1}^{s} \nu_{\alpha\beta} z_\beta \right] \frac{\partial}{\partial z_\alpha} G_{\mathbf{X}_t}(\mathbf{z}; t). \tag{3.5}$$

When (3.5) is solved, the probability $P(\mathbf{x}_\tau, \mathbf{x}_t)$ can be derived from the p.g.f. However, a simple explicit solution for (3.5) is unknown at the present time.

PROBLEMS

1. Show that the assumption of independence of individuals' transitions implies the relations

$$\nu^*_{\alpha\beta} = x_{\alpha t}\nu_{\alpha\beta} \qquad \text{and} \qquad \mu^*_{\alpha\delta} = x_{\alpha t}\mu_\alpha$$

as in (2.2).

2. *Backward differential equations.* Derive a system of backward differential equations for the probability distribution $P(\mathbf{x}_\tau, \mathbf{x}_t)$ in the emigration-immigration process using (2.2) and assuming $\lambda_\alpha(t) = \lambda_\alpha$.

3. *Continuation.* Derive the partial differential equation for the p.g.f. of the probability distribution in Problem 2. Solve the partial differential equation and compare your result with (2.65).

4. In solving a system of linear differential equations, we have either verified that the roots (eigenvalues) of the corresponding characteristic equation are real

and distinct as in Chapters 4 and 5, or we have made this assumption, as in Chapters 6, 7 and 8 [for example, (2.26)], before adding the individual solutions to obtain a general solution. Explain in detail why such a verification or assumption is necessary.

5. *Continuation.* What form will the general solution have if two of the eigenvalues are complex conjugates? Explain.

6. *Continuation.* What form will the general solution have if the characteristic equation has multiple roots? Explain.

7. Show that when the eigenvalues ρ_j are real and distinct, the vector $\mathbf{Z}$ in (2.29) satisfies the differential equation (2.13*a*).

8. Integrate (2.33) to obtain the components $\eta_j(t)$ in (2.35).

9. Show that the p.g.f. in (2.49) satisfies the partial differential equation in (2.8).

10. Show that the components ζ_α of the vector $\boldsymbol{\zeta}$ in (2.47) are given in (2.52).

11. Compute $\theta_\beta(t)$ in (2.55) from (2.54).

12. Show that the p.g.f. in (2.65) satisfies the partial differential equation in (2.8) when $\lambda_\alpha(t) = \lambda_\alpha$.

13. Find the limiting probability distribution of $\mathbf{X}_t$ as $t \to \infty$ for the case of constant immigration.

14. Consider three points on the time axis $0 < t_1 < t_2$ and let the corresponding population sizes of the illness states be denoted by

$$\mathbf{x}_0 = \begin{pmatrix} x_{10} \\ \cdot \\ \cdot \\ x_{s0} \end{pmatrix}, \quad \mathbf{X}_1 = \begin{pmatrix} X_{11} \\ \cdot \\ \cdot \\ X_{s1} \end{pmatrix}, \quad \mathbf{X}_2 = \begin{pmatrix} X_{12} \\ \cdot \\ \cdot \\ X_{s2} \end{pmatrix},$$

respectively. Derive the joint probability distribution of the random vectors $\mathbf{X}_1$ and $\mathbf{X}_2$, given $\mathbf{x}_0$, for the case of constant immigration.

15. Let the number of illness states be $s = 2$ and let $\lambda_\alpha(t) = \lambda_\alpha$, $\alpha = 1, 2$. Start with a system of differential equations corresponding to (2.3) and derive an explicit formula for the p.g.f. defined in (2.4).

16. Show that when the number of illness states $s = 1$, the partial differential equation (3.4) reduces to that in (6.7) for the linear growth described in Chapter 3.

Part 2

*CHAPTER 9

The Life Table and Its Construction

1. INTRODUCTION

Long before the development of modern probability and statistics, men were concerned with the length of life and they constructed tables to measure longevity. A crude table, credited to the Praetorian Praefect Ulpianus, was constructed as early as the middle of the third century A.D. Modern life tables date from John Graunt's *Bills of Mortality*, published in 1662, and from E. Halley's famous table for the city of Breslau, published in 1693. Halley's table already contained most of the columns in use today. During the hundred years after Halley, several life tables were constructed, including the French tables of Deparcieux (1746), of Buffon (1749), of Mourgue, and of Duvillard (both published in the 1790's), the Northampton table of Richard Price (1783), and Wigglesworth's table for Massachusetts and New Hampshire (1793). Official English tables originated during the term of William Farr as Compiler of Abstracts in the General Records Office in 1839. The first official English life table was published in 1843.

Although the life table is largely a product of actuarial science, its application is not limited to the computation of insurance premiums. Advances in theoretical statistics and stochastic processes have made it possible to study length of life from a purely statistical point of view, making the life table a valuable analytical tool for demographers, epidemiologists, physicians, zoologists, manufacturers, and investigators in other fields.

The two forms of the life table now in general use are the cohort (or generation) life table and the current life table. In its strictest form, a *cohort life table* records the actual mortality experience of a particular group of individuals (the cohort) from birth to the death of the last member

---

* This chapter may be omitted without loss of continuity.

of the group. The difficulties involved in constructing a cohort life table are apparent. Statistics covering a period of 100 years are available for only a few populations and, even when available, are likely to be unreliable. Individuals in a given cohort may emigrate or die unrecorded, and the life expectancy of a group of people already dead is of historical interest only. However, cohort life tables do have practical application in studies of animal populations, and modified cohort tables have proved useful in the analysis of patient survival in studies of treatment effectiveness.

The *current life table* considers the mortality experience of a given population during one short period of time, for example, the California population in 1960. The table projects the life span of each individual in a hypothetical cohort on the basis of the actual death rates in the given population. The current life table is then a fictitious pattern reflecting the mortality experience of a real population during a given period of time.

Cohort and current life tables may be either complete or abridged. In a *complete life table* the functions are computed for each year of life; an *abridged life table* differs only in that it deals with age intervals greater than one year, with the possible exception of the first year or the first five years of life. A typical set of intervals is: 0–1, 1–5, 5–10, 10–15, etc.

This chapter describes a general form of the life table with interpretations of its various functions, and offers a method of constructing the current life table. Theoretical aspects of life table functions are discussed in detail in Chapter 10.

2. DESCRIPTION OF THE LIFE TABLE

Cohort and current life tables are identical in appearance but different in construction. The following discussion refers to the complete current life table. The function of each column is defined and its relation to the other columns is explained; conventional symbols have been modified for the sake of simplicity. The complete current life table for the total California population in 1960, presented in Table 2 on pages 200–202, will serve as an example.

Column 1. *Age interval,* $(x, x + 1)$. Each interval in this column is defined by the two exact ages stated except for the final age interval, which is half open, with beginning age denoted by w.

Column 2. *Proportion dying in interval* $(x, x + 1)$, $\hat{q}_x$. Each $\hat{q}_x$ is an estimate of the probability that an individual alive at the exact age x will die during the interval $(x, x + 1)$. The figures in this column are derived from the corresponding age specific death rates of the current population. These proportions are the basic quantities from which figures in other columns of the table are computed. To avoid decimals, they are sometimes expressed as the number of deaths per 1000 population.

Column 3. *Number alive at age x, l_x.* The first number in this column, l_0, is the arbitrary radix, while each successive figure represents the number of survivors at the exact age x. Thus the figures in this column have meaning only in conjunction with the radix l_0. The radix is usually assigned a convenient number, such as 100,000. Table 2 shows that l_2 or 97,476 of every 100,000 persons born alive will survive to the second birthday, providing they are subject to the same mortality experience as that of the California population, 1960.

Column 4. *Number dying in interval* $(x, x + 1)$, d_x. The figures in this column are also dependent upon the radix l_0. Again using the 1960 California experience, we see that out of $l_0 = 100{,}000$ born alive, $d_0 = 2378$ will die in the first year of life. But the number 2378 is meaningless by itself, and is certainly not the number of infant deaths in California, 1960. For each age interval $(x, x + 1)$, d_x is merely the number of life table deaths.

The figures in the columns l_x and d_x are computed from the observed values of $\hat{q}_0, \hat{q}_1, \ldots, \hat{q}_w$ and the radix l_0 by using the relations

$$d_x = l_x\hat{q}_x, \qquad x = 0, 1, \ldots, w, \tag{2.1}$$

and

$$l_{x+1} = l_x - d_x, \qquad x = 0, 1, \ldots, w - 1. \tag{2.2}$$

Starting with the first age interval, we use Equation (2.1) for $x = 0$ to obtain the number d_0 dying in the interval and Equation (2.2) for $x = 0$ to obtain the number l_1 who survive to the end of the interval. With l_1 persons alive at the exact age 1 we again use the relations (2.1) and (2.2) for $x = 1$ to obtain the corresponding figures for the second interval. By repeated application of (2.1) and (2.2) we compute all the figures in columns 3 and 4.

Column 5. *Fraction of last year of life for age x, a'_x.* Each of the d_x people who die during the interval $(x, x + 1)$ has lived x complete years plus some fraction of the year $(x, x + 1)$. The average of the fractions, denoted by a'_x, plays an important role not only in the construction of life tables, but also in the stochastic studies of life table functions as presented in Chapter 10.

Column 6. *Number of years lived in interval* $(x, x + 1)$, L_x. Each member of the cohort who survives the year $(x, x + 1)$ contributes one year to L_x, while each member who dies during the year $(x, x + 1)$ contributes, on the average, a fraction a'_x of a year, so that

$$L_x = (l_x - d_x) + a'_x d_x, \qquad x = 0, 1, \ldots, w - 1, \tag{2.3}$$

where the first term on the right side is the number of years lived in the interval $(x, x + 1)$ by the $(l_x - d_x)$ survivors, and the last term is the number of years lived in $(x, x + 1)$ by the d_x persons who died during the interval. When a'_x is assumed to be $\frac{1}{2}$,

$$L_x = l_x - \tfrac{1}{2}d_x. \tag{2.4}$$

Column 7. *Total number of years lived beyond age* x, T_x. This total is equal to the sum of the number of years lived in each age interval beginning with age x, or

$$T_x = L_x + L_{x+1} + \cdots + L_w, \qquad x = 0, 1, \ldots, w, \tag{2.5}$$

with an obvious relation

$$T_x = L_x + T_{x+1}. \tag{2.6}$$

Column 8. *Observed expectation of life at age* x, $\hat{e}_x$. This is the average number of years yet to be lived by a person now aged x. Since the total number of years of life remaining to the l_x individuals is T_x,

$$\hat{e}_x = \frac{T_x}{l_x}, \qquad x = 0, 1, \ldots, w. \tag{2.7}$$

Each $\hat{e}_x$ summarizes the mortality experience of persons beyond age x in the current population under consideration, making this column the most important in the life table. Furthermore, this is the only column in the table other than $\hat{q}_x$ and a'_x that is meaningful without reference to the radix l_0. As a rule, the expectation of life $\hat{e}_x$ decreases as the age x increases, with the single exception of the first year of life, when the reverse is true due to high mortality during that year. In the 1960 California population, for example, the expectation of life at birth is $\hat{e}_0 = 70.58$ years whereas at age one $\hat{e}_1 = 71.30$. The symbol $\hat{e}_x$, denoting the observed expectation of life, is computed from the actual mortality data and is an estimate of e_x, the true unknown expectation of life at age x.[1]

Remark 1. Useful quantities which are not listed in the conventional life table are

$$\hat{p}_x = 1 - \hat{q}_x, \tag{2.8}$$

the proportion of survivors over the age interval $(x, x + 1)$, and

$$\hat{p}_{xy} = \hat{p}_x \hat{p}_{x+1} \cdots \hat{p}_{y-1} = \frac{l_y}{l_x} \tag{2.9}$$

the proportion of those living at age x who will survive to age y. When $x = 0$, $\hat{p}_{0y}$ becomes the proportion of the total born alive who survive to age y; clearly

$$\hat{p}_{0y} = \frac{l_y}{l_0}.$$

Remark 2. In a cohort table, a group of l_0 individuals born alive is followed throughout its life span. The number l_x of survivors and the number d_x of deaths recorded for each age constitute the basic variables from which, using (2.1), $\hat{q}_x$ is computed. The figures in the remaining columns are obtained by the procedures used for the current life table.

Remark 3. An abridged life table contains columns similar to those described for the complete life table. The limits of age intervals are denoted by x_i, $i = 0, 1, \ldots, w$, and the length of the interval by n_i so that $x_{i+1} - x_i = n_i$. The table contains the following columns.

Column 1. Age interval (x_i, x_{i+1}).
Column 2. Proportion dying in interval (x_i, x_{i+1}), $\hat{q}_i$.
Column 3. Number alive at age x_i, l_i.

---

[1] During the early development of the concept of expectation of life, a curtate expectation of life, defined as

$$e_x = \frac{l_{x+1} + l_{x+2} + \cdots}{l_x},$$

was first introduced. This expectation considers only the completed years of life lived by survivors, whereas the complete expectation of life takes into account also the fractional years of life lived by those who die in any year. Under the assumption that each person who dies during any year of age lives half of the year on the average, the complete expectation of life is given by

$$\mathring{e}_x = e_x + \tfrac{1}{2}.$$

Since the curtate expectation is no longer in use, in this book the symbol e_x is used to denote the true expectation of life at age x.

Column 4. Number dying in interval (x_i, x_{i+1}), d_i.

Column 5. Fraction of last age interval of life, a_i. This is the average fraction of interval (x_i, x_{i+1}) lived by an individual dying at an age included in the interval.

Column 6. Number of years lived in interval (x_i, x_{i+1}), L_i.

Column 7. Total number of years lived beyond age x_i, T_i.

Column 8. Observed expectation of life at age x_i, $\hat{e}_i$.

3. CONSTRUCTION OF THE COMPLETE LIFE TABLE

In the construction of current life tables, we are mainly concerned with the computations of $\hat{q}_x$, the proportion dying in the age interval $(x, x + 1)$, and L_x, the number of years lived by the radix in the interval $(x, x + 1)$. An important element in complete life table construction as described in this section is *the fraction of the last year of life* lived by those who die at each age; for example, a man who dies at age 30 has lived 30 complete years plus a fraction of a 31st year. The average value of this fraction is denoted by a'_x where x is the age at death. The fraction a'_x is often assigned a value of one half on the assumption that deaths occur uniformly throughout the year. But a much smaller value has been observed for the first year because of the large proportion of infant deaths occurring in the first weeks of life. Extensive studies of the fraction have been made using the 1960 California mortality data collected by the State of California, Department of Public Health (Chiang et al. [1961]) and the 1963 United States data by the National Vital Statistics Division of the National Center for Health Statistics. The results obtained thus far show that, from 5 years of age on, the fractions a'_x are invariant with respect to race, sex, and age, and that the assumed value of .5 is valid. For the ages 1 to 4 these data suggest the values $a'_1 = .43$, $a'_2 = .45$, $a'_3 = .47$ and $a'_4 = .49$ for all races and for both sexes. The value of a'_0, however, is subject to more variation and should be revised when necessary; the relevant information is often available in vital statistics publications. The data suggest $a'_0 = .10$ for the white population and $a'_0 = .14$ for the nonwhite population.

In a complete life table age intervals are one year in length. The basic quantities are the number D_x of people who died at age x in a current population and the corresponding age specific death rate M_x. To construct the life table we begin by establishing a basic relation between the proportion $\hat{q}_x$ and the rate M_x. Let N_x be the number of people alive at the exact age x, among whom D_x deaths occur in the interval $(x, x + 1)$. Then by definition $\hat{q}_x$ is given by

$$\hat{q}_x = \frac{D_x}{N_x}. \tag{3.1}$$

On the other hand, *the age specific death rate M_x is the ratio of D_x to the total number of years lived by the N_x people during the interval* $(x, x+1)$. This total number is the sum of $(N_x - D_x)$ years lived by the survivors and $a'_x D_x$ years lived by those dying during the year; hence

$$M_x = \frac{D_x}{(N_x - D_x) + a'_x D_x}. \tag{3.2}$$

When the denominator is estimated with the corresponding midyear population P_x

$$(N_x - D_x) + a'_x D_x = P_x \tag{3.3}$$

we have the familiar formula

$$M_x = \frac{D_x}{P_x}. \tag{3.4}$$

Equation (3.3) may be solved for N_x,

$$N_x = P_x + (1 - a'_x)D_x. \tag{3.5}$$

Substituting (3.5) in (3.1) gives

$$\hat{q}_x = \frac{D_x}{P_x + (1 - a'_x)D_x}, \qquad x = 0, 1, \ldots, w - 1. \tag{3.6}$$

Dividing the numerator and the denominator on the right side of (3.6) by P_x, we obtain a basic relation between q_x and M_x:

$$\hat{q}_x = \frac{M_x}{1 + (1 - a'_x)M_x}, \qquad x = 0, 1, \ldots, w - 1. \tag{3.7}$$

Formula (3.7) is fundamental in the construction of complete life tables by the present method and was given in Chiang [1960b, 1961].

To illustrate, let us consider the total California 1960 population as shown in Table 1. For the first year of life, we have $P_0 = 356{,}435$ in Column 2 and $D_0 = 8663$ in Column 3. Thus, the age specific rate for $x = 0$ is

$$M_0 = \frac{D_0}{P_0} = \frac{8663}{356{,}435} = .024305.$$

The average fraction of the year lived by an infant who dies in his first year of life is $a'_0 = .10$. Therefore, the proportion dying is

$$\hat{q}_0 = \frac{.024305}{1 + (1 - .10).024305} = .02378.$$

Table 1. Construction of complete life table for total California population, 1960

Age Interval (in years) x to $x+1$ (1)	Midyear Population[a] in Interval $(x, x+1)$ P_x (2)	Number of Deaths[b] in Interval $(x, x+1)$ D_x (3)	Death Rate in Interval $(x, x+1)$ M_x (4)	Fraction of Last Year of Life a'_x (5)	Proportion Dying in Interval $(x, x+1)$ $\hat{q}_x$ (6)
0–1	356,435	8663	.024305	.10	.02378
1–2	351,611	529	.001505	.43	.00150
2–3	351,828	341	.000969	.45	.00097
3–4	344,966	273	.000791	.47	.00079
4–5	341,712	206	.000603	.49	.00060
5–6	335,837	176	.000524	.50	.00052
6–7	331,427	142	.000428	.50	.00043
7–8	322,881	142	.000440	.50	.00044
8–9	307,669	127	.000413	.50	.00041
9–10	297,664	101	.000339	.50	.00034
10–11	299,407	88	.000294	.50	.00029
11–12	297,343	74	.000249	.50	.00025
12–13	301,886	98	.000325	.50	.00032
13–14	295,079	125	.000424	.50	.00042
14–15	228,897	116	.000507	.50	.00051
15–16	230,323	128	.000556	.50	.00056
16–17	228,391	182	.000797	.50	.00080
17–18	233,447	230	.000985	.50	.00098
18–19	208,627	250	.001198	.50	.00120
19–20	200,050	240	.001200	.50	.00120
20–21	197,460	235	.001190	.50	.00119
21–22	199,116	269	.001351	.50	.00135
22–23	198,033	226	.001141	.50	.00114
23–24	194,540	253	.001301	.50	.00130
24–25	197,324	208	.001054	.50	.00105
25–26	202,354	246	.001216	.50	.00121
26–27	194,367	246	.001266	.50	.00126
27–28	203,010	240	.001182	.50	.00118
28–29	202,512	253	.001249	.50	.00125
29–30	214,672	267	.001244	.50	.00124
30–31	217,730	267	.001226	.50	.00123
31–32	215,303	321	.001491	.50	.00149
32–33	222,225	314	.001413	.50	.00141
33–34	223,064	390	.001748	.50	.00175
34–35	235,807	388	.001645	.50	.00164

Table 1. Continued

Age Interval (in years) x to $x+1$ (1)	Midyear Population[a] in Interval $(x, x+1)$ P_x (2)	Number of Deaths[b] in Interval $(x, x+1)$ D_x (3)	Death Rate in Interval $(x, x+1)$ M_x (4)	Fraction of Last Year of Life a'_x (5)	Proportion Dying in Interval $(x, x+1)$ $\hat{q}_x$ (6)
35–36	244,497	414	.001693	.50	.00169
36–37	239,842	485	.002022	.50	.00202
37–38	237,293	509	.002145	.50	.00214
38–39	244,913	561	.002291	.50	.00229
39–40	240,599	617	.002564	.50	.00256
40–41	227,249	653	.002873	.50	.00287
41–42	221,593	653	.002947	.50	.00294
42–43	211,206	767	.003632	.50	.00362
43–44	205,589	807	.003925	.50	.00392
44–45	206,513	917	.004440	.50	.00443
45–46	208,824	975	.004669	.50	.00466
46–47	199,763	1041	.005211	.50	.00520
47–48	196,068	1077	.005493	.50	.00548
48–49	183,989	1209	.006571	.50	.00655
49–50	185,220	1255	.006776	.50	.00675
50–51	176,508	1387	.007858	.50	.00783
51–52	169,668	1381	.008139	.50	.00811
52–53	164,142	1479	.009010	.50	.00897
53–54	158,150	1541	.009744	.50	.00970
54–55	157,153	1620	.010308	.50	.01026
55–56	150,716	1714	.011372	.50	.01131
56–57	143,300	1777	.012401	.50	.01232
57–58	140,652	1849	.013146	.50	.01306
58–59	127,754	1963	.015365	.50	.01525
59–60	155,596	2044	.013137	.50	.01305
60–61	125,219	2160	.017250	.50	.01710
61–62	120,497	2053	.017038	.50	.01689
62–63	114,185	2367	.020730	.50	.02052
63–64	112,980	2347	.020774	.50	.02056
64–65	112,459	2675	.023786	.50	.02351
65–66	112,650	2956	.026241	.50	.02590
66–67	104,489	2809	.026883	.50	.02653
67–68	104,256	2985	.028631	.50	.02823
68–69	92,485	3007	.032513	.50	.03199
69–70	92,818	3132	.033743	.50	.03318

Table 1. Continued

Age Interval (in years) x to $x+1$ (1)	Midyear Population[a] in Interval $(x, x+1)$ P_x (2)	Number of Deaths[b] in Interval $(x, x+1)$ D_x (3)	Death Rate in Interval $(x, x+1)$ M_x (4)	Fraction of Last Year of Life a'_x (5)	Proportion Dying in Interval $(x, x+1)$ $\hat{q}_x$ (6)
70–71	90,139	3333	.036976	.50	.03631
71–72	87,659	3434	.039175	.50	.03842
72–73	77,284	3417	.044214	.50	.04326
73–74	70,503	3480	.049360	.50	.04817
74–75	68,341	3573	.052282	.50	.05095
75–76	64,382	3673	.057050	.50	.05547
76–77	55,582	3473	.062484	.50	.06059
77–78	49,424	3362	.068024	.50	.06579
78–79	43,301	3356	.077504	.50	.07461
79–80	41,047	3343	.081443	.50	.07826
80–81	34,864	3346	.095973	.50	.09158
81–82	28,190	2946	.104505	.50	.09932
82–83	24,539	2860	.116549	.50	.11013
83–84	22,389	2791	.124659	.50	.11735
84–85	19,728	2613	.132451	.50	.12422
85–86	15,866	2354	.148368	.50	.13812
86–87	12,358	1997	.161596	.50	.14952
87–88	10,902	1990	.182535	.50	.16727
88–89	8160	1664	.203922	.50	.18505
89–90	6929	1333	.192380	.50	.17550
90–91	4857	1280	.263537	.50	.23285
91–92	3254	999	.307007	.50	.26615
92–93	2485	802	.322736	.50	.27789
93–94	1902	611	.321241	.50	.27678
94–95	1383	472	.341287	.50	.29154
95 and over	3560	1252	.351685		1.00000

[a] U.S. Department of Commerce, Bureau of the Census, "U.S. Census of Population: 1960, Final Report PC(1)-6D. Detailed Characteristics, California," p. 6–471.

[b] Unpublished data from the Bureau of Vital Statistics, California State Department of Public Health. I express my appreciation to the Bureau for permission to use the data.

When all the values of $\hat{q}_x$ have been computed and l_0 has been selected, d_x and l_x for successive values of x are determined from Equations (2.1) and (2.2) as shown in Table 2. For the 1960 California population we determine first the number of life table infant deaths,

$$d_0 = l_0\hat{q}_0 = 100{,}000 \times .02378 = 2378,$$

and the life table survivors at age 1,

$$l_1 = l_0 - d_0 = 100{,}000 - 2378 = 97{,}622.$$

The formula for the number of years L_x lived in the age interval $(x, x + 1)$ is derived also with the aid of a'_x, the fraction of the last year of life as given in Section 2:

$$L_x = (l_x - d_x) + a'_x d_x, \qquad x = 0, 1, \ldots. \tag{2.3}$$

To take again the example of the first year of life, $a'_0 = .10$ and

$$L_0 = 97{,}622 + .10 \times 2378 = 97{,}860.$$

Remark 4. The ratio d_x/L_x is known as the *life table death rate* for age x. Since a life table is entirely based on the age specific death rates of the current population, the death rates computed from the life table should be equal to the corresponding rates of the current population; symbolically

$$\frac{d_x}{L_x} = M_x = \frac{D_x}{P_x}, \qquad x = 0, 1, \ldots. \tag{3.8}$$

To prove equation (3.8), we substitute (2.3) in the left side of (3.8) and divide the resulting expression by l_x to obtain

$$\frac{d_x}{L_x} = \frac{\hat{q}_x}{1 - (1 - a'_x)\hat{q}_x}. \tag{3.9}$$

Substitution of (3.6) for $\hat{q}_x$ in (3.9) and simplification give D_x/P_x, proving the assertion (3.8).

The final age interval in a life table is a half open interval, such as age 95 and over. The values of D_w, P_w, M_w, l_w, d_w, and T_w all refer to the open interval age w and over, and $\hat{q}_w = 1$. The length of the interval is infinite and the necessary information for determining the average number of years lived by an individual beyond age w is unavailable. We must therefore use an approach other than equation (2.3) to determine L_w. Writing the first equation in (3.8) for $x = w$, we have

$$L_w = \frac{d_w}{M_w}. \tag{3.10}$$

Table 2. Complete life table for total California population, 1960

Age Interval (in Years) x to $x+1$ (1)	Proportion Dying in Interval $(x, x+1)$ $\hat{q}_x$ (2)	Number Living at Age x l_x (3)	Number Dying in Interval $(x, x+1)$ d_x (4)	Fraction of Last Year of Life a_x' (5)	Number of Years Lived in Interval $(x, x+1)$ L_x (6)	Total Number of Years Lived Beyond Age x T_x (7)	Observed Expectation of Life at Age x $\hat{e}_x$ (8)
0–1	.02378	100,000	2378	.10	97,860	7,058,410	70.58
1–2	.00150	97,622	146	.43	97,539	6,960,550	71.30
2 3	.00097	97,476	95	.45	97,424	6,863,011	70.41
3–4	.00079	97,381	77	.47	97,340	6,765,587	69.48
4–5	.00060	97,304	58	.49	97,274	6,668,247	68.53
5–6	.00052	97,246	51	.50	97,221	6,570,973	67.57
6–7	.00043	97,195	42	.50	97,174	6,473,752	66.61
7–8	.00044	97,153	43	.50	97,132	6,376,578	65.63
8–9	.00041	97,110	40	.50	97,090	6,279,446	64.66
9–10	.00034	97,070	33	.50	97,053	6,182,356	63.69
10–11	.00029	97,037	28	.50	97,023	6,085,303	62.71
11–12	.00025	97,009	24	.50	96,997	5,988,280	61.73
12–13	.00032	96,985	31	.50	96,970	5,891,283	60.74
13–14	.00042	96,954	41	.50	96,933	5,794,313	59.76
14–15	.00051	96,913	49	.50	96,888	5,697,380	58.79
15–16	.00056	96,864	54	.50	96,837	5,600,492	57.82
16–17	.00080	96,810	77	.50	96,771	5,503,655	56.85
17–18	.00098	96,733	95	.50	96,686	5,406,884	55.89
18–19	.00120	96,638	116	.50	96,580	5,310,198	54.95
19–20	.00120	96,522	116	.50	96,464	5,213,618	54.01
20–21	.00119	96,406	115	.50	96,349	5,117,154	53.08
21–22	.00135	96,291	130	.50	96,226	5,020,805	52.14
22–23	.00114	96,161	110	.50	96,106	4,924,579	51.21
23–24	.00130	96,051	125	.50	95,988	4,828,473	50.27
24–25	.00105	95,926	101	.50	95,875	4,732,485	49.33
25–26	.00121	95,825	116	.50	95,767	4,636,610	48.39
26–27	.00126	95,709	121	.50	95,648	4,540,843	47.44
27–28	.00118	95,588	113	.50	95,531	4,445,195	46.50
28–29	.00125	95,475	119	.50	95,416	4,349,664	45.56
29–30	.00124	95,356	118	.50	95,297	4,254,248	44.61
30–31	.00123	95,238	117	.50	95,179	4,158,951	43.67
31–32	.00149	95,121	142	.50	95,050	4,063,772	42.72
32–33	.00141	94,979	134	.50	94,912	3,968,722	41.79
33–34	.00175	94,845	166	.50	94,762	3,873,810	40.84
34–35	.00164	94,679	155	.50	94,602	3,779,048	39.91
35–36	.00169	94,524	160	.50	94,444	3,684,446	38.98
36–37	.00202	94,364	191	.50	94,269	3,590,002	38.04
37–38	.00214	94,173	202	.50	94,072	3,495,733	37.12
38–39	.00229	93,971	215	.50	93,864	3,401,661	36.20
39–40	.00256	93,756	240	.50	93,636	3,307,797	35.28

Table 2. Continued

Age Interval (in Years) x to x + 1 (1)	Proportion Dying in Interval (x, x + 1) $\hat{q}_x$ (2)	Number Living at Age x l_x (3)	Number Dying in Interval (x, x + 1) d_x (4)	Fraction of Last Year of Life a_x' (5)	Number of Years Lived in Interval (x, x + 1) L_x (6)	Total Number of Years Lived Beyond Age x T_x (7)	Observed Expectation of Life at Age x $\hat{e}_x$ (8)
40–41	.00287	93,516	268	.50	93,382	3,214,161	34.37
41–42	.00294	93,248	274	.50	93,111	3,120,779	33.47
42–43	.00362	92,974	337	.50	92,805	3,027,668	32.56
43–44	.00392	92,637	363	.50	92,456	2,934,863	31.68
44–45	.00443	92,274	409	.50	92,069	2,842,407	30.80
45–46	.00466	91,865	428	.50	91,651	2,750,338	29.94
46–47	.00520	91,437	475	.50	91,200	2,658,687	29.08
47–48	.00548	90,962	498	.50	90,713	2,567,487	28.23
48–49	.00655	90,464	593	.50	90,167	2,476,774	27.38
49–50	.00675	89,871	607	.50	89,568	2,386,607	26.56
50–51	.00783	89,264	699	.50	88,915	2,297,039	25.73
51–52	.00811	88,565	718	.50	88,206	2,208,124	24.93
52–53	.00897	87,847	788	.50	87,453	2,119,918	24.13
53–54	.00970	87,059	844	.50	86,637	2,032,465	23.35
54–55	.01026	86,215	885	.50	85,772	1,945,828	22.57
55–56	.01131	85,330	965	.50	84,847	1,860,056	21.80
56–57	.01232	84,365	1039	.50	83,846	1,775,209	21.04
57–58	.01306	83,326	1088	.50	82,782	1,691,363	20.30
58–59	.01525	82,238	1254	.50	81,611	1,608,581	19.56
59–60	.01305	80,984	1057	.50	80,455	1,526,970	18.86
60–61	.01710	79,927	1367	.50	79,244	1,446,515	18.10
61–62	.01689	78,560	1327	.50	77,897	1,367,271	17.40
62–63	.02052	77,233	1585	.50	76,440	1,289,374	16.69
63–64	.02056	75,648	1555	.40	74,871	1,212,934	16.03
64–65	.02351	74,093	1742	.50	73,222	1,138,063	15.36
65–66	.02590	72,351	1874	.50	71,414	1,064,841	14.72
66–67	.02653	70,477	1870	.50	69,542	993,427	14.10
67–68	.02823	68,607	1937	.50	67,638	923,885	13.47
68–69	.03199	66,670	2133	.50	65,603	856,247	12.84
69–70	.03318	64,537	2141	.50	63,466	790,644	12.25
70–71	.03631	62,396	2266	.50	61,263	727,178	11.65
71–72	.03842	60,130	2310	.50	58,975	665,915	11.07
72–73	.04326	57,820	2501	.50	56,569	606,940	10.50
73–74	.04817	55,319	2665	.50	53,986	550,371	9.95
74–75	.05095	52,654	2683	.50	51,313	496,385	9.43
75–76	.05547	49,971	2772	.50	48,585	445,072	8.91
76–77	.06059	47,199	2860	.50	45,769	396,487	8.40
77–78	.06579	44,339	2917	.50	42,880	350,718	7.91
78–79	.07461	41,422	3090	.50	39,877	307,838	7.43
79–80	.07826	38,332	3000	.50	36,832	267,961	6.99

Table 2. Continued

Age Interval (in Years) x to $x+1$ (1)	Proportion Dying in Interval $(x, x+1)$ $\hat{q}_x$ (2)	Number Living at Age x l_x (3)	Number Dying in Interval $(x, x+1)$ d_x (4)	Fraction of Last Year of Life a_x' (5)	Number of Years Lived in Interval $(x, x+1)$ L_x (6)	Total Number of Years Lived Beyond Age T_x (7)	Observed Expectation of Life at Age x $\hat{e}_x$ (8)
80–81	.09158	35,332	3236	.50	33,714	231,129	6.54
81–82	.09932	32,096	3188	.50	30,502	197,415	6.15
82–83	.11013	28,908	3184	.50	27,316	166,913	5.77
83–84	.11735	25,724	3019	.50	24,215	139,597	5.43
84–85	.12422	22,705	2820	.50	21,295	115,382	5.08
85–86	.13812	19,885	2747	.50	18,512	94,087	4.73
86–87	.14952	17,138	2562	.50	15,857	75,575	4.41
87–88	.16727	14,576	2438	.50	13,357	59,718	4.10
88–89	.18505	12,138	2246	.50	11,015	46,361	3.82
89–90	.17550	9,892	1736	.50	9,024	35,346	3.57
90–91	.23285	8,156	1899	.50	7,207	26,322	3.23
91–92	.26615	6,257	1665	.50	5,424	19,115	3.05
92–93	.27789	4,592	1276	.50	3,954	13,691	2.98
93–94	.27678	3,316	918	.50	2,857	9,737	2.94
94–95	.29154	2,398	699	.50	2,049	6,880	2.87
95 and over	1.00000	1,699	1699		4,831	4,831	2.84

Since each one of the l_w people alive at w will eventually die, $l_w = d_w$, and from (3.10) we have the required formula

$$L_w = \frac{l_w}{M_w}. \tag{3.11}$$

Obviously

$$T_w = L_w \quad \text{and} \quad \hat{e}_w = \frac{T_w}{l_w} = \frac{L_w}{l_w}. \tag{3.12}$$

In the California 1960 life table $w = 95$, and $l_{95} = 1699$. The death rate for age 95 and over is $M_{95} = .351685$; therefore

$$L_{95} = \frac{1699}{.351685} = 4831$$

and

$$T_{95} = 4831 \quad \text{and} \quad \hat{e}_{95} = \frac{4831}{1699} = 2.84. \tag{3.11a}$$

Calculations for the remaining columns, T_x and $\hat{e}_x$, were explained adequately in Section 2.

4. CONSTRUCTION OF THE ABRIDGED LIFE TABLE

Various methods of constructing abridged life tables have been developed—among them, King's method [1914], Greville's method [1943], and the Reed-Merrell method [1939]. Each was based on intricate assumptions and involved computations; for details, see the readily available literature. The present method was suggested by Chiang [1960b, 1961].

The idea and procedure used in the construction of the abridged life table by the present method are the same as those used in the construction of the complete life table described in Section 3, except for differences that result from the length of intervals. The length of the typical ith interval (x_i, x_{i+1}) in the abridged table is $n_i = x_{i+1} - x_i$, which is greater than one year. The essential element here is the average fraction of the interval lived by each person who dies at an age included in the interval. This fraction, denoted by a_i, is conceptually a logical extension of the fraction a'_x of the last year of life. Determination and discussion of a_i will be presented in Section 4.1. We use a_i as the point of departure.

Starting with the values of a_i we can construct the abridged life table (Table 3) by following the steps outlined in Section 3. Because of its importance, however, we repeat the previous argument to derive the formula for $\hat{q}_i$, the estimate of the probability that an individual alive at age x_i will die in the interval (x_i, x_{i+1}). Let D_i be the number of deaths occurring in the age interval (x_i, x_{i+1}) during the calendar year under consideration, M_i the corresponding age specific death rate, and N_i the number of individuals alive at exact age x_i among whom D_i deaths occur in the interval. Then by definition

$$\hat{q}_i = \frac{D_i}{N_i} \tag{4.1}$$

and M_i is the ratio of D_i to the total number of years lived by the N_i individuals during the interval (x_i, x_{i+1}),

$$M_i = \frac{D_i}{(N_i - D_i)n_i + a_i n_i D_i}. \tag{4.2}$$

Solving (4.2) for N_i and substituting the resulting expression in (4.1) give the basic formula in the construction of an abridged life table

$$\hat{q}_i = \frac{n_i M_i}{1 + (1 - a_i)n_i M_i}. \tag{4.3}$$

When the age interval is a single year, (4.3) reduces to (3.7). The age specific death rate M_i may be estimated from

$$M_i = \frac{D_i}{P_i} \tag{4.4}$$

where P_i is the midyear population.

All other quantities in the table are functions of $\hat{q}_i$, a_i, and the radix l_0. The number d_i of deaths in (x_i, x_{i+1}) and the number l_{i+1} of survivals at age x_{i+1} are computed from

$$d_i = l_i \hat{q}_i, \qquad i = 0, 1, \ldots, w - 1, \tag{4.5}$$

and

$$l_{i+1} = l_i - d_i, \qquad i = 0, 1, \ldots, w - 1, \tag{4.6}$$

respectively. The number of years lived in the interval (x_i, x_{i+1}) by the l_i survivors at age x_i is

$$L_i = n_i(l_i - d_i) + a_i n_i d_i, \qquad i = 0, 1, \ldots, w - 1. \tag{4.7}$$

The final age interval is again an open interval, and L_w is computed exactly as in the complete life table:

$$L_w = \frac{l_w}{M_w} \tag{4.8}$$

where M_w is again the specific death rate for people of age x_w and over.

The total number T_i of years remaining to all the people attaining age x_i is the sum of L_j for $j = i, i + 1, \ldots, w$. The observed expectation of life $\hat{e}_i$ at age x_i is the ratio T_i/l_i, or

$$\hat{e}_i = \frac{L_i + L_{i+1} + \cdots + L_w}{l_i}, \qquad i = 0, \ldots w. \tag{4.9}$$

As an example, the construction of the abridged life table for the 1960 total United States population is given in Tables 3 and 4.

4.1. The Fraction of Last Age Interval of Life

Consider a single year $(x, x + 1)$ within the interval (x_i, x_{i+1}) so that $x_i \le x < x_{i+1}$, and let a_x' be the corresponding fraction of the last year of life. Let N_i be the number of people alive at x_i; among them a proportion $\hat{p}_{i,x}$ will survive to exact age x, and a proportion $\hat{p}_{i,x}\hat{q}_x$ will die in $(x, x + 1)$. Each of those who dies in $(x, x + 1)$ has lived, on the average, a fraction a_x' of the year $(x, x + 1)$, or a total of $(x - x_i) + a_x'$ in the entire interval (x_i, x_{i+1}). Therefore, the $N_i\hat{p}_{i,x}\hat{q}_x$ people who die in $(x, x + 1)$ live a total of $N_i\hat{p}_{i,x}\hat{q}_x[(x - x_i) + a_x']$ years in (x_i, x_{i+1}). The total number of

Table 3. Construction of abridged life table for total United States population, 1960

Age Interval (in Years) x_i to x_{i+1} (1)	Midyear Population in Interval $(x_i, x_{i+1})^a$ P_i (2)	Number of Deaths in Interval $(x_i, x_{i+1})^b$ D_i (3)	Death Rate in Interval (x_i, x_{i+1}) M_i (4)	Fraction of Last Age Interval of Life a_i (5)	Proportion Dying in Interval (x_i, x_{i+1}) $\hat{q}_i$ (6)
0–1	4,126,560	110,873	.026868	.10	.02623
1–5	16,195,304	17,682	.001092	.39	.00436
5–10	18,659,141	9,163	.000491	.46	.00245
10–15	16,815,965	7,374	.000439	.54	.00219
15–20	13,287,434	12,185	.000917	.57	.00458
20–25	10,803,165	13,348	.001236	.49	.00616
25–30	10,870,386	14,214	.001308	.50	.00652
30–35	11,951,709	19,200	.001606	.52	.00800
35–40	12,508,316	29,161	.002331	.54	.01159
40–45	11,567,216	42,942	.003712	.54	.01840
45–50	10,928,878	64,283	.005882	.54	.02902
50–55	9,696,502	90,593	.009343	.53	.04571
55–60	8,595,947	116,753	.013582	.52	.06577
60–65	7,111,897	153,444	.021576	.52	.10257
65–70	6,186,763	196,605	.031778	.52	.14763
70–75	4,661,136	223,707	.047994	.51	.21472
75–80	2,977,347	219,978	.073884	.51	.31280
80–85	1,518,206	185,231	.122006	.48	.46312
85–90	648,581	120,366	.185584	.45	.61437
90–95	170,653	50,278	.294621	.41	.78812
95+	44,551	13,882	.311598		1.00000

[a] U.S. Department of Commerce, Bureau of the Census, *United States Census of Population*, 1960. United States Summary, Detailed Characteristics, Final Report PC (1)-1D. Table 155, pp. 1–149.

[b] U.S. Department of Health, Education and Welfare, Public Health Service, National Center for Health Statistics. *Vital Statistics of the U.S.*, 1960. Vol. II, part A, Table 5-11, pp. 5-182 and 5-183.

Table 4. Abridged life table for total United States population, 1960

Age Interval (in Years) x_i to x_{i+1} (1)	Proportion Dying in Interval (x_i, x_{i+1}) $\hat{q}_i$ (2)	Number Living at Age x_i l_i (3)	Number Dying in Interval (x_i, x_{i+1}) d_i (4)	Fraction of Last Age Interval of Life a_i (5)	Number of Years Lived in Interval (x_i, x_{i+1}) L_i (6)	Total Number of Years Lived Beyond Age x_i T_i (7)	Observed Expectation of Life at Age x_i $\hat{e}_i$ (8)
0–1	.02623	100,000	2,623	.10	97,639	6,965,395	69.65
1–5	.00436	97,377	425	.39	388,471	6,867,756	70.53
5–10	.00245	96,952	238	.46	484,117	6,479,285	66.83
10–15	.00219	96,714	212	.54	483,082	5,995,168	61.99
15–20	.00458	96,502	442	.57	481,560	5,512,086	57.12
20–25	.00616	96,060	592	.49	478,790	5,030,526	52.37
25–30	.00652	95,468	622	.50	475,785	4,551,736	47.68
30–35	.00800	94,846	759	.52	472,408	4,075,951	42.97
35–40	.01159	94,087	1,090	.54	467,928	3,603,543	38.30
40–45	.01840	92,997	1,711	.54	461,050	3,135,615	33.72
45–50	.02902	91,286	2,649	.54	450,337	2,674,565	29.30
50–55	.04571	88,637	4,052	.53	433,663	2,224,228	25.09
55–60	.06577	84,585	5,563	.52	409,574	1,790,565	21.17
60–65	.10257	79,022	8,105	.52	375,658	1,380,991	17.48
65–70	.14763	70,917	10,469	.52	329,459	1,005,333	14.18
70–75	.21472	60,448	12,979	.51	270,441	675,874	11.18
75–80	.31280	47,469	14,848	.51	200,967	405,433	8.54
80–85	.46312	32,621	15,107	.48	123,827	204,466	6.27
85–90	.61437	17,514	10,760	.45	57,980	80,639	4.60
90–95	.78812	6,754	5,323	.41	18,067	22,659	3.35
95+	1.00000	1,431	1,431		4,592	4,592	3.21

years lived in (x_i, x_{i+1}) by all those who die at *any* age in the interval is of course the sum

$$\sum_{x=x_i}^{x_{i+1}-1} N_i\hat{p}_{i,x}\hat{q}_x[(x - x_i) + a'_x]. \tag{4.10}$$

Here

$$\sum_{x=x_i}^{x_{i+1}-1} N_i\hat{p}_{i,x}\hat{q}_x = N_i[1 - \hat{p}_{i,i+1}] \tag{4.11}$$

is the number of people alive at x_i who die at any age in the interval (x_i, x_{i+1}). By definition, the fraction a_i of the interval lived by an individual who dies in the interval is the ratio of (4.10) to (4.11) divided by the interval length, n_i:

$$a_i = \frac{\sum_{x=x_i}^{x_{i+1}-1} \hat{p}_{i,x}\hat{q}_x[(x - x_i) + a'_x]}{n_i[1 - \hat{p}_{i,i+1}]}, \qquad i = 0, 1, \ldots, w - 1. \tag{4.12}$$

When $a'_x = \frac{1}{2}$ for $x_i \le x < x_{i+1}$, (4.12) may be rewritten as

$$a_i = \frac{\sum_{x=x_i}^{x_{i+1}-1} \hat{p}_{i,x}\hat{q}_x(x - x_i)}{n_i[1 - \hat{p}_{i,i+1}]} + \frac{1}{2n_i}, \qquad i = 0, 1, \ldots, w - 1. \quad (4.13)$$

The proportion $\hat{q}_x$ of deaths for a single year of age is computed from (3.7), and the proportions $\hat{p}_{i,x}$ and $\hat{p}_{i,i+1}$ of survivors are computed from (2.9); therefore the fraction a_i can be determined from (4.12) or (4.13).

The values of a_i in Tables 3 and 4 were based on the mortality experience of the 1960 United States population, but it should be emphasized that they are equally applicable to other populations.[2] To see this, let us rewrite formula (4.12) as follows:

$$a_i n_i = \sum_{x=x_i}^{x_{i+1}-1} \frac{\hat{p}_{i,x}\hat{q}_x}{1 - \hat{p}_{i,i+1}} [(x - x_i) + a'_x]. \quad (4.14)$$

Here the coefficient

$$\frac{\hat{p}_{i\cdot x}\hat{q}_x}{1 - \hat{p}_{i,i+1}}, \qquad x_i \le x < x_{i+1}, \quad (4.15)$$

is the proportion of those dying in the interval (x_i, x_{i+1}) who will die in the year $(x, x + 1)$. This shows that a_i depends neither on the values of $\hat{q}_x$ or $\hat{p}_x$, nor on the specific death rate M_x, but rather on the trend of mortality within the interval. Since the trend of mortality does not vary much from one current population to another, the a_i may be regarded as constant (with the possible exception of a_0) for subpopulations of the United States and may be revised every decade. The invariant property of a_i can be verified with empirical data. In Table 5 are the fractions a_i determined from the experience of four populations: United States, 1960; United States 1959–1961; California, 1960; and England and Wales, 1960–1962. For each population, the estimated probabilities $\hat{q}_x$ for single years of age were first computed from formula (3.7); these values were then used in formula (4.12) to compute a_i. The values of the fractions a'_x for single years were those listed in Tables 1 and 2 except $a_0 = .11$ for England and Wales. Table 5 shows a remarkable agreement among the four sets of values of a_i.

Remark 5. The assumption of $a'_x = \frac{1}{2}$ for each year of age within an interval (x_i, x_{i+1}) does not necessarily imply that $a_i = \frac{1}{2}$ for the entire

---

[2] I express my appreciation to the National Center for Health Statistics, P.H.S., U.S. Department of Health, Education and Welfare for permission to use their data for the computation of a_i.

Table 5. Fraction of last age interval of life, a_i, as determined from 1960 United States data, from 1959–1961 United States data, from 1960 California data, and from 1960–1962 England and Wales data

Age Interval (in Years)	1960 United States	1959–1961 United States	1960 California	1960–1962 England and Wales
0–1	.10	.10	.10	.11
1–5	.39	.39	.39	.38
5–10	.46	.46	.46	.46
10–15	.54	.54	.57	.51
15–20	.57	.56	.57	.56
20–25	.49	.50	.49	.49
25–30	.50	.51	.50	.51
30–35	.52	.52	.53	.53
35–40	.54	.54	.54	.54
40–45	.54	.54	.54	.54
45–50	.54	.54	.54	.54
50–55	.53	.53	.53	.54
55–60	.52	.52	.51	.54
60–65	.52	.52	.53	.53
65–70	.52	.52	.52	.53
70–75	.51	.52	.52	.52
75–80	.51	.51	.51	.51
80–85	.48	.48	.49	.49
85–90	.45	.45	.46	
90–95	.41	.41	.40	

interval. As formula (4.14) shows, the value of the fraction a_i depends on the mortality pattern over an entire interval but not on the mortality rate for any single year. When the mortality rate increases with age in an interval, the fraction $a_i > \frac{1}{2}$; when the reverse pattern prevails, $a_i < \frac{1}{2}$. Consider, for example, the age intervals (5, 10) and (10, 15) in 1960 United States population. Although $a'_x = \frac{1}{2}$ for each age in the two intervals, $a_i = .46$ for interval (5, 10) and $a_i = .54$ for interval (10, 15), due to the difference in mortality pattern, as shown in Table 6.

5. SAMPLE VARIANCES OF $\hat{q}_i$, $\hat{p}_{ij}$ AND $\hat{e}_\alpha$

Statistical inference regarding life table functions requires the evaluation of random variation associated with these functions. A systematic study of the life table from a statistical point of view is made in Chapter 10.

Table 6. Computation of a_i for age intervals (5, 10) and (10, 15) based on 1960 United States population experience

Age Interval x to $x+1$ (1)	Fraction of the Last Year of Life a'_x (2)	Proportion Dying in Age Interval $\hat{q}_x$ (3)	Fraction of Last Age Interval a_i (4)
5–6	.50	.00058	
6–7	.50	.00053	
7–8	.50	.00048	.46
8–9	.50	.00044	
9–10	.50	.00041	
10–11	.50	.00037	
11–12	.50	.00040	
12–13	.50	.00040	.54
13–14	.50	.00048	
14–15	.50	.00057	

Here we briefly consider the sample variances of (i) the proportion $\hat{q}_i$ of individuals dying in an interval (x_i, x_{i+1}), (ii) the proportion $\hat{p}_{ij}$ of individuals alive at age x_i who will survive to age x_j, and (iii) the observed expectation of life $\hat{e}_\alpha$ at age x_α. The formulas for the current life table are given in Section 5.1 and those for the cohort life table in Section 5.2.

5.1. Formulas for the Current Life Table

The sample variance of the proportion of people dying in the interval is the same as the sample variance of the proportion surviving the interval,

$$S_{\hat{q}_i}^{\ 2} = S_{\hat{p}_i}^{\ 2}. \tag{5.1}$$

In the current life table, the proportion of people dying in an interval is derived from the corresponding age-specific death rate, in terms of which the sample variance of $\hat{q}_i$ should be expressed. In the formula

$$\hat{q}_i = \frac{D_i}{N_i} \tag{5.2}$$

D_i is the number of deaths in the interval (x_i, x_{i+1}) out of N_i alive at age x_i. If we assume that all individuals in the group N_i have the same probability of dying in the interval, then D_i is a binomial random variable and $\hat{q}_i$ is a binomial proportion, and their sample variances are

$$S_{D_i}{}^2 = N_i\hat{q}_i(1 - \hat{q}_i) \tag{5.3}$$

and

$$S_{\hat{q}_i}{}^2 = \frac{\hat{q}_i(1 - \hat{q}_i)}{N_i}, \tag{5.4}$$

respectively. Using (5.2) to estimate N_i in (5.4) yields

$$S_{\hat{q}_i}{}^2 = \frac{\hat{q}_i^2(1 - \hat{q}_i)}{D_i}, \tag{5.5}$$

where $\hat{q}_i$, as given in the preceding section, is

$$\hat{q}_i = \frac{n_iM_i}{1 + (1 - a_i)n_iM_i}. \tag{4.3}$$

Substituting (4.3) in (5.5), we have the required formula for the sample variance of $\hat{q}_i$,

$$S_{\hat{q}_i}{}^2 = \frac{n_i{}^2M_i(1 - a_in_iM_i)}{P_i[1 + (1 - a_i)n_iM_i]^3}. \tag{5.6}$$

When the age interval is a single year ($n_i = 1$), all the subscripts in formula (5.6) are replaced with x to obtain the corresponding formula for a single year of age:

$$S_{\hat{q}_x}{}^2 = \frac{M_x(1 - a_x'M_x)}{P_x[1 + (1 - a_x')M_x]^3}. \tag{5.7}$$

If a_x' is assumed to be $\frac{1}{2}$, (5.7) becomes

$$S_{\hat{q}_x}{}^2 = \frac{4M_x(2 - M_x)}{P_x(2 + M_x)^3}. \tag{5.8}$$

The proportion $\hat{p}_{ij}$ surviving the interval (x_i, x_j) is computed from

$$\hat{p}_{ij} = \hat{p}_i\hat{p}_{i+1}\cdots\hat{p}_{j-1}, \qquad i < j;\ i, j = 0, 1, \ldots \tag{5.9}$$

where the individual proportions

$$\hat{p}_i = 1 - \hat{q}_i, \qquad i = 0, 1, \ldots \tag{5.10}$$

are based on the corresponding age-specific death rates and are independent of one another. Therefore the sample variance of $\hat{p}_{ij}$ may be determined from (cf. Equation (5.30), Chapter 1):

$$S_{\hat{p}_i}^2 = \sum_{h=i}^{j-1}\left(\frac{\partial}{\partial \hat{p}}\hat{p}_{ij}\right)^2 S_{\hat{p}_h}{}^2. \tag{5.11}$$

A straightforward computation gives the required formula

$$S_{\hat{p}_{ij}}^2 = \hat{p}_{ij}^2 \sum_{h=i}^{j-1} \hat{p}_h^{-2} S_{\hat{p}_h}^2 \tag{5.12}$$

with the sample variance of $\hat{p}_i$ given in (5.6).

The sample variance of the observed expectation of life also can be expressed in terms of the sample variance of $\hat{p}_i$. From formula (5.31) in Chapter 10 we have

$$S_{\hat{e}_\alpha}^2 = \sum_{i=\alpha}^{w-1} \hat{p}_{\alpha i}^2 [\hat{e}_{i+1} + (1 - a_i)n_i]^2 S_{\hat{p}_i}^2, \qquad \alpha = 0, 1, \ldots, w - 1. \tag{5.13}$$

Formula (5.13) holds true, in fact, for both the current and the cohort life tables.

Using formula (5.13) and referring to Table 7, the essential steps in the computation of the sample variance of $\hat{e}_\alpha$ are as follows.

1. Record the length of age interval n_i in Column 2, and the fraction of last age interval of life a_i in Column 3.
2. Compute the sample variance of $\hat{p}_i$ (or $\hat{q}_i$) from (5.6) and record it in Column 4.
3. Compute the product $l_i^2[\hat{e}_{i+1} + (1 - a_i)]^2 S_{\hat{p}_i}^2$ and enter it in Column 5.
4. Sum the products in Column 5 from the bottom of the table up to x_i and enter the sum in Column 6.
5. Divide the sum in Column 6 by l_i^2 to obtain the sample variance of the observed expectation of life in Column 7.

By taking the square root of the sample variance, we obtain the sample standard error of the observed expectation of life, as shown in Column 8.

The main life table functions and the corresponding standard errors for the 1960 total United States population is given in Table 8.

5.2. Formulas for the Cohort Life Table

In a cohort life table the proportions $\hat{q}_i$ and $\hat{p}_i$ computed directly from

$$\hat{q}_i = \frac{d_i}{l_i} \qquad \text{and} \qquad \hat{p}_i = \frac{l_{i+1}}{l_i} \tag{5.14}$$

are ordinary binomial proportions with the sample variance

$$S_{\hat{q}_i}^2 = S_{\hat{p}_i}^2 = \frac{1}{l_i} \hat{p}_i \hat{q}_i. \tag{5.15}$$

Table 7. Computation of the sample variance of the observed expectation of life total United States population, 1960

Age Interval (in Years) x_i to x_{i+1} (1)	Length of Interval n_i (2)	Fraction of Last Age Interval of Life a_i (3)	Sample Variance of $\hat{p}_i$ $10^8 S_{\hat{p}_i}^2$ (4)	$l_i^2[\hat{e}_{i+1} + (1 - a_i)]^2 S_{\hat{p}_i}^2$ (5)	$\sum_{j \geq i} l_j^2[\hat{e}_{j+1} + (1 - a_j)]^2 S_{\hat{p}_j}^2$ (6)	Sample Variance of $\hat{e}_i$ $10^4 S_{\hat{e}_i}^2$ (7)	Sample Standard Error of $\hat{e}_i$ $S_{\hat{e}_i}$ (8)
0–1	1	.10	.6043	308,328.66	1,384,473.52	1.3845	.012
1–5	4	.39	.1070	48,684.08	1,076,144.86	1.1349	.011
5–10	5	.46	.0653	25,686.27	1,027,460.78	1.0931	.010
10–15	5	.54	.0649	21,433.28	1,001,774.51	1.0710	.010
15–20	5	.57	.1714	47,445.51	980,341.23	1.0527	.010
20–25	5	.49	.2825	65,770.32	932,895.72	1.0110	.010
25–30	5	.50	.2971	55,984.55	867,125.40	.9514	.010
30–35	5	.52	.3307	49,278.91	811,140.85	.9017	.009
35–40	5	.54	.4553	52,293.09	761,861.94	.8606	.009
40–45	5	.54	.7739	66,833.91	709,568.85	.8205	.009
45–50	5	.54	1.2721	79,526.83	642,734.94	.7713	.009
50–55	5	.53	2.2009	95,654.42	563,208.11	.7169	.008
55–60	5	.52	3.4613	97,872.06	467,553.69	.6535	.008
60–65	5	.52	6.1531	105,623.14	369,681.63	.5920	.008
65–70	5	.52	9.4489	87,635.79	264,058.49	.5250	.007
70–75	5	.51	16.1842	71,425.04	176,422.70	.4828	.007
75–80	5	.51	30.5659	52,370.93	104,997.66	.4660	.007
80–85	5	.48	62.1657	34,293.39	52,626.73	.4946	.007
85–90	5	.45	120.9280	13,802.48	18,333.34	.5977	.008
90–95	5	.41	261.7560	4,530.86	4,530.86	.9932	.010

Table 8. Main life table functions and the corresponding standard errors total United States population, 1960

Age Interval (in Years) x_i to x_{i+1}	Proportion Dying in Interval (x_i, x_{i+1})		Proportion Surviving to Age x_i		Observed Expectation of Life at Age x_i	
	$1{,}000\hat{q}_i$	$1{,}000 S_{\hat{q}_i}$	$100{,}000\hat{p}_{0i}$	$100{,}000 S_{p_{0i}}$	$\hat{e}_i$	$S_{\hat{e}_i}$
(1)	(2)	(3)	(4)	(5)	(6)	(7)
0–1	26.23	0.08	100,000	0	69.65	.012
1–5	4.36	0.03	97,377	8	70.53	.011
5–10	2.45	0.03	96,952	8	66.83	.010
10–15	2.19	0.03	96,714	9	61.99	.010
15–20	4.58	0.04	96,502	9	57.12	.010
20–25	6.16	0.05	96,060	10	52.37	.010
25–30	6.52	0.06	95,468	11	47.68	.010
30–35	8.00	0.06	94,846	12	42.97	.009
35–40	11.59	0.07	94,087	13	38.30	.009
40–45	18.40	0.09	92,997	15	33.72	.009
45–50	29.02	0.11	91,286	16	29.30	.009
50–55	45.71	0.15	88,637	19	25.09	.008
55–60	65.77	0.19	84,585	22	21.17	.008
60–65	102.57	0.25	79,022	26	17.48	.008
65–70	147.63	0.31	70,917	31	14.18	.007
70–75	214.72	0.40	60,448	34	11.18	.007
75–80	312.80	0.55	47,469	36	8.54	.007
80–85	463.12	0.79	32,621	36	6.27	.007
85–90	614.37	1.10	17,514	32	4.60	.008
90–95	788.12	1.62	6,754	23	3.35	.010
95+	1,000.00		1,431	12	3.21	

The proportion $\hat{p}_{ij}$ of survivors is a general form of $\hat{p}_i$ and is equal to $\hat{p}_i$ when $j = i + 1$. Corresponding to (5.15) the sample variance of $\hat{p}_{ij}$ is

$$S^2_{\hat{p}_{ij}} = \frac{1}{l_i}\hat{p}_{ij}(1 - \hat{p}_{ij}), \qquad i < j;\ i, j = 0, 1, \ldots. \tag{5.16}$$

Obviously, $\hat{p}_{ij}$ can be written also in the form of a product of $\hat{p}_i$,

$$\hat{p}_{ij} = \hat{p}_i\hat{p}_{i+1}\cdots\hat{p}_{j-1}, \tag{5.17}$$

as in (5.9) for the current life table. It is proved in Chapter 10 that the $\hat{p}_i$'s are linearly uncorrelated. Therefore the sample variance of $\hat{p}_{ij}$ also has the form as given in (5.12),

$$S^2_{\hat{p}_{ij}} = \hat{p}_{ij}^{\ 2}\sum_{h=i}^{j-1}\hat{p}_h^{-2}S_{\hat{p}_h}^{\ 2}. \tag{5.12}$$

Substituting (5.15) in (5.12) gives the formula

$$S^2_{\hat{p}_{ij}} = \hat{p}_{ij}{}^2 \sum_{h=i}^{j-1} \hat{p}_h{}^{-2} \frac{1}{l_h} \hat{p}_h(1 - \hat{p}_h). \tag{5.18}$$

It is easily shown that the right side of (5.18) is equal to the right side of (5.16).

Let Y_α denote the future lifetime of an individual at age x_α; the observed expectation of life $\hat{e}_\alpha$ is simply the sample mean $\bar{Y}_\alpha$ of the l_α values of Y_α, or

$$\hat{e}_\alpha = \bar{Y}_\alpha. \tag{5.19}$$

These values of Y_α are recorded in the life table in the form of a frequency distribution in which d_i is the frequency in the interval (x_i, x_{i+1}), $i = \alpha, \alpha + 1, \ldots, w$. On the average, each of the d_i people lives x_i years plus a fraction a_i of the interval (x_i, x_{i+1}), or $x_i - x_\alpha + a_i n_i$ years beyond x_α; that is, for each of the d_i individuals,

$$Y_\alpha = x_i - x_\alpha + a_i n_i, \qquad i = \alpha, \alpha + 1, \ldots, w. \tag{5.20}$$

The sample mean of Y_α can then be expressed as

$$\hat{e}_\alpha = \bar{Y}_\alpha = \frac{1}{l_\alpha} \sum_{i=\alpha}^{w} (x_i - x_\alpha + a_i n_i) d_i \tag{5.21}$$

and the sample variance of Y_α as

$$S^2_{Y_\alpha} = \frac{1}{l_\alpha} \sum_{i=\alpha}^{w} [(x_i - x_\alpha + a_i n_i) - \hat{e}_\alpha]^2 d_i. \tag{5.22}$$

Therefore, the sample variance of the sample mean $\bar{Y}_\alpha$, or $\hat{e}_\alpha$, is given by

$$S_{\hat{e}_\alpha}{}^2 = \frac{1}{l_\alpha} S^2_{Y_\alpha}, \tag{5.23}$$

or

$$S_{\hat{e}_\alpha}{}^2 = \frac{1}{l_\alpha^2} \sum_{i=\alpha}^{w} [(x_i - x_\alpha + a_i n_i) - \hat{e}_\alpha]^2 d_i. \tag{5.24}$$

It was noted at the end of Section 5.1 that formula (5.13) holds true also for the cohort life table. This implies the equality

$$\sum_{i=\alpha}^{w-1} \hat{p}_{\alpha i}{}^2 [\hat{e}_{i+1} + (1 - a_i) n_i]^2 \frac{1}{l_i} p_i q_i = \frac{1}{l_\alpha^2} \sum_{i=\alpha}^{w} [(x_i - x_\alpha + a_i n_i) - \hat{e}_\alpha]^2 d_i. \tag{5.25}$$

For proof of (5.25) the reader is referred to Chiang [1960b].

PROBLEMS

1. Show that

$$\sum_{x=x_i}^{x_{i+1}} \frac{p_{i,x}q_x}{1 - p_{i,i+1}} = 1.$$

2. Verify Equation (5.25).

3. *Life table based on a sample of deaths.* One series of United States life tables is based on a sample of deaths. The actual number D_i of deaths in the age interval (x_i, x_{i+1}) is not observed directly but is estimated by a sampling procedure. The estimated values are then used as the basis for the computation of the mortality rate and the life table functions (see Sirken [1966] and Chiang [1967]). We have then a sample of size δ taken without replacement from the total of D death certificates so that $\delta/D = f$ is a preassigned sampling fraction. A number δ_i in the sample falls into the age interval (x_i, x_{i+1}) with

$$\delta_0 + \delta_1 + \cdots + \delta_w = \delta.$$

Derive the formula for the variance of δ_i and the covariance between δ_i and δ_j.

4. *Continuation.* In a life table described in Problem 3, the number D_i is estimated from $\hat{D}_i = \delta_i/f$ and the corresponding probability of dying from

$$\hat{q}_i = \frac{n_i\hat{D}_i}{P_i + (1 - a_i)n_i\hat{D}_i}. \tag{4.3a}$$

(a) Derive the formula for the variance of $\hat{q}_i$.

(b) In this case there is a covariance between $\hat{q}_i$ and $\hat{q}_j$. Why? Derive the formula for the covariance.

(c) Derive the formula for the observed expectation of life $\hat{e}_\alpha$.

5. The material in Table 9 is taken from Miller and Thomas [1958] in their study of the effects of larval crowding and body size on the longevity of adult *Drosophila Melanogaster*. Construct a life table for males and one for females, and compare their longevity, assuming $a_i = \frac{1}{2}$.

6. *England and Wales life tables.* Table 10 gives the population size and number of deaths by age and sex for England and Wales, 1961. Construct one life table for males and one for females. Comment briefly.

7. *Continuation.* Use formula (5.5) to compute the sample variance of $\hat{q}_i$ and formula (5.13) to compute the sample variance of $\hat{e}_i$ for the total population of England and Wales, 1961.

8. *Continuation.* Compare and discuss the mortality and survival experience of the populations of the United States in 1960 and England and Wales in 1961 as summarized in the life tables in Table 4 and Table 11.

Table 9. Number of adult Drosophila Melanogaster living and number dying by age and sex

	Males		Females	
Age Interval (in Days) x_i to x_{i+1}	Number Living at x_i	Number Dying in Interval (x_i, x_{i+1})	Number Living at x_i	Number Dying in Interval (x_i, x_{i+1})
0–5	270	2	275	4
5–10	268	4	271	7
10–15	264	3	264	3
15–20	261	7	261	7
20–25	254	3	254	13
25–30	251	3	241	22
30–35	248	16	219	31
35–40	232	66	188	68
40–45	166	36	120	51
45–50	130	54	69	38
50–55	76	42	31	26
55–60	34	21	5	5
60 and over	13	13		

Table 10. Population and deaths by age and sex, and fraction of last age interval of life, England and Wales, 1961

Age Interval	Populations[a] P_i			Deaths[a] D_i			Fraction of Last Age Interval[b]
x_i to x_{i+1}	Total	Males	Females	Total	Males	Females	a_i
0–1	785,000	403,000	382,000	17,393	9,988	7,405	.11
1–5	2,880,000	1,479,000	1,401,000	2,662	1,536	1,126	.38
5–10	3,254,000	1,668,000	1,586,000	1,320	803	517	.46
10–15	3,667,000	1,876,000	1,791,000	1,186	745	441	.51
15–20	3,216,000	1,640,000	1,576,000	2,137	1,517	620	.56
20–25	2,933,000	1,494,000	1,439,000	2,335	1,589	746	.49
25–30	2,860,000	1,460,000	1,400,000	2,323	1,519	804	.51
30–35	2,977,000	1,499,000	1,478,000	3,168	1,862	1,306	.53
35–40	3,225,000	1,603,000	1,622,000	5,180	2,997	2,183	.54
40–45	3,076,000	1,522,000	1,554,000	7,932	4,542	3,390	.54
45–50	3,237,000	1,596,000	1,641,000	14,052	8,351	5,701	.54
50–55	3,211,000	1,572,000	1,639,000	23,489	14,596	8,893	.54
55–60	2,962,000	1,415,000	1,547,000	36,370	23,830	12,540	.54
60–65	2,465,000	1,101,000	1,364,000	50,152	31,461	18,691	.53
65–70	1,996,000	830,000	1,166,000	63,345	36,553	26,792	.53
70–75	1,540,000	601,000	939,000	79,373	41,180	38,193	.52
75–80	1,087,000	397,000	690,000	88,968	41,320	47,648	.51
80–85	601,000	200,000	401,000	80,126	32,708	47,418	.49
85+	297,000	92,000	205,000	70,241	23,685	46,556	

[a] The Registrar General's Statistical Review of England and Wales for the year 1961, Table 1, page 2 (population) and Table 14, page 118 (deaths).

[b] The values of a_i were computed on the basis of 1960–62 England and Wales mortality data using formula (4.12).

Table 11. Abridged life table for total England and Wales population, 1961

Age Interval (in Years) x_i to x_{i+1} (1)	Proportion Dying in Interval (x_i, x_{i+1}) $\hat{q}_i$ (2)	Number Living at Age x_i l_i (3)	Number Dying in Interval (x_i, x_{i+1}) d_i (4)	Fraction of Last Age Interval of Life a_i (5)	Number of Years Lived in Interval (x_i, x_{i+1}) L_i (6)	Total Number of Years Lived Beyond Age x_i T_i (7)	Observed Expectation of Life at Age x_i $\hat{e}_i$ (8)
0–1	.02173	100,000	2,173	.11	98,066	7,100,874	71.01
1–5	.00369	97,827	361	.38	390,413	7,002,808	71.58
5–10	.00203	97,466	198	.46	486,795	6,612,395	67.84
10–15	.00162	97,268	158	.51	485,953	6,125,600	62.98
15–20	.00332	97,110	322	.56	484,842	5,639,647	58.07
20–25	.00397	96,788	384	.49	482,961	5,154,805	53.26
25–30	.00405	96,404	390	.51	481,064	4,671,844	48.46
30–35	.00531	96,014	510	.53	478,872	4,190,780	43.65
35–40	.00800	95,504	764	.54	475,763	3,711,908	38.87
40–45	.01282	94,740	1,215	.54	470,906	3,236,145	34.16
45–50	.02149	93,525	2,010	.54	463,002	2,765,239	29.57
50–55	.03597	91,515	3,292	.54	450,003	2,302,237	25.16
55–60	.05971	88,223	5,268	.54	428,999	1,852,234	20.99
60–65	.09709	82,955	8,054	.53	395,848	1,423,235	17.16
65–70	.14767	74,901	11,061	.53	348,512	1,027,387	13.72
70–75	.22934	63,840	14,641	.52	284,062	678,875	10.63
75–80	.34088	49,199	16,771	.51	204,906	394,813	8.02
80–85	.49748	32,428	16,132	.49	121,003	189,907	5.86
85+	1.00000	16,296	16,296		68,904	68,904	4.23

CHAPTER 10

Probability Distributions of Life Table Functions

1. INTRODUCTION

The concept of the life table originated in longevity studies of man; it was always presented as a subject peculiar to public health, demography, and actuarial science. As a result, its development has not received sufficient attention in the field of statistics. Actually, the problems of mortality studies are similar to those of reliability theory and life testing, and they may be described in terms familiar to the statistically oriented mind. From a statistical point of view, human life is a random experiment and its outcome, survival or death, is subject to chance. The life table systematically records the outcomes of many such experiments for a large number of individuals over a period of time. Thus the quantities in the table are random variables subject to established statistical analysis. The purpose of this chapter is to derive probability distributions of life table functions and to discuss some optimum properties of these functions when they are used as estimates of the corresponding unknown quantities. The presentation will focus on the cohort life table but, whenever necessary, clarification will be made for application to the current life table. A typical abridged life table is reproduced on page 219.

The following symbols are also used in the text:

$$p_{ij} = \Pr\{\text{an individual alive at age } x_i \text{ will survive to age } x_j\},\quad i \le j;\ i, j = 0, 1, \ldots, \tag{1.1}$$

and

$$1 - p_{ij} = \Pr\{\text{an individual alive at age } x_i \text{ will die before age } x_j\},\quad i \le j;\ i, j = 0, 1, \ldots. \tag{1.2}$$

Table 1. Life table

Age Interval (in Years) x_i to x_{i+1}	Number Living at Age x_i l_i	Proportion Dying in Interval (x_i, x_{i+1}) $\hat{q}_i$	Fraction of Last Age Interval of Life a_i	Number Dying in Interval (x_i, x_{i+1}) d_i	Number of Years Lived in Interval (x_i, x_{i+1}) L_i	Total Number of Years Lived Beyond Age x_i T_i	Observed Expectation of Life at Age x_i $\hat{e}_i$
x_0 to x_1	l_0	$\hat{q}_0$	a_0	d_0	L_0	T_0	$\hat{e}_0$
.	.	.	.	.	.	.	.
.	.	.	.	.	.	.	.
.	.	.	.	.	.	.	.
.	.	.	.	.	.	.	.
x_w and over	l_w	$\hat{q}_w$		d_w	L_w	T_w	$\hat{e}_w$

When $x_j = x_{i+1}$, we drop the second subscript and write p_i for $p_{i,i+1}$. No particular symbol is introduced for the probability $1 - p_{ij}$ except when $x_j = x_{i+1}$, in which case we let $1 - p_i = q_i$. Finally, the symbol e_i is used to denote the true, unknown expectation of life at age x_i, estimated by the "observed expectation of life," $\hat{e}_i$.

All the quantities in the life table, with the exception of l_0 and a_i, are treated as random variables in this chapter. The radix l_0 is conventionally set equal to a convenient number, such as $l_0 = 100{,}000$, so that the value of l_i clearly indicates the proportion of survivors to age x_i. We adopt the convention and consider l_0 a constant in deriving the probability distributions of other life table functions. The distributions of the quantities in columns L_i and T_i are not discussed because of their limited use. One remark should be made regarding the final age interval (x_w and over): In a conventional table the last interval is usually an open interval, for example 95 and over; statistically speaking, x_w is a random variable and is treated accordingly. This point is discussed in Section 2.1. Throughout this chapter we shall assume a homogeneous population in which all individuals are subjected to the same force of mortality and in which one individual's survival is independent of the survival of any other individual in the group.

1.1. Probability Distribution of the Number of Survivors

The various functions of the life table are usually given for integral ages or for other discrete intervals. In the derivation of the distribution of survivors, however, age is more conveniently treated as a continuous variable with formulas derived for l_x, the number of individuals surviving the age interval $(0, x)$, for all possible values of x.

The probability distribution of l_x depends on the force of mortality, or the intensity of risk of death, $\mu(x)$, defined as follows:

$$\mu(x)\Delta + o(\Delta) = \Pr\{\text{an individual alive at age } x \text{ will die in interval } (x, x + \Delta)\}. \tag{1.3}$$

It has been shown in the pure death process in Section 5, Chapter 3, that l_x has a binomial distribution with

$$\Pr\{l_x = k \mid l_0\} = \binom{l_0}{k} p_{0x}^k (1 - p_{0x})^{l_0 - k} \tag{1.4}$$

where

$$p_{0x} = \exp\left\{-\int_0^x \mu(\tau)\, d\tau\right\} \tag{1.5}$$

is the probability that an individual alive at age 0 will survive to age x, and the corresponding p.g.f. is given by

$$G_{l_x \mid l_0}(s) = (1 - p_{0x} + p_{0x}s)^{l_0}. \tag{1.6}$$

For $x = x_i$, the probability that an individual will survive the age interval $(0, x_i)$ is

$$p_{0i} = \exp\left\{-\int_0^{x_i} \mu(\tau)\, d\tau\right\} \tag{1.7}$$

and the p.g.f. of l_i is

$$G_{l_i \mid l_0}(s) = (1 - p_{0i} + p_{0i}s)^{l_0}. \tag{1.8}$$

Therefore the expectation and variance of l_i given l_0 are

$$E[l_i \mid l_0] = l_0 p_{0i} \tag{1.9}$$

and

$$\sigma^2_{l_i \mid l_0} = l_0 p_{0i}(1 - p_{0i}). \tag{1.10}$$

In general, the probability of surviving the age interval (x_i, x_j) is

$$p_{ij} = \exp\left\{-\int_{x_i}^{x_j} \mu(\tau)\, d\tau\right\} \tag{1.11}$$

with the obvious relation

$$p_{\alpha j} = p_{\alpha i} p_{ij} \tag{1.12}$$

for $\alpha \le i \le j$. The p.g.f. for the conditional distribution of l_j given $l_i > 0$ is

$$G_{l_j \mid l_i}(s_j) = E(s_j^{l_j} \mid l_i) = (1 - p_{ij} + p_{ij}s_j)^{l_i}. \tag{1.13}$$

When $j = i + 1$, (1.13) becomes

$$G_{l_{i+1}|l_i}(s_{i+1}) = E(s_{i+1}^{l_{i+1}} \mid l_i) = (1 - p_i + p_i s_{i+1})^{l_i}. \tag{1.14}$$

Although formula (1.13) holds true whatever $x_i < x_j$ may be, the conditional probabilities of l_j relative to $l_0, l_1, \ldots, l_i$ are the same as those relative to l_i in the sense that for each k

$$\Pr\{l_j = k \mid l_0, \ldots, l_i\} = \Pr\{l_j = k \mid l_i\}. \tag{1.15}$$

In other words, for each u the sequence $l_0, l_1, \ldots, l_u$ is a Markov process. Thus

$$E(l_j \mid l_0, \ldots, l_i) = E(l_j \mid l_i) \tag{1.16}$$

and

$$E(s_j^{l_j} \mid l_0, \ldots, l_i) = E(s_j^{l_j} \mid l_i), \qquad i < j; i, j = 0, 1, \ldots, u. \tag{1.17}$$

2. JOINT PROBABILITY DISTRIBUTION OF THE NUMBERS OF SURVIVORS

Let us introduce, for a given u, the p.g.f. of the joint probability distribution of $l_1, \ldots, l_u$,

$$G_{l_1,\ldots,l_u|l_0}(s_1, \ldots, s_u) = E(s_1^{l_1} \cdots s_u^{l_u} \mid l_0) \tag{2.1}$$

where $|s_i| < 1$ for $i = 1, \ldots, u$.

To derive an explicit formula for the p.g.f. (2.1), we use the identity (see Equation (4.22) in Chapter 1):

$$E[s_1^{l_1} \cdots s_{i+1}^{l_{i+1}} \mid l_0] = E[s_1^{l_1} \cdots s_i^{l_i} E\{s_{i+1}^{l_{i+1}} \mid l_0, \ldots, l_i\} \mid l_0] \tag{2.2}$$

and write

$$E[s_1^{l_1} \cdots s_{i+1}^{l_{i+1}} \mid l_0] = E[s_1^{l_1} \cdots s_i^{l_i} E\{s_{i+1}^{l_{i+1}} \mid l_i\} \mid l_0]$$
$$i = 0, 1, \ldots, u - 1, \tag{2.3}$$

where the conditional expectation of the quantity inside the braces is the p.g.f. of the conditional distribution of l_{i+1} given l_i, with the explicit function presented in (1.14). Now we use the identity (2.3) for $i = u - 1$ and rewrite the p.g.f. (2.1) as

$$G_{l_1,\ldots,l_u|l_0}(s_1, \ldots, s_u) = E[s_1^{l_1} \cdots s_{u-1}^{l_{u-1}} E\{s_u^{l_u} \mid l_{u-1}\} \mid l_0]. \tag{2.4}$$

In view of (1.14), $E\{s_u^{l_u} \mid l_{u-1}\} = (1 - p_{u-1} + p_{u-1}s_u)^{l_{u-1}}$, hence

$$\begin{aligned} G_{l_1,\ldots,l_u|l_0}(s_1, \ldots, s_u) &= E[s_1^{l_1} \cdots s_{u-1}^{l_{u-1}}(1 - p_{u-1} + p_{u-1}s_u)^{l_{u-1}} \mid l_0] \\ &= E[s_1^{l_1} \cdots s_{u-2}^{l_{u-2}} z_{u-1}^{l_{u-1}} \mid l_0], \end{aligned} \tag{2.5}$$

where

$$z_{u-1} = s_{u-1}(1 - p_{u-1} + p_{u-1}s_u) \tag{2.6}$$

is less than unity in absolute value, and

$$1 - z_{u-1} = (1 - s_{u-1}) + p_{u-1}s_{u-1}(1 - s_u). \tag{2.7}$$

Formula (2.5) contains only $u - 1$ random variables, $l_1, \ldots, l_{u-1}$, or one less than we had originally. Using formulas (2.3) and (1.14) for $i = u - 2$, the random variable l_{u-1} drops out, and the p.g.f. becomes

$$G_{l_1,\ldots,l_u|l_0}(s_1, \ldots, s_u) = E[s_1^{l_1} \cdots s_{u-3}^{l_{u-3}} z_{u-2}^{l_{u-2}} \mid l_0]. \tag{2.8}$$

Here

$$z_{u-2} = s_{u-2} - p_{u-2}s_{u-2}(1 - s_{u-1}) - p_{u-2}p_{u-1}s_{u-2}s_{u-1}(1 - s_u) \tag{2.9}$$

is also less than unity in absolute value, and

$$1 - z_{u-2} = (1 - s_{u-2}) + p_{u-2}s_{u-2}(1 - s_{u-1}) + p_{u-2}p_{u-1}s_{u-2}s_{u-1}(1 - s_u). \tag{2.10}$$

By repeated application of the same process, all the random variables are removed from the expression of the p.g.f., and the following formula is finally reached:

$$G_{l_1,\ldots,l_u|l_0}(s_1, \ldots, s_u) = [1 - \{p_{01}(1 - s_1) + p_{02}s_1(1 - s_2) + p_{03}s_1s_2(1 - s_3) + \cdots + p_{0u}s_1s_2 \cdots s_{u-1}(1 - s_u)\}]^{l_0}. \tag{2.11}$$

The p.g.f. (2.11) is then used to derive the joint probability function and moments of the random variables $l_1, \ldots, l_u$ by differentiating (2.11) with respect to the arguments. The joint probability function turns out to be

$$\Pr\{l_1 = k_1, \ldots, l_u = k_u \mid l_0\} = \prod_{i=0}^{u-1} \binom{k_i}{k_{i+1}} p_i^{k_{i+1}}(1 - p_i)^{k_i - k_{i+1}}$$
$$k_{i+1} = 0, 1, \ldots, k_i, \qquad \text{with} \qquad k_0 = l_0. \tag{2.12}$$

The expected values and covariances are

$$E(l_i \mid l_0) = l_0 p_{0i} \tag{2.13}$$

and

$$\sigma_{l_i,l_j|l_0} = l_0 p_{0j}(1 - p_{0i}), \qquad i \le j;\ i, j = 1, \ldots, u. \tag{2.14}$$

When $i = j$, (2.14) reduces to the variance of l_i (Equation (1.10)). These results show that *for a given u, $l_1, \ldots, l_u$ in the life table form a chain of binomial distributions; the p.g.f., the joint probability distribution, the expected values, and covariances are given in* (2.11), (2.12), (2.13) *and* (2.14), *respectively.* ▶

2.1. An Urn Scheme

The life table functions can be generated from an entirely different approach. As an example, consider an experiment in which balls are drawn with replacement from an infinite sequence of urns, numbered $0, 1, \ldots$. In the ith urn there is a proportion p_i of white balls and a proportion q_i of black balls with $0 < p_i < 1$ and $p_i + q_i = 1$. Beginning with the 0th urn a number l_0 of balls is drawn of which l_1 are white; then a total of l_1 balls is drawn from the first urn of which l_2 are white; l_2 balls are then drawn from the second urn of which l_3 are white, and so on. In general, the number l_{i+1} of white balls drawn from the ith urn is the number of balls to be drawn from the next or the $(i + 1)$th urn. The experiment terminates as soon as no white balls are drawn. Let the last urn from which balls are drawn be the Wth urn, so that $l_i > 0$ for $i \leq W$ and $l_i = 0$ for $i > W$.

The correspondence between the urn scheme and the problem of the life table is evident. For example l_0 is the initial size of the cohort, p_i (or q_i) is the probability of surviving (or dying in) an age interval, and l_i is the number of survivors at age x_i, $i = 0, 1, \ldots$. The number W corresponding to the beginning of the last age interval is also a random variable, which we shall now discuss.

To derive the probability distribution of W, we note that for $W = w$ there must be $l_w = k_w$ drawings from the wth urn for $1 \leq k_w \leq l_0$ and all k_w balls drawn must be black balls. Therefore we have the probability

$$\Pr\{W = w\} = \sum_{k_w=1}^{l_0} \binom{l_0}{k_w} p_{0w}^{k_w}(1 - p_{0w})^{l_0-k_w}(1 - p_w)^{k_w}, \qquad w = 0, 1, \ldots, \tag{2.15}$$

where, for convenience,

$$p_{0w} = p_0 p_1 \cdots p_{w-1}. \tag{2.16}$$

The expectation of W is more conveniently obtained if the probability in (2.15) is rewritten as

$$\sum_{k_w=1}^{l_0} \binom{l_0}{k_w} p_{0w}^{k_w}(1 - p_{0w})^{l_0-k_w}(1 - p_w)^{k_w} = (1 - p_{0,w+1})^{l_0} - (1 - p_{0w})^{l_0} \tag{2.17}$$

which can be verified by direct computation. The expectation of W is now given by

$$E(W) = \sum_{w=0}^{\infty} w[(1 - p_{0,w+1})^{l_0} - (1 - p_{0w})^{l_0}]. \tag{2.18}$$

For a given v, we write the partial sum

$$\sum_{w=0}^{v} w[(1 - p_{0,w+1})^{l_0} - (1 - p_{0w})^{l_0}] = \sum_{w=1}^{v} [(1 - p_{0,v+1})^{l_0} - (1 - p_{0w})^{l_0}]. \tag{2.19}$$

Letting $v \to \infty$ and $p_{0,v+1} \to 0$, we have from (2.19)

$$E(W) = \sum_{w=1}^{\infty} [1 - (1 - p_{0w})^{l_0}]. \tag{2.20}$$

For $l_0 = 1$,

$$E(W) = p_{01} + p_{02} + p_{03} + \cdots \tag{2.21}$$

which is closely related to the expectation of life e_0 (see footnote 2 on page 237).

If the force of mortality were independent of age with $\mu(\tau) = \mu$ for $0 \le \tau < \infty$, then the proportion of white balls in each urn would be constant with $p_i = p$. In this case $p_{0i} = p^i$,

$$\begin{aligned} \Pr\{W = w\} &= (1 - p^{w+1})^{l_0} - (1 - p^w)^{l_0} \\ &= \sum_{k=1}^{l_0} (-1)^{k+1} \binom{l_0}{k} (1 - p^k) p^{wk}, \end{aligned} \tag{2.22}$$

and the expectation, the variance, and the p.g.f. of W all have closed forms. Using (2.22) we compute the p.g.f. of W

$$\begin{aligned} G_W(s) &= \sum_{w=0}^{\infty} s^w \sum_{k=1}^{l_0} (-1)^{k+1} \binom{l_0}{k} (1 - p^k) p^{wk} \\ &= \sum_{k=1}^{l_0} (-1)^{k+1} \binom{l_0}{k} (1 - p^k)(1 - p^k s)^{-1}, \end{aligned} \tag{2.23}$$

the expectation

$$E(W) = \sum_{k=1}^{l_0} (-1)^{k+1} \binom{l_0}{k} (1 - p^k)^{-1} p^k \tag{2.24}$$

and the variance

$$\sigma_W^2 = \sum_{k=1}^{l_0} (-1)^{k+1} \binom{l_0}{k} (1 + p^k)(1 - p^k)^{-2} p^k - [E(W)]^2. \tag{2.25}$$

When $l_0 = 1$, W has the geometric distribution [see Equation (4.13), Chapter 2]. ▶

3. JOINT PROBABILITY DISTRIBUTION OF THE NUMBERS OF DEATHS

In a life table covering the entire life span of each individual in a given population, the sum of the deaths at all ages is equal to the size of the original cohort. Symbolically,

$$d_0 + d_1 + \cdots + d_w = l_0, \tag{3.1}$$

where d_w is the number of deaths in the age interval (x_w and over). Each individual in the original cohort has a probability $p_{0i}q_i$ of dying in the interval (x_i, x_{i+1}), $i = 0, 1, \ldots, w$. Since an individual dies once and only once in the span covered by the life table,

$$p_{00}q_0 + \cdots + p_{0w}q_w = 1, \tag{3.2}$$

where $p_{00} = 1$ and $q_w = 1$. Thus we have the well-known results: *The numbers of deaths, $d_0, \ldots, d_w$, in a life table have a multinomial distribution with the joint probability distribution*

$$\Pr\{d_0 = \delta_0, \ldots, d_w = \delta_w \mid l_0\} = \frac{l_0!}{\delta_0! \cdots \delta_w!}(p_{00}q_0)^{\delta_0} \cdots (p_{0w}q_w)^{\delta_w}; \tag{3.3}$$

the expectation, variance and covariance are given, respectively, by

$$E(d_i \mid l_0) = l_0 p_{0i} q_i, \tag{3.4}$$

$$\sigma^2_{d_i \mid l_0} = l_0 p_{0i} q_i (1 - p_{0i} q_i), \tag{3.5}$$

and

$$\sigma_{d_i, d_j \mid l_0} = -l_0 p_{0i} q_i p_{0j} q_j \qquad \textit{for} \qquad i \neq j;\ i, j = 0, 1, \ldots, w. \quad ►(3.6)$$

In the discussion above, age 0 was chosen only for simplicity. For any given age, say x_α, the probability that an individual alive at age x_α will die in the interval (x_i, x_{i+1}) subsequent to x_α is $p_{\alpha i}q_i$ and the sum

$$\sum_{i=\alpha}^{w} p_{\alpha i} q_i = 1, \tag{3.7}$$

and thus the numbers of deaths in intervals beyond x_α also have a multinomial distribution.

4. OPTIMUM PROPERTIES OF $\hat{p}_j$ AND $\hat{q}_j$

The quantity $\hat{q}_j$ (or $\hat{p}_j$) is an estimator of the probability that an individual alive at age x_i will die in (or survive) the interval (x_i, x_{i+1}), with

$$\hat{p}_j + \hat{q}_j = 1, \qquad i = 0, 1, \ldots. \tag{4.1}$$

Therefore, $\hat{p}_j$ and $\hat{q}_j$ have the same optimum properties. For convenience, we consider $\hat{p}_j$ in the following discussion.

4.1. Maximum Likelihood Estimator of p_j

Let us introduce, for each individual in the cohort l_0, a sequence of random variables $\{\varepsilon_i\}$ defined as follows:

$$\begin{aligned} \varepsilon_i &= 1 \quad \text{if the individual dies in } (x_i, x_{i+1}) \\ &= 0 \quad \text{otherwise} \end{aligned} \tag{4.2}$$

and

$$\sum_{i=0}^{\infty} \varepsilon_i = 1$$

with the corresponding probabilities

$$\begin{aligned} \Pr\{\varepsilon_i = 1\} &= p_{0i}q_i = p_{0i}(1 - p_i) \\ \Pr\{\varepsilon_i = 0\} &= 1 - p_{0i}(1 - p_i). \end{aligned} \tag{4.3}$$

Easy computation shows that

$$\Pr\{\varepsilon_0 = 1\} + \Pr\{\varepsilon_1 = 1\} + \cdots = q_0 + p_0q_1 + \cdots = 1. \tag{4.4}$$

This means the probability is one that an individual alive at age x_0 will eventually die. The joint probability of the random variables in the sequence is

$$\prod_{i=0}^{\infty} [p_{0i}(1 - p_i)]^{\varepsilon_i}. \tag{4.5}$$

For the entire cohort, there are l_0 sequences of random variables $\{\varepsilon_{i\beta}\}$, $\beta = 1, \ldots, l_0$. For each β, let $f_\beta(\varepsilon_{i\beta}; p_i)$ be the corresponding probability function,

$$\begin{aligned} f_\beta(\varepsilon_{i\beta}; p_i) &= \prod_{i=0}^{\infty} [p_{0i}(1 - p_i)]^{\varepsilon_{i\beta}} \qquad \text{for} \quad \sum_{i=0}^{\infty} \varepsilon_{i\beta} = 1 \\ &= 0 \qquad \text{otherwise.} \end{aligned} \tag{4.6}$$

Assuming that the survival of an individual is independent of the survival of other individuals, the l_0 sequences are stochastically independent; therefore the joint probability distribution of the l_0 sequences is given by

$$\begin{aligned} f(\varepsilon_{i\beta}; p_i) &= \prod_{\beta=1}^{l_0} f_\beta(\varepsilon_{i\beta}; p_i) \\ &= \prod_{\beta=1}^{l_0} \prod_{i=0}^{\infty} [p_{0i}(1 - p_i)]^{\varepsilon_{i\beta}}, \end{aligned} \tag{4.7}$$

which is known as the likelihood function of the random variables $\varepsilon_{i\beta}$ for $i = 0, \ldots$ and $\beta = 1, \ldots, l_0$. Making the substitution of

$$\sum_{\beta=1}^{l_0} \varepsilon_{i\beta} = d_i \tag{4.8}$$

we rewrite (4.7) as

$$L = f(\varepsilon_{i\beta}; p_i) = \prod_{i=0}^{\infty} [p_{0i}(1 - p_i)]^{d_i}. \tag{4.9}$$

Now the maximum-likelihood estimators, say $\hat{p}_j$, are the values of p_j for which the likelihood function (4.9) attains a maximum. In this case, the maximizing values $\hat{p}_j$ can be obtained by differentiation. Letting

$$\log L = \log f(\varepsilon_{i\beta}; p_i) = \sum_{i=0}^{\infty} d_i \log [p_{0i}(1 - p_i)] \tag{4.10}$$

setting

$$\frac{\partial}{\partial p_j} \log L = 0, \qquad j = 0, 1, \ldots, \tag{4.11}$$

and solving the equations in (4.11), we have the maximum-likelihood estimators

$$\hat{p}_j = \frac{\sum_{i=j+1}^{\infty} d_i}{\sum_{i=j}^{\infty} d_i} = \frac{l_{j+1}}{l_j}, \tag{4.12}$$

where

$$l_j = \sum_{i=j}^{\infty} d_i \tag{4.13}$$

is the number of survivors at age x_j. It should be noted that if for some age x_w all the l_w individuals alive at x_w die within the interval (x_w, x_{w+1}), that is, $\varepsilon_{w\beta} = 1$ for all β, then $\varepsilon_{i\beta} = 0$, $d_i = 0$, and $l_i = 0$ for all $i > w$, so that there is no contribution to the likelihood function beyond the wth factor. Consequently, the maximum-likelihood estimator in (4.12) is defined only for $l_j > 0$. With this understanding, let us compute the first two moments.

We have shown in Section 2 that, given $l_j > 0$, the number l_{j+1} has the binomial distribution; therefore

$$E[\hat{p}_j] = E\left(\frac{l_{j+1}}{l_j}\right) = E\left[\frac{1}{l_j} E(l_{j+1} \mid l_j)\right] = p_j, \tag{4.14}$$

and $\hat{p}_j$ and hence $\hat{q}_j$ are unbiased estimators of the corresponding probabilities. Direct computation gives also

$$E[\hat{p}_j^2] = E\left(\frac{1}{l_j}\right) p_j(1 - p_j) + p_j^2 \tag{4.15}$$

and the variance

$$\sigma_{\hat{p}_j}^2 = E\left(\frac{1}{l_j}\right) p_j(1 - p_j) = \sigma_{\hat{q}_j}^2. \tag{4.16}$$

When l_0 is large, (4.16) may be approximated by

$$\sigma_{\hat{p}_j}^2 = \frac{1}{E(l_j)} p_j(1 - p_j). \tag{4.17}$$

Justification of (4.17) is left to the reader.

For the covariance between $\hat{p}_j$ and $\hat{p}_k$ for $j < k$, we require that l_k and hence l_j and l_{j+1} be positive and compute the conditional expectation

$$E[\hat{p}_k \mid \hat{p}_j] = E\left[\frac{l_{k+1}}{l_k} \mid \hat{p}_j\right] = E\left[\frac{1}{l_k} E(l_{k+1} \mid l_k) \mid \hat{p}_j\right] = p_k = E(\hat{p}_k), \tag{4.18}$$

from which it follows that

$$E[\hat{p}_j\hat{p}_k] = E[\hat{p}_j E(\hat{p}_k \mid \hat{p}_j)] = E[\hat{p}_j]E[\hat{p}_k]$$

and that

$$\sigma_{\hat{p}_j, \hat{p}_k} = 0. \tag{4.19}$$

Observe that formula (4.19) of zero covariance holds true only for proportions in two nonoverlapping age intervals. If the two intervals considered both begin with age x_α but extend to the ages x_j and x_k, respectively, the covariance between the proportions $\hat{p}_{\alpha j}$ and $\hat{p}_{\alpha k}$ is not equal to zero. Easy computation shows that

$$\sigma_{\hat{p}_{\alpha j}, \hat{p}_{\alpha k}} = E\left(\frac{1}{l_\alpha}\right) p_{\alpha k}(1 - p_{\alpha j}), \qquad \alpha < j \le k. \tag{4.20}$$

When $k = j$, (4.20) becomes the variance of $\hat{p}_{\alpha j}$. ▶

Although $\hat{p}_j$ and $\hat{p}_k$ have zero covariance, they are not independently distributed. The following proof is for the case where $j = 0$ and $k = 1$. To prove that $\hat{p}_0$ and $\hat{p}_1$ are not independently distributed, it is sufficient to prove the inequality

$$E(\hat{p}_0^2)E(\hat{p}_1^2) > E(\hat{p}_0^2\hat{p}_1^2). \tag{4.21}$$

In view of (4.15), the left side of (4.21) is

$$E\left(\frac{l_1^2}{l_0^2}\right)\left[E\left(\frac{1}{l_1}\right)p_1q_1 + p_1^2\right] \tag{4.22}$$

and the right side of (4.21) can be written

$$E\left[\frac{l_1^2}{l_0^2}\left(\frac{1}{l_1}\,p_1q_1 + p_1^2\right)\right]; \tag{4.23}$$

therefore inequality (4.21) holds true if and only if

$$E(l_1^2)E\left(\frac{1}{l_1}\right) > E(l_1). \tag{4.24}$$

According to Theorem 2 in Chapter 1 [see Equation (4.11), Chapter 1],

$$E\left(\frac{1}{l_1}\right) > \frac{1}{E(l_1)}.$$

Hence (4.21) holds true if

$$E(l_1^2)\frac{1}{E(l_1)} > E(l_1) \tag{4.25}$$

or if

$$E(l_1^2) - [E(l_1)]^2 > 0 \tag{4.26}$$

which is always true since the left side is the variance of l_1, proving (4.21) and the independence of $\hat{p}_0$ and $\hat{p}_1$ (see Problem 16 in Chapter 1). ▶

4.2. Cramér-Rao Lower Bound for the Variance of an Unbiased Estimator of p_j

In (4.16) we have the exact formula for the variance of $\hat{p}_j$ given in (4.12). We now determine the lower limit for the variance of an unbiased estimator. Let $\tilde{p}_j$ be any unbiased estimator of p_j. According to a theorem by Cramér and Rao, the variance of $\tilde{p}_j$ satisfies the inequality[1]

$$\text{Var}\,(\tilde{p}_j) \geq \frac{1}{-E\left(\dfrac{\partial^2}{\partial p_j^2}\log L\right)} \tag{4.27}$$

[1] It should be pointed out that when the parameters p_j, $j = 0, 1, \ldots$, are estimated jointly, the lower bound is a function of the expectations

$$E[(\partial^2 \log L)/\partial p_j^2] \qquad \text{and} \qquad E[(\partial^2 \log L)/\partial p_j\, \partial p_k],$$

$j, k = 0, 1, \ldots$. Formula (4.27), however, is correct, because in the present case the expectations of the "mixed" partial derivatives vanish whatever $j \neq k$ may be.

where log L is the logarithm of the likelihood function defined in (4.10). First let us sketch a proof of (4.27).

The derivative $(\partial \log L)/\partial p_j$ obviously is a random variable; we compute the expectations

$$E\left[\frac{\partial}{\partial p_j}\log L\right] = E\left[\frac{1}{L}\frac{\partial}{\partial p_j}L\right] = \frac{\partial}{\partial p_j}1 = 0 \tag{4.28}$$

and

$$E\left[\frac{\partial^2}{\partial p_j^2}\log L\right] = E\left[\frac{1}{L}\frac{\partial^2}{\partial p_j^2}L - \left(\frac{1}{L}\frac{\partial}{\partial p_j}L\right)^2\right]$$

$$= \frac{\partial^2}{\partial p_j^2}1 - E\left[\left(\frac{\partial}{\partial p_j}\log L\right)^2\right] = -E\left[\left(\frac{\partial}{\partial p_j}\log L\right)^2\right], \tag{4.29}$$

from which we have the variance of $(\partial \log L)/\partial p_j$

$$\text{Var}\left(\frac{\partial}{\partial p_j}\log L\right) = E\left[\left(\frac{\partial}{\partial p_j}\log L\right)^2\right] = -E\left[\frac{\partial^2}{\partial p_j^2}\log L\right]. \tag{4.30}$$

Let $\tilde{p}_j$ be any unbiased estimator of p_j, then the covariance between $\tilde{p}_j$ and $(\partial \log L)/\partial p_j$ is unity, as shown by the following computation

$$\text{Cov}\left(\tilde{p}_j, \frac{\partial}{\partial p_j}\log L\right) = E\left[\tilde{p}_j\frac{\partial}{\partial p_j}\log L\right] = E\left[\tilde{p}_j\frac{1}{L}\frac{\partial}{\partial p_j}L\right]$$

$$= \frac{\partial}{\partial p_j}E[\tilde{p}_j] = \frac{\partial}{\partial p_j}p_j = 1. \tag{4.31}$$

Since the square of the covariance between two random variables cannot exceed the product of the two variances, we have

$$\text{Var}(\tilde{p}_j)\,\text{Var}\left(\frac{\partial}{\partial p_j}\log L\right) \geq 1. \tag{4.32}$$

Substituting (4.30) in (4.32) gives (4.27) and the proof is complete. ▶

In the present case

$$-E\left[\frac{\partial^2}{\partial p_j^2}\log L\right] = \frac{l_0 p_{0j}}{p_j(1-p_j)} \tag{4.33}$$

and the lower bound is

$$\frac{1}{l_0 p_{0j}}\,p_j(1-p_j). \tag{4.34}$$

The difference between the lower bound in (4.34) and the exact formula (4.16) lies in the difference between $1/l_0 p_{0j}$ and $E(1/l_j)$. Therefore, relative to the lower bound, the efficiency of $\hat{p}_j$ is $l_0 p_{0j}/E(1/l_j)$. However, we shall show in the next section that the maximum-likelihood estimator $\hat{p}_j$ given in (4.12) has the minimum variance of all the unbiased estimators of p_j; thus the lower bound cannot be attained in this case.

4.3. Sufficiency and Efficiency of $\hat{p}_j$

We first define *sufficient statistics* in general terms. Let $X_1, \ldots, X_n$ be a sample of random variables with the joint probability (density) function $f(x_1, \ldots, x_n; \theta_1, \ldots, \theta_r)$ depending upon unknown parameters $\theta_1, \ldots, \theta_r$. Functions $T_k = T_k(X_1, \ldots, X_n)$, $k = 1, \ldots, r$, are called joint sufficient statistics for $\theta_1, \ldots, \theta_r$ if and only if the joint probability (density) function $f(x_1, \ldots, x_n; \theta_1, \ldots, \theta_r)$ can be factorized as follows:

$$f(x_1, \ldots, x_n; \theta_1, \ldots, \theta_r) = g(t_1, \ldots, t_r; \theta_1, \ldots, \theta_r) \times h(x_1, \ldots, x_n; t_1, \ldots, t_r) \quad (4.35)$$

where g is a function of both the sufficient statistics and the parameters, whereas h is independent of the parameters. This factorization shows that, given the sufficient statistics $T_1, \ldots, T_r$, any function of $x_1, \ldots, x_n$ adds no further information regarding the parameters $\theta_1, \ldots, \theta_r$. In other words, sufficient statistics exhaust all the information about the parameters contained in the sample. The concept of sufficient statistics was introduced by Fisher [1922], and the above factorization was developed by Neyman [1935]. For a detailed discussion on sufficiency and other optimum properties of estimators, see Lehmann [1959], Hogg and Craig [1965], Kendall and Stuart [1961], and other standard textbooks in statistics.

In the present case, the random variables are $\varepsilon_{i\beta}$ and the parameters are $p_0, p_1, \ldots$; the joint probability of $\varepsilon_{i\beta}$ is given in (4.7) or (4.9),

$$f(\varepsilon_{i\beta}; p_i) = \prod_{i=0}^{\infty} [p_{0i}(1 - p_i)]^{d_i}. \quad (4.9)$$

Using the relation $l_i = d_i + d_{i+1} + \ldots$, we rewrite (4.9) as

$$f(\varepsilon_{i\beta}; p_i) = \prod_{i=0}^{\infty} (1 - p_i)^{d_i} p_i^{l_{i+1}} \quad (4.36)$$

which is easily factorized:

$$f(\varepsilon_{i\beta}; p_i) = g(l_i; p_i) h(\varepsilon_{i\beta}; l_i) \quad (4.37)$$

where

$$g(l_i; p_i) = \prod_{i=0}^{\infty} \frac{l_i!}{(l_i - l_{i+1})!\, l_{i+1}!} (1 - p_i)^{l_i - l_{i+1}} p_i^{\,l_{i+1}} \tag{4.38}$$

and

$$h(\varepsilon_{i\beta}; l_i) = \prod_{i=0}^{\infty} \frac{\left[\sum_{\beta=1}^{l_0} \varepsilon_{i\beta}\right]! \left(l_i - \sum_{\beta=1}^{l_0} \varepsilon_{i\beta}\right)!}{l_i!}. \tag{4.39}$$

According to the Fisher-Neyman factorization criterion, the statistics l_i are jointly sufficient for p_i, $i = 0, 1, \ldots$. ▶

We want to show that for any unbiased estimator $\tilde{p}_j$ of p_j

$$\sigma_{\tilde{p}_j}^{\;2} \geq \sigma_{\hat{p}_j}^{\;2} \tag{4.40}$$

and the equality sign holds true if and only if $\tilde{p}_j = \hat{p}_j$. To prove (4.40) we note that, since l_j and l_{j+1} are sufficient statistics for p_j, the conditional expectation

$$E(\tilde{p}_j \mid l_j, l_{j+1}) = V(l_j, l_{j+1}) \tag{4.41}$$

is independent of p_j. Since $\tilde{p}_j$ is unbiased, we have

$$p_j = E(\tilde{p}_j) = E[E(\tilde{p}_j \mid l_j, l_{j+1})] = E[V(l_j, l_{j+1})] \tag{4.42}$$

whatever may be $0 \leq p_j \leq 1$. Now if we restrict l_j to positive values only, then the joint probability distribution of l_j and l_{j+1} is

$$\frac{1}{1 - (1 - p_{0j})^{l_0}} \left[\binom{l_0}{l_j} p_{0j}^{\;l_0} (1 - p_{0j})^{l_0 - l_j}\right]\left[\binom{l_j}{l_{j+1}} p_j^{l_{j+1}} (1 - p_j)^{l_j - l_{j+1}}\right] \tag{4.43}$$

for $l_j = 1, 2, \ldots, l_0$ and $l_{j+1} = 0, 1, \ldots, l_j$. Substituting (4.43) in (4.42) yields the identity

$$p_j \equiv \sum_{l_j=1}^{l_0} \sum_{l_{j+1}=0}^{l_j} \frac{V(l_j, l_{j+1})}{1 - (1 - p_{0j})^{l_0}} \left[\binom{l_0}{l_j} p_{0j}^{\;l_0} (1 - p_{0j})^{l_0 - l_j}\right]$$
$$\times \left[\binom{l_j}{l_{j+1}} p_j^{l_{j+1}} (1 - p_j)^{l_j - l_{j+1}}\right]. \tag{4.44}$$

Since (4.44) is an identity in p_j, it has the unique solution

$$V(l_j, l_{j+1}) = \frac{l_{j+1}}{l_j} \tag{4.45}$$

meaning that

$$E[\tilde{p}_j \mid l_j, l_{j+1}] = \hat{p}_j. \tag{4.46}$$

The variance of $\tilde{p}_j$ can now be computed:

$$\sigma_{\tilde{p}_j}{}^2 = E(\tilde{p}_j - p_j)^2 = E[(\tilde{p}_j - \hat{p}_j) + (\hat{p}_j - p_j)]^2$$
$$= E(\tilde{p}_j - p_j)^2 + E(\hat{p}_j - p_j)^2 + 2E(\tilde{p}_j - \hat{p}_j)(\hat{p}_j - p_j) \quad (4.47)$$

where the expectation of the cross product

$$E(\tilde{p}_j - \hat{p}_j)(\hat{p}_j - p_j) = E[E\{(\tilde{p}_j - \hat{p}_j)(\hat{p}_j - p_j) \mid l_j, l_{j+1}\}]$$
$$= E[E\{(\tilde{p}_j - \hat{p}_j) \mid l_j, l_{j+1}\}(\hat{p}_j - p_j)] = 0. \quad (4.48)$$

Therefore

$$\sigma_{\tilde{p}_j}{}^2 = E(\tilde{p}_j - \hat{p}_j)^2 + \sigma_{\hat{p}_j}{}^2 \quad (4.49)$$

and

$$\sigma_{\tilde{p}_j}{}^2 > \sigma_{\hat{p}_j}{}^2 \quad (4.50)$$

which was to be shown. The two variances are equal only in the trivial case when $\tilde{p}_j = \hat{p}_j$.

Thus we have shown that *the quantities $\hat{p}_j$ and $\hat{q}_j$ in the life table are the unique, unbiased, efficient estimators of the corresponding probabilities p_j and q_j.* ▶

5. DISTRIBUTION OF THE OBSERVED EXPECTATION OF LIFE

The observed expectation of life summarizes the mortality experience of a population from a given age to the end of the life span. At age x_i the expectation expresses the average number of years remaining to each individual living at that age if all individuals are subjected to the estimated probabilities of death $\hat{q}_j$ for $j \geq i$. This is certainly the most useful column in the life table.

To avoid confusion in notation, let α denote a fixed number and x_α a particular age. We are interested in the distribution of $\hat{e}_\alpha$, the observed expectation of life at the age x_α. Consider l_α, the number of survivors to age x_α, and let Y_α denote the future lifetime beyond age x_α of a particular individual. Clearly, Y_α is a continuous random variable that can take on any non-negative real value. Let y_α be the value that the random variable Y_α assumes, then $x_\alpha + y_\alpha$ is the entire life span of the individual. Let $f(y_\alpha)$ be the probability density function of the random variable Y_α, and let dy_α be an infinitesimal time interval. Since Y_α can assume values between y_α and $y_\alpha + dy_\alpha$ if and only if the individual survives the age interval $(x_\alpha, x_\alpha + y_\alpha)$ and dies in the interval $(x_\alpha + y_\alpha, x_\alpha + y_\alpha + dy_\alpha)$, we have

$$f(y_\alpha)\, dy_\alpha = \exp\left\{-\int_{x_\alpha}^{x_\alpha+y_\alpha} \mu(\tau)\, d\tau\right\}\mu(x_\alpha + y_\alpha)\, dy_\alpha, \qquad y_\alpha \geq 0. \quad (5.1)$$

Function $f(y_\alpha)$ in (5.1) is a proper probability density function since it is never negative and since the integral of the function from $y_\alpha = 0$ to $y_\alpha = \infty$ is equal to unity. Clearly, $f(y_\alpha)$ can never be negative whatever the value of y_α. To evaluate the integral

$$\int_0^\infty f(y_\alpha)\,dy_\alpha = \int_0^\infty \exp\left\{-\int_{x_\alpha}^{x_\alpha+y_\alpha} \mu(\tau)\,d\tau\right\}\mu(x_\alpha + y_\alpha)\,dy_\alpha \tag{5.2}$$

we define a quantity Φ

$$\Phi = \int_{x_\alpha}^{x_\alpha+y_\alpha} \mu(\tau)\,d\tau = \int_0^{y_\alpha} \mu(x_\alpha + t)\,dt \tag{5.3}$$

and substitute the differential

$$d\Phi = \mu(x_\alpha + y_\alpha)\,dy_\alpha \tag{5.4}$$

in the integral to give the solution

$$\int_0^\infty f(y_\alpha)\,dy_\alpha = \int_0^\infty e^{-\Phi}\,d\Phi = 1. \tag{5.5}$$

The mathematical expectation of the random variable Y_α is the expected length of life beyond age x_α. In accordance with the definition given the symbol e_α, we may write

$$e_\alpha = \int_0^\infty y_\alpha f(y_\alpha)\,dy_\alpha = \int_0^\infty y_\alpha \exp\left\{-\int_{x_\alpha}^{x_\alpha+y_\alpha} \mu(\tau)\,d\tau\right\}\mu(x_\alpha + y_\alpha)\,dy_\alpha. \tag{5.6}$$

Thus the expectation e_α and the variance

$$\sigma_{Y_\alpha}^2 = \int_0^\infty (y_\alpha - e_\alpha)^2 f(y_\alpha)\,dy_\alpha \tag{5.7}$$

both depend on the intensity of risk of death $\mu(\tau)$.

Remark. The expectation of life at age x_α is conventionally defined as

$$e_\alpha = \int_0^\infty \exp\left\{-\int_{x_\alpha}^{x_\alpha+y_\alpha} \mu(\tau)\,d\tau\right\}dy_\alpha. \tag{5.8}$$

It is instructive to prove that the two alternative definitions (5.6) and (5.8) are identical. Let $u = y_\alpha$, $du = dy_\alpha$,

$$v = -\exp\left\{-\int_{x_\alpha}^{x_\alpha+y_\alpha} \mu(\tau)\,d\tau\right\} \tag{5.9}$$

and

$$dv = \exp\left\{-\int_{x_\alpha}^{x_\alpha+y_\alpha} \mu(\tau)\,d\tau\right\}\mu(x_\alpha + y_\alpha)\,dy_\alpha. \tag{5.10}$$

Integrating (5.6) by parts gives

$$\int_0^\infty y_\alpha \exp\left\{-\int_{x_\alpha}^{x_\alpha+y_\alpha} \mu(\tau)\,d\tau\right\}\mu(x_\alpha + y_\alpha)\,dy_\alpha$$

$$= -y_\alpha \exp\left\{-\int_{x_\alpha}^{x_\alpha+y_\alpha} \mu(\tau)\,d\tau\right\}\Bigg|_0^\infty + \int_0^\infty \exp\left\{-\int_{x_\alpha}^{x_\alpha+y_\alpha} \mu(\tau)\,d\tau\right\} dy_\alpha. \tag{5.11}$$

The first term on the right vanishes and the second term is the same as (5.8), proving the identity.

5.1. Observed Expectation of Life and Sample Mean Length of Life

The future lifetimes of l_α survivors may be regarded as a sample of l_α independent and identically distributed random variables, $Y_{\alpha k}, k = 1, \ldots, l_\alpha$, each of which has the probability density function (5.1), the expectation (5.6) and the variance (5.7). According to the central limit theorem, for large l_α the distribution of the sample mean

$$\bar{Y}_\alpha = \frac{1}{l_\alpha}\sum_{k=1}^{l_\alpha} Y_{\alpha k} \tag{5.12}$$

is approximately normal with an expectation as given in (5.6) and a variance $\sigma_{Y_\alpha}{}^2/l_\alpha$. Naturally one would suggest that the sample mean $\bar{Y}_\alpha$ be used as an estimate of the true expectation of life e_α, or

$$\bar{Y}_\alpha = \hat{e}_\alpha. \tag{5.13}$$

We now show that Equation (5.13) is indeed true.

Like any continuous random variable, Y_α is not accurately measured. In fact, the values of the l_α random variables are not individually recorded in the life table, but grouped in the form of a frequency distribution in which the ages x_i and x_{i+1} are the lower and upper limits for the interval i, and the deaths, d_i, are the corresponding frequencies for $i = \alpha, \alpha + 1, \ldots, w$. The sum of the frequencies equals the number of survivors at age x_α, or

$$d_\alpha + \cdots + d_w = l_\alpha. \tag{5.14}$$

The total number of years remaining to the l_α survivors depends on the exact age at which death occurs, that is, on the distribution of deaths within each age interval. Suppose that the distribution of deaths in the interval (x_i, x_{i+1}) is such that, on the average, each of the d_i individuals lives $a_i n_i$ years in the interval. Each thus lives $x_i + a_i n_i$ years, or $x_i - x_\alpha + a_i n_i$ years after age x_α, and the sample mean is given by

$$\bar{Y}_\alpha = \frac{1}{l_\alpha}\sum_{i=\alpha}^{w}(x_i - x_\alpha + a_i n_i)d_i = \frac{1}{l_\alpha}\left[\sum_{i=\alpha}^{w}(x_i - x_\alpha)d_i + \sum_{i=1}^{w} a_i n_i d_i\right]. \tag{5.15}$$

By definition

$$x_i - x_\alpha = n_\alpha + n_{\alpha+1} + \cdots + n_{i-1} = \sum_{j=\alpha}^{i-1} n_j;$$

hence

$$\begin{aligned}\sum_{i=\alpha}^{w}(x_i - x_\alpha)d_i &= \sum_{i=\alpha}^{w}\left(\sum_{j=\alpha}^{i-1} n_j\right)d_i \\ &= \sum_{j=\alpha}^{w-1} n_j \sum_{i=j+1}^{w} d_i = \sum_{j=\alpha}^{w-1} n_j(l_j - d_j)\end{aligned} \tag{5.16}$$

since $l_j = d_j + d_{j+1} + \cdots + d_w$. Substituting (5.16) in (5.15) gives

$$\begin{aligned}\bar{Y}_\alpha &= \frac{1}{l_\alpha}\left[\sum_{j=\alpha}^{w-1} n_j(l_j - d_j) + \sum_{i=\alpha}^{w} a_i n_i d_i\right] \\ &= \frac{1}{l_\alpha}\left[\sum_{i=\alpha}^{w-1}\{n_i(l_i - d_i) + a_i n_i d_i\} + a_w n_w d_w\right].\end{aligned} \tag{5.17}$$

We let [cf. Equation (4.7) in Chapter 9]

$$L_i = n_i(l_i - d_i) + a_i n_i d_i, \qquad i = \alpha, \cdots, w - 1, \tag{5.18}$$

be the number of years lived by the l_α individuals in the interval (x_i, x_{i+1}), and let

$$L_w = a_w n_w d_w = a_w n_w l_w \tag{5.19}$$

be the number of years lived by l_α beyond age x_w. Using (5.18) and (5.19) we rewrite (5.17) as

$$\bar{Y}_\alpha = \frac{1}{l_\alpha}\sum_{i=\alpha}^{w} L_i \tag{5.20}$$

which, of course, is $\hat{e}_\alpha$, the observed expectation of life at age x_α as given in (4.9) in Chapter 9, proving (5.13).

5.2. Variance of the Observed Expectation of Life

We know from (5.13) that the variance of $\hat{e}_\alpha$ can be obtained from the variance of Y_α. For practical purposes we need to have a formula for the variance of $\hat{e}_\alpha$ which can be estimated for the cohort and the current life tables. Using the relation $d_i = l_i - l_{i+1}$, we rewrite (5.15) as

$$\hat{e}_\alpha = a_\alpha n_\alpha + \sum_{i=\alpha+1}^{w} c_i \frac{l_i}{l_\alpha} = a_\alpha n_\alpha + \sum_{i=\alpha+1}^{w} c_i \hat{p}_{\alpha i} \tag{5.21}$$

where $c_i = (1 - a_{i-1})n_{i-1} + a_i n_i$. Because the proportion $\hat{p}_{\alpha i}$ in (5.21) is an unbiased estimate of $p_{\alpha i}$, the expectation of $\hat{e}_\alpha$ as given by (5.6) can be rewritten[2]

$$e_\alpha = a_\alpha n_\alpha + \sum_{i=\alpha+1}^{w} c_i p_{\alpha i}, \qquad \alpha = 0, 1, \ldots, w. \tag{5.22}$$

We also note a relation between the expectations of life at the beginning of two consecutive intervals,

$$e_i - a_i n_i = [e_{i+1} + (1 - a_i)n_i]p_i. \tag{5.23}$$

Now the observed expectation of life as given in (5.21) is a linear function of $p_{\alpha i}$; therefore, its variance is

$$\text{Var}(\hat{e}_\alpha) = \sum_{i=\alpha+1}^{w} c_i^2 \,\text{Var}(\hat{p}_{\alpha i}) + 2 \sum_{i=\alpha+1}^{w-1} \sum_{j=i+1}^{w} c_i c_j \,\text{Cov}(\hat{p}_{\alpha i}, \hat{p}_{\alpha j}). \tag{5.24}$$

Substituting formula (4.20) in (5.24), we have

$$\text{Var}(\hat{e}_\alpha) = E\left(\frac{1}{l_\alpha}\right)\left[\sum_{i=\alpha+1}^{w} c_i^2 p_{\alpha i}(1 - p_{\alpha i}) + 2 \sum_{i=\alpha+1}^{w-1} \sum_{j=i+1}^{w} c_i c_j p_{\alpha j}(1 - p_{\alpha i})\right]$$

$$= E\left(\frac{1}{l_\alpha}\right)\left[\sum_{i=\alpha+1}^{w} c_i^2 p_{\alpha i} + 2 \sum_{i=\alpha+1}^{w-1} c_i \sum_{j=i+1}^{w} c_j p_{\alpha j} - \left(\sum_{i=\alpha+1}^{w} c_i p_{\alpha i}\right)^2\right]. \tag{5.25}$$

Using the relation $p_{\alpha j} = p_{\alpha i} p_{ij}$ and formula (5.22)

$$\text{Var}(\hat{e}_\alpha) = E\left(\frac{1}{l_\alpha}\right)\left[\sum_{i=\alpha+1}^{w} c_i^2 p_{\alpha i} + 2 \sum_{i=\alpha+1}^{w} c_i p_{\alpha i}(e_i - a_i n_i) - (e_\alpha - a_\alpha n_\alpha)^2\right]$$

$$= E\left(\frac{1}{l_\alpha}\right)\left[\sum_{i=\alpha+1}^{w} \{c_i(c_i - 2a_i n_i) + 2c_i e_i\} p_{\alpha i} - (e_\alpha - a_\alpha n_\alpha)^2\right]. \tag{5.26}$$

---

[2] If $n_i = 1$ and $a_i = \frac{1}{2}$ for all i, then $c_i = 1$ and (5.21) and (5.22) become

$$\hat{e}_\alpha = \frac{1}{2} + \sum_{i=\alpha+1}^{w} \hat{p}_{\alpha i} \qquad \text{and} \qquad e_\alpha = \frac{1}{2} + \sum_{i=\alpha+1}^{w} p_{\alpha i},$$

respectively.

Since $c_i = (1 - a_{i-1})n_{i-1} + a_i n_i$, the quantity inside the braces may be rewritten as

$$\begin{aligned} c_i(c_i - 2a_i n_i) + 2c_i e_i &\\ &= \{(1 - a_{i-1})^2 n_{i-1}^2 - a_i^2 n_i^2\} + 2\{(1 - a_{i-1})n_{i-1} + a_i n_i\}e_i \\ &= \{e_i + (1 - a_{i-1})n_{i-1}\}^2 - \{e_i - a_i n_i\}^2 \\ &= \{e_i + (1 - a_{i-1})n_{i-1}\}^2 - \{e_{i+1} + (1 - a_i)n_i\}^2 p_i^2. \end{aligned} \tag{5.27}$$

Substitution of the last expression in (5.26) gives

$$\begin{aligned} \operatorname{Var}(\hat{e}_\alpha) = E\left(\frac{1}{l_\alpha}\right)\Bigg[\sum_{i=\alpha+1}^{w} (\{e_i + (1 - a_{i-1})n_{i-1}\}^2 p_{\alpha i} &\\ - \{e_{i+1} + (1 - a_i)n_i\}^2 p_i^2 p_{\alpha i}) - (e_\alpha - a_\alpha n_\alpha)^2\Bigg]. \end{aligned} \tag{5.28}$$

Making the substitutions of $p_w = 0$ and

$$(e_\alpha - a_\alpha n_\alpha) = \{e_{\alpha+1} + (1 - a_\alpha)n_\alpha\}p_\alpha$$

in (5.28) and combining terms, we find that

$$\begin{aligned} \operatorname{Var}(\hat{e}_\alpha) = E\left(\frac{1}{l_\alpha}\right)\Bigg[\sum_{i=\alpha}^{w-1} (\{e_{i+1} + (1 - a_i)n_i\}^2 p_{\alpha,i+1} &\\ - \{e_{i+1} + (1 - a_i)n_i\}^2 p_i^2 p_{\alpha i})\Bigg]. \end{aligned} \tag{5.29}$$

Since $p_{\alpha,i+1} = p_{\alpha i}p_i$, we have the final formula for the variance of the observed expectation of life at age x_α,

$$\operatorname{Var}(\hat{e}_\alpha) = E\left(\frac{1}{l_\alpha}\right)\left[\sum_{i=\alpha}^{w-1} \{e_{i+1} + (1 - a_i)n_i\}^2 p_{\alpha i} p_i (1 - p_i)\right], \quad \alpha = 0, 1, \ldots, w - 1. \tag{5.30}$$

Formula (5.30) is reduced to

$$\operatorname{Var}(\hat{e}_\alpha) = \sum_{i=\alpha}^{w-1} p_{\alpha i}^2 \{e_{i+1} + (1 - a_i)n_i\}^2 \sigma_{\hat{q}_i}^2, \qquad \alpha = 0, 1, \ldots, w - 1, \tag{5.31}$$

when $E(1/l_\alpha)$ is approximated by $1/E(l_\alpha)$, $p_{\alpha i}$ is written for $E(l_i)/E(l_\alpha)$, and (4.17) is used for the variance of $\hat{q}_i$.

Thus we have

Theorem. *If the distribution of deaths in the age interval* (x_i, x_{i+1}) *is such that, on the average, each of the* d_i *individuals lives* $a_i n_i$ *years in the interval, for* $i = \alpha, \alpha + 1, \ldots, w$, *then as* l_α *approaches infinity, the* *probability distribution of the observed expectation of life at age x_α, as given by* (5.21), *is asymptotically normal and has the mean and variance as given by* (5.22) *and* (5.31), *respectively.* ▶

It should be noted that although (5.31) is an approximation of the exact formula (5.30) for the variance of $\hat{e}_\alpha$ when l_α is a random variable, it becomes identical to (5.30) when l_α is a given constant, such as l_0.

Although the theorem concerning the asymptotic distribution of the observed expectation of life is true for the cohort and the current life tables, it is not immediately clear that formula (5.31) holds also for the current life table. In the current life table the basic random variable $\hat{q}_i$ is computed from actual mortality experience and, in general, its variance is not given by either formula (4.16) or formula (4.17). We must therefore prove (5.31) for the current life table.

The observed expectation of life as given in (5.21) is a linear function of $\hat{p}_{\alpha j}$, which in the current life table is computed from

$$\hat{p}_{\alpha j} = \hat{p}_\alpha \hat{p}_{\alpha+1} \cdots \hat{p}_{j-1}, \qquad j = \alpha+1, \ldots, w. \tag{5.32}$$

Clearly, the derivatives taken at the true point $(p_\alpha, p_{\alpha+1}, \ldots, p_{j-1})$ are

$$\begin{aligned} \frac{\partial}{\partial \hat{p}_i} \hat{p}_{\alpha j} &= p_{\alpha i} p_{i+1,j}, \qquad \text{for} \quad \alpha \le i < j; \\ &= 0, \qquad \text{otherwise.} \end{aligned} \tag{5.33}$$

Hence

$$\begin{aligned} \frac{\partial}{\partial \hat{p}_i} \hat{e}_\alpha &= \sum_{j=i+1}^{w} c_j p_{\alpha i} p_{i+1,j} \\ &= p_{\alpha i}\left(c_{i+1} + \sum_{j=i+2}^{w} c_j p_{i+1,j} \right) \\ &= p_{\alpha i}[e_{i+1} + (1 - a_i)n_i]. \end{aligned} \tag{5.34}$$

Because of (5.33), the derivative (5.34) vanishes when $i = w$. Since it has been shown in Section 4 that the covariance between proportions of survivors of two nonoverlapping age intervals is zero, the variance of the observed expectation of life may be computed from the following:

$$\operatorname{Var}(\hat{e}_\alpha) = \sum_{i=\alpha}^{w-1} \left\{ \frac{\partial}{\partial \hat{p}_i} \hat{e}_\alpha \right\}^2 \operatorname{Var}(\hat{p}_i). \tag{5.35}$$

Substitution of (5.34) in (5.35) gives formula (5.31). ▶

PROBLEMS

1. Derive the probability distribution for the number l_x of survivors at age x given l_0 at age $x = 0$ by the following methods.

(a) By establishing and solving the differential equations for the probabilities $\Pr\{l_x = k \mid l_0\}$.

(b) By the method of probability generating function.

2. Verify by induction that the p.g.f. of the joint probability distribution of $l_1, \ldots, l_u$ given l_0 is that given in (2.11).

3. Derive the joint probability distribution of l_1 and l_2 given l_0 using the relation

$$\Pr\{l_1 = k_1 \text{ and } l_2 = k_2 \mid l_0\} = \Pr\{l_1 = k_1 \mid l_0\} \Pr\{l_2 = k_2 \mid k_1\}.$$

Generalize the result to the distribution of $l_1, \ldots, l_u$.

4. Use the rule for expectations in Equation (4.22) in Chapter 1 to compute $E(l_i\, l_j \mid l_0)$ and use the result to derive the covariance between l_i and l_j.

5. *Distribution of W.* Show that the sum of the probabilities

$$\Pr\{W = w\} = \sum_{k_w=1}^{l_0} \binom{l_0}{k_w} p_{0w}^{k_w}(1 - p_{0w})^{l_0-k_w}(1 - p_w)^{k_w}$$

from $w = 0$ to $w = \infty$ is unity.

6. *Continuation.* Verify the equality in (2.17).

7. *Continuation.* In dealing with infinite sums (or integrals), strange results may appear if caution is not observed. The following is an example. The p.g.f. of the distribution of W as given in (2.22) is defined by

$$G_W(s) = \sum_{w=0}^{\infty} s^w[(1 - p^{w+1})^{l_0} - (1 - p^w)^{l_0}].$$

Collecting terms according to $(1 - p^w)^{l_0}$ we have

$$G_W(s) = \sum_{w=1}^{\infty} (s^{w-1} - s^w)(1 - p^w)^{l_0}.$$

Substitution of $s = 1$ in the last expression gives $G_W(1) = 0$ (!) and taking the derivative of $G_W(s)$ at $s = 1$ yields the expectation

$$E(W) = \sum_{w=1}^{\infty} [(w - 1) - w](1 - p^w)^{l_0} = -\sum_{w=1}^{\infty} (1 - p^w)^{l_0} < 0 \text{ (!)}$$

What went wrong in the above argument?

8. *The estimator $\hat{p}_j$.* Show that the estimator $\hat{p}_j = l_{j+1}/l_j$ as given in (4.12) is a maximizing value of the likelihood function L in (4.9).

9. Justify the following approximation

$$E\left(\frac{1}{l_j}\right) \doteq \frac{1}{E(l_j)}.$$

10. Derive the covariance between the proportions $\hat{p}_{\alpha j}$ and $\hat{p}_{\alpha k}$.

11. Show that whatever may be $j \neq k$,

$$E\left(\frac{\partial^2}{\partial p_j \partial p_k} \log L\right) = 0$$

where $\log L$ is given in (4.10). Evaluate the expectation for $k = j$, and use the result to determine the lower bound of the variance of an unbiased estimator of p_j.

12. Show that the function $V(l_j, l_{j+1}) = l_{j+1}/l_j$ is the unique solution of the identity (4.44).

13. *Completeness.* Let $V(l_j, l_{j+1})$ be a continuous function of l_j and l_{j+1} (but not a function of the parameter p_j). The family of the probability density functions of $V(l_j, l_{j+1})$ for all p_j, $0 < p_j < 1$, is said to be complete if

$$E[V(l_j, l_{j+1})] = 0$$

for every p_j implies that $V(l_j, l_{j+1}) = 0$ for all possible values of l_j and l_{j+1}.

Show that the family of the probability density functions of $V(l_j, l_{j+1}) = l_{j+1}/l_j$ is complete.

In general, if a function $V(l_j, l_{j+1})$ of sufficient statistics l_j and l_{j+1} is an unbiased estimator of p_j and the probability density function of $V(l_j, l_{j+1})$ is complete, then, according to a theorem of Blackwell [1947] and Rao [1949], $V(l_j, l_{j+1})$ is the unique best estimator of p_j (see also Lehmann and Scheffe [1950]).

14. *Expectation of life.* Suppose the intensity of risk of death (force of mortality) is independent of age so that $\mu(x) = \mu$. Find e_α, the expectation of life at age x_α, and the variance of Y_α, the future length of life beyond x_α, as given in formulas (5.6) and (5.7).

15. *Continuation.* Compute the expectation of life e_α and the variance of Y_α for $\mu(x) = \mu a x^{a-1}$ (see Equation (5.14) in Chapter 3, the Weibull distribution).

16. Show that the quantities $\bar{Y}_\alpha$ in (5.15) and $\hat{e}_\alpha$ in (5.21) are equal, or

$$\frac{1}{l_\alpha} \sum_{i=\alpha}^{w} (x_i - x_\alpha + a_i n_i) d_i = a_\alpha n_\alpha + \sum_{i=\alpha+1}^{w} c_i \hat{p}_{\alpha i}$$

where $c_i = (1 - a_{i-1}) n_{i-1} + a_i n_i$ and $\hat{p}_{\alpha i} = l_i / l_\alpha$.

CHAPTER 11

Competing Risks

1. INTRODUCTION

Death is not a repetitive event and it is usually attributed to a single cause; however, various risks competing for the life of an individual must be considered in cause-specific mortality studies. In an investigation of congenital malformation as a cause of infant death, for example, some subjects will die from other causes such as tuberculosis. These infants have no chance either of dying from congenital malformation or of surviving the first year of life. What then is the contribution of their survival experience to such a mortality study, and what adjustment must be made for the competing effect of tuberculosis in the assessment of congenital malformation as a cause of death? Competing risks also should be taken into account in studies of the relative susceptibility of people in different illness states to other diseases. Are people suffering from arteriosclerotic heart disease more likely to die from pneumonia than people without a heart condition? A meaningful comparison between the two groups with respect to their susceptibility to pneumonia would have to evaluate the effect of arteriosclerosis as a competing risk.

The study of survival and the application of life table methodology to these problems require an understanding of the following three types of probability of death from a specific cause.

1. The crude probability. The probability of death from a specific cause in the presence of all other risks acting in a population, or

$Q_{i\delta}$ = Pr{an individual alive at time x_i will die in the interval (x_i, x_{i+1}) from cause R_δ in the presence of all other risks in the population}.

2. The net probability. The probability of death if a specific risk is the only risk in effect in the population or, conversely, the probability of death if a specific risk is eliminated from the population.

$$q_{i\delta} = \Pr\{\text{ an individual alive at } x_i \text{ will die in the interval } (x_i, x_{i+1}) \text{ if } R_\delta \text{ is the only risk acting on the population}\};$$

$$q_{i.\delta} = \Pr\{\text{an individual alive at } x_i \text{ will die in the interval } (x_i, x_{i+1}) \text{ if } R_\delta \text{ is eliminated as a risk of death}\}.$$

3. The partial crude probability. The probability of death from a specific cause when another risk (or risks) is eliminated from the population.

$$Q_{i\delta.1} = \Pr\{\text{an individual alive at } x_i \text{ will die in the interval } (x_i, x_{i+1}) \text{ from } R_\delta \text{ if } R_1 \text{ is eliminated as a risk of death}\};$$

$$Q_{i\delta.12} = \Pr\{\text{an individual alive at } x_i \text{ will die in the interval } (x_i, x_{i+1}) \text{ from } R_\delta \text{ if } R_1 \text{ and } R_2 \text{ are eliminated as risks of death}\}.$$

When the cause of death is not specified, we have the probabilities

$$p_i = \Pr\{\text{an individual alive at } x_i \text{ will survive the interval } (x_i, x_{i+1})\}$$

and

$$q_i = \Pr\{\text{an individual alive at } x_i \text{ will die in the interval } (x_i, x_{i+1})\},$$

with $p_i + q_i = 1$.

The use of the terms "risk" and "cause" needs clarification. Both terms may refer to the same condition, but are distinguished here by their position in time relative to the occurrence of death. Prior to death the condition referred to is a risk; after death the same condition is the cause. For example, tuberculosis is a risk of dying to which an individual is exposed, but tuberculosis is also the cause of death if it is the disease from which the individual eventually dies.

In the human population the net and partial crude probabilities cannot be estimated directly, but only through their relations with the crude probability. The study of these relations is part of the problem of "competing risks," or "multiple decrement." The concept of competing risks originated in a controversy over the value of vaccination. Bernoulli, D'Alembert, and Laplace each derived a method of determining the change in population composition that would take place if smallpox were eliminated as a cause of death. It was Makeham [1874], however, who first formulated a theory of decremental forces and explored its practical applications. An interesting account of the development may be found in

Todhunter [1949]. Greville [1948] discussed deterministically multiple decrement tables; Fix and Neyman [1951] studied the problem of competing risks for cancer patients; and Chiang [1961] approached the problem from a stochastic viewpoint. Other interesting studies include those by Berkson and Elvebeck [1960], Berman [1963], Cornfield [1957], and Kimball [1958].

Formulas expressing relations between net and crude probabilities have been developed by assuming either a constant intensity of risk of death (force of mortality) or a uniform distribution of deaths. We shall review these formulas in this chapter, assuming a constant relative intensity. Partial crude probabilities have not received sufficient attention in view of their often indispensable role in studies of cause-specific mortality. Their relations with the corresponding crude probabilities will also be discussed.

2. RELATIONS BETWEEN CRUDE, NET, AND PARTIAL CRUDE PROBABILITIES

Suppose that r risks of death are acting simultaneously on each individual in a population, and let these risks be denoted by $R_1, \ldots, R_r$. For each risk, R_δ, there is a corresponding intensity function (or force of mortality) $\mu(t; \delta)$ such that,

$$\mu(t; \delta)\Delta + o(\Delta) = \Pr\{\text{an individual alive at time } t \text{ will die in interval } (t, t + \Delta) \text{ from cause } R_\delta\}, \qquad \delta = 1, \ldots, r, \tag{2.1}$$

and the sum

$$\mu(t; 1) + \cdots + \mu(t; r) = \mu(t) \tag{2.2}$$

is the total intensity (or the total force of mortality). Although for each risk R_δ the intensity $\mu(t; \delta)$ is a function of time t, we assume that within the time interval (x_i, x_{i+1}) the ratio

$$\frac{\mu(t; \delta)}{\mu(t)} = c_{i\delta} \tag{2.3}$$

is independent of time t, but is a function of the interval (x_i, x_{i+1}) and risk R_δ. Assumption (2.3) permits the risk-specific intensity $\mu(t; \delta)$ to vary in absolute magnitude, but requires that it remain a constant proportion of the total intensity in an interval.

Consider death without specification of cause. The probability that an individual alive at x_i will survive the interval (x_i, x_{i+1}) is

$$p_i = \exp\left\{-\int_{x_i}^{x_{i+1}} \mu(t)\, dt\right\}, \qquad i = 0, 1, \ldots, \tag{2.4}$$

and the probability of his dying in the interval is $q_i = 1 - p_i$ (see Section 1, Chapter 10).

To derive the crude probability of dying from risk R_δ, we consider a point t within the interval (x_i, x_{i+1}). The probability that an individual alive at x_i will die from R_δ in interval $(t, t + dt)$ is

$$\exp\left\{-\int_{x_i}^{t} \mu(\tau)\, d\tau\right\}\mu(t; \delta)\, dt \tag{2.5}$$

where the exponential function is the probability of surviving from x_i to t when all risks are acting, and the factor $\mu(t; \delta)\, dt$ is the instantaneous death probability from cause R_δ in time interval $(t, t + dt)$. Summing (2.5) over all possible values of t, for $x_i < t \le x_{i+1}$, gives the crude probability

$$Q_{i\delta} = \int_{x_i}^{x_{i+1}} \exp\left\{-\int_{x_i}^{t} \mu(\tau)\, d\tau\right\}\mu(t; \delta)\, dt. \tag{2.6}$$

Under assumption (2.3) of a constant relative intensity, (2.6) may be rewritten as

$$Q_{i\delta} = \frac{\mu(t; \delta)}{\mu(t)} \int_{x_i}^{x_{i+1}} \exp\left\{-\int_{x_i}^{t} \mu(\tau)\, d\tau\right\}\mu(t)\, dt. \tag{2.7}$$

Integrating gives

$$Q_{i\delta} = \frac{\mu(t; \delta)}{\mu(t)}\left[1 - \exp\left\{\int_{x_i}^{x_{i+1}} \mu(t)\, dt\right\}\right] = \frac{\mu(t; \delta)}{\mu(t)} q_i; \tag{2.8}$$

hence

$$\frac{\mu(t; \delta)}{\mu(t)} = \frac{Q_{i\delta}}{q_i}, \qquad x_i < t \le x_{i+1}; \qquad \delta = 1, \ldots, r. \tag{2.9}$$

Equation (2.9) is obvious, for if the ratio of the risk-specific intensity to the total intensity is constant throughout an interval, this constant should also be equal to the ratio of the corresponding probabilities of dying over the entire interval. Equations (2.2) and (2.9) imply a trivial equality,

$$Q_{i1} + \cdots + Q_{ir} = q_i, \qquad i = 0, 1, \ldots . \tag{2.10}$$

2.1. Relations Between Crude and Net Probabilities

The net probability of death in the interval (x_i, x_{i+1}) when R_δ is the only operating risk is obviously

$$q_{i\delta} = 1 - \exp\left\{-\int_{x_i}^{x_{i+1}} \mu(t; \delta)\, dt\right\} \tag{2.11}$$

which, in view of (2.3), can be written as

$$q_{i\delta} = 1 - \exp\left\{-\frac{\mu(t; \delta)}{\mu(t)} \int_{x_i}^{x_{i+1}} \mu(t)\, dt\right\} = 1 - p_i^{\mu(t;\delta)/\mu(t)}. \tag{2.12}$$

With Equation (2.9), (2.12) gives the relation between the net and the crude probabilities,

$$q_{i\delta} = 1 - p_i^{Q_{i\delta}/q_i}, \qquad \delta = 1, \ldots, r. \tag{2.13}$$

The net probability of death when risk R_δ is eliminated can be derived in the same way:

$$q_{i\cdot\delta} = 1 - \exp\left\{-\int_{x_i}^{x_{i+1}} [\mu(t) - \mu(t, \delta)]\, dt\right\}$$

$$= 1 - p_i^{(q_i - Q_{i\delta})/q_i}. \tag{2.14}$$

Because of the absence of competing risks, the net probability is always greater than the corresponding crude probability, or

$$q_{i\delta} > Q_{i\delta}. \tag{2.15}$$

Furthermore, if two risks R_δ and R_ε are such that

$$Q_{i\delta} > Q_{i\varepsilon},$$

then

$$q_{i\delta} > q_{i\varepsilon} \quad \text{and} \quad q_{i\cdot\delta} < q_{i\cdot\varepsilon}. \tag{2.16}$$

Verification of (2.15) and (2.16) is left to the reader.

2.2. Relations between Crude and Partial Crude Probabilities

Suppose now that risk R_1 is eliminated from the population. In the presence of all other risks, let $Q_{i\delta\cdot1}$ be the partial crude probability that an individual alive at time x_i will die in the interval (x_i, x_{i+1}) from cause R_δ for $\delta = 2, \ldots, r$. We shall express $Q_{i\delta\cdot1}$ in terms of the probabilities p_i and q_i and of the crude probabilities Q_{i1} and $Q_{i\delta}$. Using the multiplication and addition theorems as in (2.6) we have

$$Q_{i\delta\cdot1} = \int_{x_i}^{x_{i+1}} \exp\left\{-\int_{x_i}^{t} [\mu(\tau) - \mu(\tau; 1)]\, d\tau\right\} \mu(t; \delta)\, dt. \tag{2.17}$$

To simplify (2.17), we note from (2.9) that the ratio $\mu(t;\delta)/[\mu(t)-\mu(t;1)]$ is equal to $Q_{i\delta}/(q_i - Q_{i1})$ and is independent of time t. The partial crude probability may then be rewritten as

$$Q_{i\delta\cdot 1} = \frac{\mu(t;\delta)}{\mu(t)-\mu(t;1)}\int_{x_i}^{x_{i+1}} \exp\left\{-\int_{x_i}^{t}[\mu(\tau)-\mu(\tau;1)]\,d\tau\right\}[\mu(t)-\mu(t;1)]\,dt$$

$$= \frac{Q_{i\delta}}{q_i - Q_{i1}}\left[1-\exp\left\{-\int_{x_i}^{x_{i+1}}[\mu(t)-\mu(t;1)]\,dt\right\}\right]$$

$$= \frac{Q_{i\delta}}{q_i - Q_{i1}}\,q_{i\cdot 1}. \tag{2.18}$$

Substituting (2.14) for $\delta = 1$ in (2.18) gives the final form

$$Q_{i\delta\cdot 1} = \frac{Q_{i\delta}}{q_i - Q_{i1}}\,[1 - p_i^{(q_i - Q_{i1})/q_i}], \qquad \delta = 2, \ldots, r. \tag{2.19}$$

The sum of $Q_{i\delta\cdot 1}$ for $\delta = 2, \ldots, r$ is equal to the net probability of death when risk R_1 is eliminated from the population, and we have

$$\sum_{k=2}^{r} Q_{i\delta\cdot 1} = \sum_{k=2}^{r}\frac{Q_{i\delta}}{q_i - Q_{i1}}\,[1 - p_i^{(q_i - Q_{i1})/q_i}] = 1 - p_i^{(q_i - Q_{i1})/q_i} = q_{i\cdot 1}, \tag{2.20}$$

as might have been anticipated.

Formula (2.19) can be easily generalized to other cases where more than one risk is eliminated. If risks R_1 and R_2 are eliminated, the partial crude probability that an individual alive at time x_i will die from cause R_δ in the interval (x_i, x_{i+1}) is

$$Q_{i\delta\cdot 12} = \frac{Q_{i\delta}}{q_i - Q_{i1} - Q_{i2}}\,[1 - p_i^{(q_i - Q_{i1} - Q_{i2})/q_i}], \qquad \delta = 3, \ldots, r. \tag{2.21}$$

In the discussion of these three types of probability, both q_i and p_i are assumed to be greater than zero but less than unity. If q_i were zero ($p_i = 1$), then $Q_{i\delta}$ would also be zero for $\delta = 1, \ldots, r$. Then the ratios $Q_{i\delta}/q_i$, $Q_{i\delta}/(q_i - Q_{i1})$, and $(q_i - Q_{i1})/q_i$ and, consequently, formulas (2.13), (2.14), (2.19), and (2.21) would become meaningless. In other words, if an individual were certain to survive an interval, it would be meaningless to speak of his chance of dying from a specific risk. On the other hand, if p_i were zero ($q_i = 1$), formula (2.4) shows that the integral

$$\int_{x_i}^{x_{i+1}} \mu(t)\,dt$$

would approach infinity; this, fortunately, is extremely unrealistic.

3. JOINT PROBABILITY DISTRIBUTION OF THE NUMBERS OF DEATHS AND THE NUMBERS OF SURVIVORS

In studies of competing risks in a given population, deaths are classified according to cause, and the number of deaths from each specific cause is the basic random variable for estimating the corresponding probability. As in the cohort life table, we let l_0 be the initial population size at age x_0, and l_i the number of survivors at the beginning of the interval (x_i, x_{i+1}). The number d_i of deaths occurring in each interval is grouped by cause with $d_{i\delta}$ deaths from R_δ so that

$$d_i = d_{i1} + \cdots + d_{ir} \tag{3.1}$$

and

$$l_i = d_{i1} + \cdots + d_{ir} + l_{i+1}, \qquad i = 0, 1, \ldots . \tag{3.2}$$

The probability distribution of the random variables in (3.2) can be readily derived. Given an individual alive at x_i, his probability $Q_{i\delta}$ of dying in (x_i, x_{i+1}) from R_δ, $\delta = 1, \ldots, r$, and his probability p_i of surviving the interval satisfy the condition

$$1 = Q_{i1} + \cdots + Q_{ir} + p_i. \tag{3.3}$$

Therefore, given l_i individuals alive at x_i, the conditional distribution of $d_{i\delta}$ and l_{i+1} is multinomial with the probability distribution[1]

$$\frac{l_i!}{\prod_{\delta=1}^{r} d_{i\delta}!\, l_{i+1}!} \prod_{\delta=1}^{r} Q_{i\delta}^{d_{i\delta}} p_i^{l_{i+1}}, \tag{3.4}$$

and the corresponding p.g.f.

$$g_{d_{i\delta}, l_{i+1}|l_i}(s_{i\delta}, s_{i+1}) = E\left(\prod_{\delta=1}^{r} s_{i\delta}^{d_{i\delta}} s_{i+1}^{l_{i+1}} \,\middle|\, l_i\right) = \left(\sum_{\delta=1}^{r} Q_{i\delta} s_{i\delta} + p_i s_{i+1}\right)^{l_i}, \tag{3.5}$$

where $|s_{i\delta}| \le 1$ and $|s_i| \le 1$.

We can see from (3.4) and (3.5) that for any positive integer u the joint probability distribution of all the random variables $d_{i1}, \ldots, d_{ir}, l_{i+1}$, for $i = 0, 1, \ldots, u$ is

$$\prod_{i=0}^{u} \frac{l_i!}{\prod_{\delta=1}^{r} d_{i\delta}!\, l_{i+1}!} Q_{i1}^{d_{i1}} \cdots Q_{ir}^{d_{ir}} p_i^{l_{i+1}} \tag{3.6}$$

---

[1] For simplicity of formulas, no symbols for the values that the random variables l_i and $d_{i\delta}$ assume are introduced in this chapter.

with $d_{i\delta}$ and l_{i+1} satisfying Equation (3.2). The corresponding p.g.f., defined as

$$G_{d_{i\delta},l_{i+1}|l_0}(s_{i\delta}, s_{i+1}) = E\left(\prod_{i=0}^{u} s_{i1}^{d_{i1}} \cdots s_{ir}^{d_{ir}} s_{i+1}^{l_{i+1}} \,\middle|\, l_0\right), \tag{3.7}$$

can be derived by the same approach used in Chapter 10 to obtain the p.g.f. (2.11). Direct computation gives

$$G_{d_{i\delta},l_{i+1}|l_0}(s_{i\delta}, s_{i+1}) = \left[\sum_{\delta=1}^{r} Q_{0\delta}s_{0\delta} + \sum_{j=1}^{u}\left(\prod_{i=0}^{j-1} p_i s_{i+1}\right)\left(\sum_{\delta=1}^{r} Q_{j\delta}s_{j\delta}\right) + \prod_{i=0}^{u} p_i s_{i+1}\right]^{l_0}. \tag{3.8}$$

We shall not reproduce the derivation of (3.8) but, instead, we shall prove it by induction.

Clearly (3.8) holds true for $u = 0$ since, in this case, (3.8) becomes

$$\left(\sum_{\delta=1}^{r} Q_{0\delta}s_{0\delta} + p_0 s_1\right)^{l_0}, \tag{3.9}$$

which is identical to (3.5) for $i = 0$. Suppose that (3.8) is true for $u - 1$; we want to prove that (3.8) is true also for u. Under this inductive assumption for $u - 1$, we have

$$E\left(\prod_{i=0}^{u-1}\prod_{\delta=1}^{r} s_{i\delta}^{d_{i\delta}} s_{i+1}^{l_{i+1}} \mid l_0\right)$$
$$= \left[\sum_{\delta=1}^{r} Q_{0\delta}s_{0\delta} + \sum_{j=1}^{u-1}\left(\prod_{i=0}^{j-1} p_i s_{i+1}\right)\left(\sum_{\delta=1}^{r} Q_{j\delta}s_{j\delta}\right) + \prod_{i=0}^{u-1} p_i s_{i+1}\right]^{l_0}. \tag{3.10}$$

The p.g.f. (3.7) may be written as

$$G_{d_{i\delta},l_{i+1}|l_0}(s_{i\delta}, s_{i+1}) = E\left[\left\{\prod_{i=0}^{u-1}\left(\prod_{\delta=1}^{r} s_{i\delta}^{d_{i\delta}} s_{i+1}^{l_{i+1}}\right)\right\} E\left\{\prod_{\delta=1}^{r} s_{u\delta}^{d_{u\delta}} s_{u+1}^{l_{u+1}} \,\middle|\, l_u\right\} \,\middle|\, l_0\right] \tag{3.11}$$

since the distribution of $d_{u1}, \ldots, d_{ur}$ and l_{u+1} is dependent only on the number of survivors (l_u) at the beginning of the interval (x_u, x_{u+1}). The conditional expectation inside the brackets in (3.11) is the p.g.f. of the conditional probability distribution of $d_{u1}, \ldots, d_{ur}$ and l_{u+1} given l_u. Therefore, (3.5) may be substituted in (3.11) to give

$$G_{d_{i\delta},l_{i+1}|l_0}(s_{i\delta}, s_{i+1}) = E\left[\left\{\prod_{i=0}^{u-1}\left(\prod_{\delta=1}^{r} s_{i\delta}^{d_{i\delta}} s_{i+1}^{l_{i+1}}\right)\right\}\left\{\sum_{\delta=1}^{r} Q_{u\delta}s_{u\delta} + p_u s_{u+1}\right\} \,\middle|\, l_0\right]$$
$$= E\left[\left\{\prod_{i=0}^{u-2}\left(\prod_{\delta=1}^{r} s_{i\delta}^{d_{i\delta}} s_{i+1}^{l_{i+1}}\right)\right\}\left\{\prod_{\delta=1}^{r} s_{u-1,\delta}^{d_{u-1}}{}^{i+1} z_u^{l_u}\right\} \,\middle|\, l_0\right] \tag{3.12}$$

where

$$z_u = s_u\left\{\sum_{\delta=1}^{r} Q_{u\delta}s_{u\delta} + p_u s_{u+1}\right\} \tag{3.13}$$

is less than unity in absolute value. Because of formula (3.10), (3.12) becomes

$$G_{d_{i\delta},l_{i+1}|l_0}(s_{i\delta}, s_{i+1}) = \left[\sum_{\delta=1}^{r} Q_{0\delta}s_{0\delta} + \sum_{j=1}^{u-1}\left(\prod_{i=0}^{j-1} p_i s_{i+1}\right)\left(\sum_{\delta=1}^{r} Q_{j\delta}s_{j\delta}\right) + \left(\prod_{i=0}^{u-2} p_i s_{i+1}\right) p_{u-1}z_u\right]^{l_0}. \tag{3.14}$$

Using (3.13), we rewrite the last term inside the brackets in the form

$$\begin{aligned}\left(\prod_{i=0}^{u-2} p_i s_{i+1}\right) p_{u-1}z_u &= \left(\prod_{i=0}^{u-2} p_i s_{i+1}\right) p_{u-1}s_u\left(\sum_{\delta=1}^{r} Q_{u\delta}s_{u\delta} + p_u s_{u+1}\right)\\ &= \left(\prod_{i=0}^{u-1} p_i s_{i+1}\right)\left(\sum_{\delta=1}^{r} Q_{u\delta}s_{u\delta}\right) + \prod_{i=0}^{u} p_i s_{i+1}.\end{aligned} \tag{3.15}$$

When (3.15) is substituted in (3.14), the resulting expression becomes identical to (3.8) and the proof is complete.

The generating function (3.8) assumes a value of unity at the point $(s_{i1}, \ldots, s_{ir}, s_{i+1}) = (1, \ldots, 1, 1)$ for $i = 0, 1, \ldots, u$,

$$\begin{aligned}G_{d_{i\delta},l_{i+1}|l_0}(1, 1) &= \left[\sum_{\delta=1}^{r} Q_{0\delta} + \sum_{j=1}^{u}\left(\prod_{i=0}^{j-1} p_i\right)\left(\sum_{\delta=1}^{r} Q_{j\delta}\right) + \prod_{i=0}^{u} p_i\right]^{l_0}\\ &= (q_0 + p_0q_1 + \cdots + p_0p_1\cdots p_{u-1}q_u + p_0p_1\cdots p_u)^{l_0} = 1.\end{aligned} \tag{3.16}$$

The sum of the probabilities (3.6) over all possible values of the random variables is thus unity, as it should be.

The moments of the random variables $d_{i\delta}$ and l_i are easily computed from the p.g.f. (3.8). Taking the first derivative of (3.8), we have the expectations

$$E(d_{i\delta} \mid l_0) = l_0 p_{0i}Q_{i\delta}, \qquad \delta = 1, \ldots, r, \tag{3.17}$$

and

$$E(l_i \mid l_0) = l_0 p_{0i}, \qquad i = 0, 1, \ldots. \tag{3.18}$$

Here $p_{0i} = p_0p_1\cdots p_{i-1}$ is the probability that an individual alive at age x_0 will survive to x_i, and $p_{0i}Q_{i\delta}$ is the probability that the individual will die in interval (x_i, x_{i+1}) from cause R_δ. From the second derivatives we have the variances and covariances

$$\sigma^2_{d_{i\delta}|l_0} = l_0 p_{0i}Q_{i\delta}(1 - p_{0i}Q_{i\delta}) \tag{3.19}$$

and

$$\sigma_{d_{i\delta},d_{j\varepsilon}|l_0} = -l_0 p_{0i} Q_{i\delta} p_{0j} Q_{j\varepsilon}, \qquad \delta \neq \varepsilon;\ \delta, \varepsilon = 1, \ldots, r;$$
$$i, j = 0, 1, \ldots. \qquad (3.20)$$

The covariance between the number dying and the number surviving can be derived in a similar way:

$$\sigma_{d_{i\delta},l_j|l_0} = -l_0 p_{0i} Q_{i\delta} p_{0j} \qquad (3.21)$$

and

$$\sigma_{l_i,d_{j\delta}|l_0} = l_0(1 - p_{0i}) p_{0j} Q_{j\delta}, \qquad \delta = 1, \ldots, r;$$
$$i < j;\ i, j = 0, 1, \ldots. \qquad (3.22)$$

It is interesting to see that the covariance between l_i and $d_{j\delta}$ in (3.22) is of a different form from those in (3.20) and (3.21). The positive covariance in (3.22) indicates that the larger the number of survivors at age x_i, the greater the probability that a large number of deaths from R_δ will occur in a subsequent interval (x_j, x_{j+1}). The covariance between l_i and l_j for $i \leq j$

$$\sigma_{l_i,l_j|l_0} = l_0(1 - p_{0i}) p_{0j} \qquad (3.23)$$

has been given in (2.14) of Chapter 10. These results show that, *for each u, the random variables $d_{i\delta}$, l_{i+1} for $i = 0, 1, \ldots, u$; $\delta = 1, \ldots, r$ form a chain of multinomial distributions with the probability distribution and the* p.g.f. *given in* (3.6) *and* (3.8), *respectively.* ▶

4. ESTIMATION OF THE CRUDE, NET, AND PARTIAL CRUDE PROBABILITIES

Having discussed in preceding sections the relations among the crude, net, and partial crude probabilities and the joint probability distributions of the basic random variables $d_{i\delta}$ and l_i, it is now possible to derive estimators of the probabilities. First we estimate the crude probabilities; we then use these estimators and the relations among the different probabilities to estimate the net and partial crude probabilities.

The estimators of the crude probabilities $Q_{i\delta}$ and p_i can be derived directly from the joint probability function

$$\prod_{i=0}^{u} \frac{l_i!}{d_{i1}! \cdots d_{ir}!\, l_{i+1}!} Q_{i1}^{d_{i1}} \cdots Q_{ir}^{d_{ir}} p_i^{l_{i+1}} \qquad (3.6)$$

by using the maximum likelihood principle. To familiarize the reader with another important method of estimation, however, we shall use Neyman's reduced χ^2 method (Neyman [1949]).

The random variables in the present case are $d_{i\delta}$ and l_{i+1}, with the expectations

$$E(d_{i\delta} \mid l_0) = l_0 p_{0i} Q_{i\delta}, \qquad \delta = 1, \ldots, r; \quad i = 0, 1, \ldots, u, \tag{3.17}$$

and

$$E(l_{i+1} \mid l_0) = l_0 p_{0i} p_i = l_0 p_{0i}\left(1 - \sum_{\delta=1}^{r} Q_{i\delta}\right), \qquad \delta = 1, \ldots, r; \quad i = 0, 1, \ldots, u. \tag{3.18}$$

According to Neyman's theory, "best" estimators of $Q_{i\delta}$ minimize the reduced form of the χ^2:

$$\chi_0^{\,2} = \sum_{i=0}^{u}\left[\sum_{\delta=1}^{r}\frac{(d_{i\delta} - l_0 p_{0i} Q_{i\delta})^2}{d_{i\delta}} + \frac{\left\{l_{i+1} - l_0 p_{0i}\left(1 - \sum_{\delta=1}^{r} Q_{i\delta}\right)\right\}^2}{l_{i+1}}\right]. \tag{4.1}$$

The reduced χ^2 (4.1) differs from its ordinary form in that the random variable, rather than its expectation, appears in the denominator. The substitution of

$$p_i = 1 - \sum_{\delta=1}^{r} Q_{i\delta} \tag{4.2}$$

in (4.1) makes it necessary to derive only the minimizing values $\hat{Q}_{i\delta}$. The estimator $\hat{p}_i$ is the complement of the sum of $\hat{Q}_{i\delta}$, and will appear automatically as a by-product. Taking the derivative of $\chi_0^{\,2}$ with respect to $Q_{i\delta}$ and setting the derivatives equal to zero lead to the equations

$$\frac{\hat{Q}_{i\delta}}{d_{i\delta}} = \frac{\hat{p}_i}{l_{i+1}}, \qquad \delta = 1, \ldots, r. \tag{4.3}$$

For each i, we sum (4.3) over δ to get

$$\frac{\hat{p}_i}{l_{i+1}} = \frac{\sum_{\delta=1}^{r}\hat{Q}_{i\delta} + \hat{p}_i}{\sum_{\delta=1}^{r} d_{i\delta} + l_{i+1}} = \frac{1}{l_i}. \tag{4.4}$$

Substituting (4.4) in (4.3) gives the desired estimators

$$\hat{p}_i = \frac{l_{i+1}}{l_i} \tag{4.5}$$

and

$$\hat{Q}_{i\delta} = \frac{d_{i\delta}}{l_i}. \tag{4.6}$$

In writing the reduced χ^2 (4.1), we understand that if for some age x_w all the l_w individuals alive at x_w die during the interval (x_w, x_{w+1}), then $d_{i\delta} = 0$ and $l_i = 0$ for all $i > w$, so that there is no contribution to the reduced χ^2 beyond the wth term. Consequently the estimators (4.5) and (4.6) are defined only for positive l_i.

The estimators in (4.5) and (4.6) are recognized as maximizing values of the joint probability (3.6) (see Problem 7). Thus, in this case, Neyman's reduced χ^2 method yields estimators identical to those obtained by the maximum likelihood principle. Using the same argument as that in Section 4 of Chapter 10, we can show that the estimators $\hat{p}_i$ and $\hat{Q}_{i\delta}$ in (4.5) and (4.6) are the unique, unbiased, efficient estimators of the corresponding probabilities.

The exact formulas for variances and covariances of the estimators in (4.5) and (4.6) may be obtained by direct computation. They are

$$\operatorname{Var}(\hat{Q}_{i\delta} \mid l_0) = E\left(\frac{1}{l_i}\right) Q_{i\delta}(1 - Q_{i\delta}), \tag{4.7}$$

$$\sigma^2_{\hat{p}_i \mid l_0} = \sigma^2_{\hat{q}_i \mid l_0} = E\left(\frac{1}{l_i}\right) p_i q_i, \tag{4.8}$$

$$\begin{aligned} \operatorname{Cov}(\hat{Q}_{i\delta}, \hat{Q}_{j\varepsilon} \mid l_0) &= -E\left(\frac{1}{l_i}\right) Q_{i\delta} Q_{i\varepsilon} \qquad & j = i \\ &= 0 & j \neq i \end{aligned} \tag{4.9}$$

$$\operatorname{Cov}(\hat{p}_i, \hat{p}_j \mid l_0) = \operatorname{Cov}(\hat{q}_i, \hat{q}_j \mid l_0) = 0 \qquad j \neq i \tag{4.10}$$

and

$$\begin{aligned} \operatorname{Cov}(\hat{p}_i, \hat{Q}_{j\delta} \mid l_0) &= -E\left(\frac{1}{l_i}\right) p_i Q_{i\delta} \qquad & j = i \\ &= 0 & j \neq i. \end{aligned} \tag{4.11}$$

When the original cohort l_0 is large, the expectation of the reciprocal of l_i may be approximated by the reciprocal of the expectation $E(l_i)$ [see Equation (4.17) in Chapter 10].

Formulas for the estimators of the net and partial crude probabilities may be derived by using their relations with the crude probabilities in Section 2. Substituting (4.5) and (4.6) in the formulas (2.13), (2.14),

(2.19), and (2.21) gives the following estimators:

$$\hat{q}_{i\delta} = 1 - \left(\frac{l_{i+1}}{l_i}\right)^{d_{i\delta}/d_i}, \qquad \delta = 1, \ldots, r; \tag{4.12}$$

$$\hat{q}_{i\cdot\delta} = 1 - \left(\frac{l_{i+1}}{l_i}\right)^{(d_i - d_{i\delta})/d_i}, \qquad \delta = 1, \ldots, r; \tag{4.13}$$

$$\hat{Q}_{i\delta\cdot 1} = \frac{d_{i\delta}}{d_i - d_{i1}}\left[1 - \left(\frac{l_{i+1}}{l_i}\right)^{(d_i - d_{i1})/d_i}\right], \qquad \delta = 2, \ldots, r; \tag{4.14}$$

$$\hat{Q}_{i\delta\cdot 12} = \frac{d_{i\delta}}{d_i - d_{i1} - d_{i2}}\left[1 - \left(\frac{l_{i+1}}{l_i}\right)^{(d_i - d_{i1} - d_{i2})/d_i}\right], \qquad \begin{array}{l}\delta = 3, \ldots, r; \\ i = 0, 1, \ldots, u.\end{array} \tag{4.15}$$

For each i, the sum of the estimators $\hat{Q}_{i\delta\cdot 1}$ over δ is equal to $\hat{q}_{i\cdot 1}$:

$$\sum_{\delta=2}^{r} \hat{Q}_{i\delta\cdot 1} = \sum_{\delta=2}^{r} \frac{d_{i\delta}}{d_i - d_{i1}}\left[1 - \left(\frac{l_{i+1}}{l_i}\right)^{(d_i - d_{i1})/d_i}\right]$$

$$= 1 - \left(\frac{l_{i+1}}{l_i}\right)^{(d_i - d_{i1})/d_i} = \hat{q}_{i\cdot 1}. \tag{4.16}$$

Although formulas (4.12) to (4.15) for the estimators of the net and partial crude probabilities are quite simple, the exact formulas for their variances and covariances are difficult to obtain. Approximate formulas, however, can be developed [see Equation (5.30) in Chapter1]. For example, for the variance of the estimator $\hat{q}_{i\delta}$, we write

$$\text{Var}(\hat{q}_{i\delta} \mid l_0) = \left\{\frac{\partial}{\partial Q_{i\delta}} q_{i\delta}\right\}^2 \text{Var}(\hat{Q}_{i\delta} \mid l_0) + \left\{\frac{\partial}{\partial \hat{p}_i} \hat{q}_{i\delta}\right\}^2 \text{Var}(\hat{p}_i \mid l_0)$$

$$+ 2\left\{\frac{\partial}{\partial \hat{Q}_{i\delta}} \hat{q}_{i\delta}\right\}\left\{\frac{\partial}{\partial \hat{p}_i} \hat{q}_{i\delta}\right\} \text{Cov}(\hat{Q}_{i\delta}, \hat{p}_i \mid l_0) \tag{4.17}$$

where the partial derivatives are taken at the true point $(Q_{i\delta}, p_i)$. Using formulas (4.7), (4.8), and (4.11) for the variances and covariances and simplifying, we have

$$\text{Var}(\hat{q}_{i\delta} \mid l_0) = E\left(\frac{1}{l_i}\right)\frac{(1 - q_{i\delta})^2}{p_i q_i}[p_i \log(1 - q_{i\delta}) \log(1 - q_{i.\delta}) + Q_{i\delta}^2]. \tag{4.18}$$

When $E(1/l_i)$ is approximated by $1/E(l_i)$, (4.18) becomes

$$\operatorname{Var}(\hat{q}_{i\delta} \mid l_0) = \frac{(1-q_{i\delta})^2}{l_0 p_{0i} p_i q_i} [p_i \log(1-q_{i\delta}) \log(1-q_{i.\delta}) + Q_{i\delta}^2]. \quad (4.19)$$

Using the same approach, we obtain the covariances

$$\operatorname{Cov}(\hat{q}_{i\delta}, \hat{q}_{i\varepsilon} \mid l_0) = \frac{(1-q_{i\delta})(1-q_{i\varepsilon})}{l_0 p_{0i} p_i q_i} [p_i \log(1-q_{i\delta}) \log(1-q_{i\varepsilon}) - Q_{i\delta} Q_{i\varepsilon}] \quad (4.20)$$

and

$$\operatorname{Cov}(\hat{q}_{i\delta}, \hat{p}_i \mid l_0) = -\frac{(1-q_{i\delta}) Q_{i\delta}}{l_0 p_{0i}}. \quad (4.21)$$

Approximate formulas for the variances and covariances of $\hat{q}_{x\cdot\delta}$ and $\hat{Q}_{x\delta\cdot 1}$ may be obtained in the same way. For completeness they are listed below.

$$\operatorname{Var}(\hat{q}_{i.\delta} \mid l_0) = \frac{(1-q_{i\cdot\delta})^2}{l_0 p_{0i} p_i q_i} [p_i \log(1-q_{i\delta}) \log(1-q_{i\cdot\delta}) + (q_i - Q_{i\delta})^2], \quad (4.22)$$

$$\operatorname{Cov}(\hat{q}_{i.\delta}, \hat{q}_{i\cdot\varepsilon} \mid l_0) = -\frac{(1-q_{i\cdot\delta})(1-q_{i\cdot\varepsilon})}{l_0 p_{0i} p_i q_i} \times [p_i \log(1-q_{i\delta}) \log(1-q_{i\varepsilon}) - (q_i - Q_{i\delta})(q_i - Q_{i\varepsilon})], \quad \delta \neq \varepsilon; \quad (4.23)$$

$$\operatorname{Cov}(\hat{q}_{i.\delta}, \hat{p}_i \mid l_0) = -\frac{(1-q_{i\cdot\delta})(q_i - Q_{i\delta})}{l_0 p_{0i}}, \quad (4.24)$$

$$\operatorname{Var}(\hat{Q}_{i\delta.1} \mid l_0) = \frac{q_i - Q_{i1} - Q_{i\delta}}{l_0 p_{0i}(q_i - Q_{i1}) Q_{i\delta}} Q_{i\delta\cdot 1}^2 + \frac{[Q_{i\delta\cdot 1}(q_i - Q_{i1}) - Q_{i\delta}]^2}{l_0 p_{0i} p_i q_i (q_i - Q_{i1})} \left[(q_i - Q_{i1}) + Q_{i1} p_i \left(\frac{\log p_i}{q_i} \right)^2 \right], \quad (4.25)$$

$$\operatorname{Cov}(\hat{Q}_{i\delta.1}, \hat{Q}_{i\varepsilon.1} \mid l_0) = -\frac{Q_{i\delta\cdot 1} Q_{i\varepsilon\cdot 1}}{l_0 p_{0i}(q_i - Q_{i1})} + \frac{[Q_{i\delta\cdot 1}(q_i - Q_{i1}) - Q_{i\delta}][Q_{i\varepsilon\cdot 1}(q_i - Q_{i1}) - Q_{i\varepsilon}]}{l_0 p_{0i} p_i q_i (q_i - Q_{i1})} \times \left[(q_i - Q_{i1}) + Q_{i1} p_i \left(\frac{\log p_i}{q_i} \right)^2 \right], \quad \delta \neq \varepsilon; \quad (4.26)$$

$$\operatorname{Cov}(\hat{Q}_{i\delta.1}, \hat{p}_i \mid l_0) = -\frac{[Q_{i\delta} - Q_{i\delta\cdot 1}(q_i - Q_{i1})]}{l_0 p_{0i}}, \quad \delta = 2, \ldots, r, \quad i = 0, 1, \ldots. \quad (4.27)$$

Since the sum of the estimators $\hat{Q}_{i\delta\cdot 1}$ for $\delta = 2, \ldots, r$ is equal to $\hat{q}_{i\cdot 1}$ [Equation (4.16)], the following relations hold true for the variances and covariances:

$$\sum_{\delta=2}^{r} \sigma^2_{\hat{Q}_{i\delta\cdot 1} \mid l_0} + 2 \sum_{\delta=2}^{r-1} \sum_{\varepsilon=\delta+1}^{r} \operatorname{Cov}(\hat{Q}_{i\delta\cdot 1}, \hat{Q}_{i\varepsilon\cdot 1} \mid l_0) = \sigma^2_{\hat{q}_{i\cdot 1} \mid l_0} \tag{4.28}$$

and

$$\sum_{\delta=2}^{r} \operatorname{Cov}(\hat{Q}_{i\delta\cdot 1}, \hat{p}_i \mid l_0) = \sigma_{\hat{q}_{i\cdot 1}, \hat{p}_i \mid l_0}. \tag{4.29}$$

When (4.22), (4.24), (4.25), (4.26), and (4.27) are substituted into (4.28) and (4.29), we have two identities which can be verified by direct computation.

5. APPLICATION TO CURRENT MORTALITY DATA

When current population data on causes of death are used to study mortality, the estimators of the various probabilities are somewhat different from those presented in the last section. To derive these estimators, let us first reintroduce the symbols used in Chapter 9. For age interval (x_i, x_{i+1}), we let $n_i = x_{i+1} - x_i$ be the length of the interval, P_i the midyear population, D_i the number of deaths occurring during the calendar year, a_i the average fraction of the interval lived by each of the D_i individuals, and N_i the number of people alive at x_i among whom D_i deaths occur. The age-specific death rate is given by

$$M_i = \frac{D_i}{P_i}, \tag{5.1}$$

and the probability of dying in the interval (x_i, x_{i+1}) is estimated from

$$\hat{q}_i = \frac{D_i}{N_i}. \tag{5.2}$$

When N_i is expressed in terms of P_i and D_i [cf. Equation (3.5) in Chapter 9]

$$N_i = [P_i + (1 - a_i)n_i D_i]/n_i, \tag{5.3}$$

equation (5.2) becomes

$$\hat{q}_i = \frac{n_i M_i}{1 + (1 - a_i)n_i M_i}, \tag{5.4}$$

as given in equation (4.3) in Chapter 9, and its complement is

$$\hat{p}_i = \frac{1 - a_i n_i M_i}{1 + (1 - a_i)n_i M_i}. \tag{5.5}$$

The D_i deaths are now further divided according to cause with $D_{i\delta}$ dying from cause R_δ, $\delta = 1, \ldots, r$, and

$$D_i = D_{i1} + \cdots + D_{ir} \tag{5.6}$$

so that

$$M_{i\delta} = \frac{D_{i\delta}}{P_i} \tag{5.7}$$

are age-cause-specific death rates. The estimator of the crude probability of dying from R_δ in the presence of competing risks is obviously

$$\hat{Q}_{i\delta} = \frac{D_{i\delta}}{N_i} \tag{5.8}$$

which is the analogue to Equation (4.6). Substituting (5.3) in (5.8) gives the required formula

$$\hat{Q}_{i\delta} = \frac{n_i M_{i\delta}}{1 + (1 - a_i) n_i M_i}, \qquad \begin{matrix} \delta = 1, \ldots, r; \\ i = 0, 1, \ldots . \end{matrix} \tag{5.9}$$

The net and partial probabilities can be estimated by using formulas (5.4), (5.5), (5.9), and the relations (2.13), (2.14), (2.19), and (2.21) established in Section 2. Simple computations give the estimators of the net probabilities

$$\hat{q}_{i\delta} = 1 - \hat{p}_i^{D_{i\delta}/D_i}, \tag{5.10}$$

$$\hat{q}_{i\cdot\delta} = 1 - \hat{p}_i^{(D_i - D_{i\delta})/D_i}, \qquad \delta = 1, \ldots, r; \tag{5.11}$$

and those of the partial crude probabilities

$$\hat{Q}_{i\delta\cdot 1} = \frac{D_{i\delta}}{D_i - D_{i1}} [1 - \hat{p}_i^{(D_i - D_{i1})/D_i}], \qquad \delta = 2, \ldots, r; \tag{5.12}$$

and

$$\hat{Q}_{i\delta\cdot 12} = \frac{D_{i\delta}}{D_i - D_{i1} - D_{i2}} [1 - \hat{p}_i^{(D_i - D_{i1} - D_{i2})/D_i}], \qquad \begin{matrix} \delta = 3, \ldots, r; \\ i = 0, 1, \ldots . \end{matrix} \tag{5.13}$$

Here $\hat{p}_i$ is given in (5.5) and the ratio $D_{i\delta}/D_i$ is the proportionate mortality.

The approximate formulas for sample variance and covariance of the estimators $\hat{Q}_{i\delta}$, $\hat{q}_{i\delta}$, $\hat{q}_{i\cdot\delta}$, $\hat{Q}_{i\delta\cdot 1}$, and $\hat{Q}_{i\delta\cdot 12}$ can be derived from formulas (2.19) to (4.27) by replacing the probabilities with the corresponding estimators and by the substitution of $l_0 p_{0i} = N_i$ or, in view of (5.3),

$$\frac{1}{l_0 p_{0i}} = \frac{n_i}{P_i + (1 - a_i) n_i D_i}. \tag{5.14}$$

For example, the sample variance of the estimator $\hat{q}_{i\cdot\delta}$ is

$$S^2_{\hat{q}_{i.\delta}} = \frac{n_i(1-\hat{q}_{i\cdot\delta})^2}{[P_i + (1-a_i)n_iD_i]\hat{p}_i\hat{q}_i} \times [\hat{p}_i \log(1-\hat{q}_{i\delta})\log(1-\hat{q}_{i\cdot\delta}) + (\hat{q}_i - \hat{Q}_{i\delta})^2] \qquad \delta = 1, \ldots, r; \quad i = 0, 1, \ldots. \tag{5.15}$$

Using the estimators of the probabilities derived in this chapter, we can construct cohort or current life tables for a population with respect to a particular cause of death. For example, we consider the case where risk R_δ is eliminated from a population and give formulas for two main life table functions, the proportion of survivors and the observed expectation of life. Here the basic quantity is the estimator $\hat{q}_{i\cdot\delta}$ of the net probability of dying in (x_i, x_{i+1}) when R_δ is eliminated. The proportion of people alive at an age x_α who will survive to age x_j is computed from

$$\hat{p}_{\alpha j\cdot\delta} = \prod_{i=\alpha}^{j-1}(1-\hat{q}_{i\cdot\delta}) \tag{5.16}$$

with the sample variance

$$S^2_{\hat{p}_{\alpha j.\delta}} = \hat{p}^2_{\alpha j.\delta}\sum_{i=\alpha}^{j-1}(1-\hat{q}_{i\cdot\delta})^{-2}S^2_{\hat{q}_{i.\delta}}, \qquad \alpha < j; \quad \alpha, j = 0, 1, \ldots. \tag{5.17}$$

When $\alpha = 0$ the quantity in formula (5.16) is equivalent to l_j/l_0.

The formula for the observed expectation of life at age x_α when risk R_δ is eliminated is given by

$$\hat{e}_{\alpha\cdot\delta} = a_\alpha n_\alpha + \sum_{j\geqq\alpha} c_j\hat{p}_{\alpha j\cdot\delta}, \tag{5.18}$$

where $c_j = (1-a_{j-1})n_{j-1} + a_jn_j$, and the sample variance is

$$S^2_{\hat{e}_{\alpha\cdot\delta}} = \sum_{j\geqq\alpha}\hat{p}^2_{\alpha j\cdot\delta}[\hat{e}_{j+1\cdot\delta} + (1-a_j)n_j]^2S^2_{\hat{q}_{j\cdot\delta}}, \qquad \alpha = 0, 1, \ldots. \tag{5.19}$$

Since $\hat{q}_{i\cdot\delta}$ is less than $\hat{q}_i$, the quantity $\hat{p}_{\alpha j\cdot\delta}$ in (5.16) is greater than the corresponding proportion $\hat{p}_{\alpha j}$ of survivors when R_δ is operating in the population; and $\hat{e}_{\alpha\cdot\delta}$ in (5.18) is greater than the corresponding observed expectation of life $\hat{e}_\alpha$ when all risks are acting, or

$$\hat{e}_{\alpha\cdot\delta} - \hat{e}_\alpha > 0. \tag{5.20}$$

The difference between $\hat{e}_{\alpha\cdot\delta}$ and $\hat{e}_\alpha$ in (5.20) is the additional years of life that an individual of age x_α could expect to live if risk R_δ were eliminated, or the years of life lost to an individual of age x_α due to the presence of risk R_δ.

Example. Cardiovascular-renal (CVR) diseases have caused more deaths in the human population than any other disease. As a group, they were responsible for over 55 per cent of all deaths in the United States in recent years. To evaluate the impact of these diseases on human longevity, we can compare the mortality and survival experience of the current population with the hypothetical experience of the same population under conditions that would exist if CVR diseases were removed as causes of death. The theory of competing risks provides convenient methods for the analysis of such a problem. The basic quantity needed for this purpose is the net probability ($q_{i.1}$) that an individual alive at age x_i would die before age x_{i+1} if CVR diseases (R_1) were eliminated as a risk of death.

Table 1 contains the required data for illustration: population sizes, number of deaths from all causes, and deaths from CVR diseases by age for United States white males and females, in the year 1960. From these data two life tables were constructed: one based on $\hat{q}_i$ using formula (5.4), and the other on $\hat{q}_{i.1}$ using formula (5.11). The main life table columns are reproduced in Tables 2 to 4, each reflecting in a different way the effect of eliminating CVR diseases as a cause of death.

Table 2 gives a comparison between $\hat{q}_i$ and $\hat{q}_{i.1}$. The difference $\hat{q}_i - \hat{q}_{i.1}$ is the reduction in the probability of dying in age interval (x_i, x_{i+1}) if CVR diseases were eliminated as a risk of death or, alternatively, the excess probability of dying due to the presence of these diseases. This difference, while not pronounced below age 30, advances with age at an accelerated rate. If the effect of CVR diseases were removed, the reduction in the probability of dying for white males is 32% of the existing probability for age interval 35 to 40, over 50% for interval 50 to 55, and about 60% for interval 70 to 75. The estimated probabilities $\hat{q}_i$ and $\hat{q}_{i.1}$ and their difference $\hat{q}_i - \hat{q}_{i.1}$ are lower for females than for males. From age 30 to age 70 the relative reduction in the probability of dying is also lower for females than for males, but the reverse is true for other ages.

The impact of CVR diseases on the probability of survival and the expectation of life are shown in Tables 3 and 4. The quantities $\hat{p}_{0i.1}$ were computed from formula (5.16) and $\hat{e}_{i.1}$ from (5.18). Table 4 shows that these diseases cause an average loss of 12 years in the expectation of life for white males under age 50 and 13 years for females under age 65. At older ages, the loss in the expectation of life decreases slightly in absolute

Table 1. Population, deaths from all causes, and deaths from major cardiovascular-renal diseases[c] by age (white males and females, United States 1960)

Age Interval (in years) x_i to x_{i+1}	White Males: Midyear Population[a] P_i	White Males: Deaths from All Causes[b] D_i	White Males: Deaths from Cardio-vascular-Renal Diseases[c b] D_{i1}	White Females: Midyear Population[a] P_i	White Females: Deaths from All Causes[b] D_i	White Females: Deaths from Cardio-vascular-Renal Diseases[c b] D_{i1}
0–1	1,794,784	48,063	228	1,719,014	34,416	132
1–5	7,063,044	7,409	153	6,788,716	5,791	108
5–10	8,191,158	4,408	177	7,876,239	3,019	121
10–15	7,488,562	3,847	208	7,188,766	2,214	196
15–20	5,893,946	7,308	355	5,772,421	2,901	273
20–25	4,657,470	7,755	481	4,822,377	2,916	414
25–30	4,725,480	7,182	768	4,839,982	3,459	602
30–35	5,216,424	9,039	1,808	5,379,640	5,215	955
35–40	5,461,528	13,803	4,444	5,708,902	8,396	1,727
40–45	5,094,821	21,336	9,125	5,298,273	12,622	3,252
45–50	4,850,486	34,247	16,796	4,988,493	18,268	5,405
50–55	4,314,976	50,716	26,812	4,462,148	24,696	8,726
55–60	3,774,623	66,540	36,907	3,986,830	32,340	13,913
60–65	3,100,045	85,890	49,649	3,419,584	46,710	24,216
65–70	2,637,044	108,726	65,609	3,031,276	65,834	38,539
70–75	1,972,947	119,269	75,371	2,339,916	85,018	55,304
75–80	1,214,577	109,193	73,057	1,545,378	96,133	67,579
80–85	591,251	83,885	58,713	829,736	91,771	68,073
85–90	235,566	49,502	36,133	367,538	65,009	49,817
90–95	56,704	18,253	13,604	100,350	29,407	22,826
95+	12,333	4,219	3,136	24,331	8,173	6,356
Not stated		267	106		170	99
Total	78,347,769	860,857	473,640	80,489,910	644,478	368,633

[a] U.S. Census of Population 1960, U.S. Summary, Detailed Characteristics, Table 156. Bureau of the Census, U.S. Department of Commerce.

[b] Vital Statistics of the U.S. 1960, Vol. II, Part A, Table 5-11. National Center for Health Statistics, U.S. Department of Health, Education and Welfare.

[c] Category Numbers 330–334, 400–468, and 592–594 of the 1955 Revision of the *International Classification of Diseases, Injuries, and Causes of Death*, World Health Organization, 1957.

value but increases spectacularly relative to the existing life expectancy. If CVR diseases were eliminated as a risk of death, a white male could expect an additional length of life of 30% over the present life expectancy at 30 years of age, 50% at age 50, and 100% at age 70.

Further inference using the life table quantities may be made by computing the corresponding sample variances. But we leave the details to the reader.

Table 2. The probability of dying and the effect of eliminating CVR disease as a risk of death in each age interval (white males and females, United States 1960)

Age Interval (in years) x_i to x_{i+1}	White Males				White Females			
	CVR Present	CVR Eliminated	Difference		CVR Present	CVR Eliminated	Difference	
	$\hat{q}_i$	$\hat{q}_{i\cdot 1}$	$\hat{q}_i - \hat{q}_{i\cdot 1}$	$\frac{\hat{q}_i - \hat{q}_{i\cdot 1}}{\hat{q}_i}$	$\hat{q}_i$	$\hat{q}_{i\cdot 1}$	$\hat{q}_i - \hat{q}_{i\cdot 1}$	$\frac{\hat{q}_i - \hat{q}_{i\cdot 1}}{\hat{q}_i}$
(1)	(2)	(3)	(4)	(5)	(6)	(7)	(8)	(9)
0–1	.02615	.02603	.00012	0.5%	.01967	.01959	.00008	0.4%
1–5	.00419	.00410	.00009	2.1	.00341	.00334	.00007	2.1
5–10	.00269	.00258	.00011	4.1	.00191	.00184	.00007	3.7
10–15	.00257	.00243	.00014	5.4	.00154	.00140	.00014	9.1
15–20	.00618	.00588	.00030	4.9	.00251	.00227	.00024	9.6
20–25	.00829	.00778	.00051	6.2	.00302	.00259	.00043	14.2
25–30	.00757	.00676	.00081	10.7	.00357	.00295	.00062	17.4
30–35	.00863	.00691	.00172	19.9	.00484	.00395	.00089	18.4
35–40	.01256	.00854	.00402	32.0	.00733	.00583	.00150	20.5
40–45	.02074	.01192	.00882	42.5	.01185	.00881	.00304	25.7
45–50	.03474	.01785	.01689	48.6	.01816	.01282	.00534	29.4
50–55	.05719	.02737	.02982	52.1	.02732	.01775	.00957	35.0
55–60	.08456	.03858	.04598	54.4	.03978	.02287	.01691	42.5
60–65	.12989	.05702	.07287	56.1	.06613	.03241	.03372	51.0
65–70	.18759	.07908	.10851	57.8	.10321	.04416	.05905	57.2
70–75	.26327	.10636	.15691	59.6	.16682	.06179	.10503	63.0
75–80	.36837	.14106	.22731	61.7	.26990	.08920	.18070	67.0
80–85	.51822	.19679	.32143	62.0	.42950	.13492	.29458	68.6
85–90	.66589	.25627	.40962	61.5	.59498	.19040	.40458	68.0
90–95	.82555	.35901	.46654	56.5	.78586	.29170	.49416	62.9
95+	1.00000	1.00000	0	0	1.00000	1.00000	0	0

Table 3. The probability of survival and the effect of eliminating CVR disease as a risk of death (white males and females, United States 1960)

Age Interval in (years) x_i to x_{i+1}	White Males				White Females			
	CVR Present	CVR Eliminated	Difference		CVR Present	CVR Eliminated	Difference	
	$\hat{p}_{0i}$	$\hat{p}_{0i\cdot1}$	$\hat{p}_{0i\cdot1} - \hat{p}_{0i}$	$\frac{\hat{p}_{0i\cdot1} - \hat{p}_{0i}}{\hat{p}_{0i}}$	$\hat{p}_{0i}$	$\hat{p}_{0i\cdot1}$	$\hat{p}_{0i\cdot1} - \hat{p}_{0i}$	$\frac{\hat{p}_{0i\cdot1} - \hat{p}_{0i}}{\hat{p}_{0i}}$
(1)	(2)	(3)	(4)	(5)	(6)	(7)	(8)	(9)
0–1	1.00000	1.00000	.00000	0.0%	1.00000	1.00000	1.00000	0.0%
1–5	.97385	.97397	.00012	0.0	.98033	.98041	.00008	0.0
5–10	.96977	.96998	.00021	0.0	.97699	.97714	.00015	0.0
10–15	.96716	.96748	.00032	0.0	.97512	.97534	.00022	0.0
15–20	.96467	.96513	.00046	0.0	.97362	.97397	.00035	0.0
20–25	.95871	.95946	.00075	0.1	.97118	.97176	.00058	0.1
25–30	.95076	.95200	.00124	0.1	.96825	.96924	.00099	0.1
30–35	.94356	.94556	.00200	0.2	.96479	.96638	.00159	0.2
35–40	.93542	.93903	.00361	0.4	.96012	.96256	.00244	0.3
40–45	.92367	.93101	.00734	0.8	.95308	.95695	.00387	0.4
45–50	.90451	.91991	.01540	1.7	.94179	.94852	.00673	0.7
50–55	.87309	.90349	.03040	3.5	.92469	.93636	.01167	1.3
55–60	.82316	.87876	.05560	6.8	.89943	.91974	.02031	2.3
60–65	.75355	.84486	.09131	12.1	.86365	.89871	.03506	4.1
65–70	.65567	.79669	.14102	21.5	.80654	.86958	.06304	7.8
70–75	.53267	.73369	.20102	37.7	.72330	.83118	.10788	14.9
75–80	.39243	.65565	.26322	67.1	.60264	.77982	.17718	29.4
80–85	.24787	.56316	.31529	127.2	.43999	.71026	.27027	61.4
85–90	.11942	.45234	.33292	278.8	.25101	.61443	.36342	144.8
90–95	.03990	.33642	.29652	743.2	.10166	.49744	.39578	389.3
95+	.00696	.21564	.20868	2998.3	.02177	.35234	.33057	1518.5

Table 4. The expectation of life and the effect of eliminating CVR disease as a risk of death (white males and females, United States, 1960)

Age Interval (in years) x_i to x_{i+1}	White Males				White Females			
	CVR Present	CVR Eliminated	Difference		CVR Present	CVR Eliminated	Difference	
	$\hat{e}_i$	$\hat{e}_{i\cdot 1}$	$\hat{e}_{i\cdot 1} - \hat{e}_i$	$\frac{\hat{e}_{i\cdot 1} - \hat{e}_i}{\hat{e}_i}$	$\hat{e}_i$	$\hat{e}_{i\cdot 1}$	$\hat{e}_{i\cdot 1} - \hat{e}_i$	$\frac{\hat{e}_{i\cdot 1} - \hat{e}_i}{\hat{e}_i}$
(1)	(2)	(3)	(4)	(5)	(6)	(7)	(8)	(9)
0–1	67.27	78.95	11.68	17.4%	74.01	86.77	12.76	17.2%
1–5	68.08	80.05	11.97	17.6	74.50	87.50	13.00	17.4
5–10	64.36	76.38	12.02	18.7	70.75	83.79	13.04	18.4
10–15	59.52	71.57	12.05	20.2	65.88	78.94	13.06	19.8
15–20	54.67	66.74	12.07	22.1	60.97	74.05	13.08	21.5
20–25	49.99	62.11	12.12	24.2	56.12	69.21	13.09	23.3
25–30	45.39	57.58	12.19	26.9	51.28	64.38	13.10	25.5
30–35	40.72	52.96	12.24	30.1	46.46	59.56	13.10	28.2
35–40	36.05	48.31	12.26	34.0	41.67	54.79	13.12	31.5
40–45	31.47	43.70	12.23	38.9	36.96	50.10	13.14	35.6
45–50	27.08	39.19	12.11	44.7	32.37	45.52	13.15	40.6
50–55	22.96	34.86	11.90	51.8	27.92	41.07	13.15	47.1
55–60	19.19	30.76	11.57	60.3	23.63	36.77	13.14	55.6
60–65	15.73	26.89	11.16	70.9	19.50	32.57	13.07	67.0
65–70	12.69	23.36	10.67	84.1	15.70	28.57	12.87	82.0
70–75	10.01	20.15	10.14	101.3	12.20	24.77	12.57	103.0
75–80	7.68	17.24	9.56	124.5	9.13	21.23	12.10	132.5
80–85	5.67	14.65	8.98	158.4	6.57	18.06	11.49	174.9
85–90	4.20	12.66	8.46	201.4	4.71	15.51	10.80	229.3
90–95	3.07	11.24	8.17	266.1	3.32	13.62	10.30	310.2
95+	2.92	11.39	8.47	290.1	2.98	13.39	10.41	349.3

PROBLEMS

1. The net probability $q_{i\delta}$ of dying from R_δ in the time interval (x_i, x_{i+1}) may be defined as

$$q_{i\delta} = \int_{x_i}^{x_{i+1}} \exp\left[-\int_{x_i}^{t} \mu(\tau, \delta)\, d\tau\right] \mu(t, \delta)\, dt.$$

Justify this formula and show that it is identical to the one given in (2.11).

2. Derive the formula in (2.21) for the partial crude probability $Q_{i\delta.12}$.

3. Use the method in Section 2, Chapter 10, to derive the p.g.f. (3.8) for the joint probability distribution of $d_{i\delta}$ and l_i.

4. Derive the formulas for the covariances $\text{Cov}(d_{i\delta}, l_j)$ and $\text{Cov}(l_i, d_{j\delta})$ for $i < j$.

5. Since
$$\sum_{\delta=1}^{r} d_{i\delta} = l_i - l_{i+1}$$

the covariances have the relations

$$\sum_{\delta=1}^{r} \text{Cov}(d_{i\delta}, l_j) = \text{Cov}(l_i, l_j) - \text{Cov}(l_{i+1}, l_j)$$

and

$$\sum_{\delta=1}^{r} \text{Cov}(l_i, d_{j\delta}) = \text{Cov}(l_i, l_j) - \text{Cov}(l_i, l_{j+1}).$$

Justify these relations and use (3.21), (3.22), and (3.23) to verify them.

6. *An urn scheme.* Modify the urn scheme in Section 2.1 in Chapter 10 to accommodate the different risks of death, and describe the crude probability $Q_{i\delta}$, the net probabilities $q_{i\delta}$, $q_{i\cdot\delta}$ and the partial crude probability $Q_{i\delta\cdot 1}$ for a specific risk R_δ in this scheme.

7. Derive the maximum-likelihood estimators $\hat{Q}_{i\delta}$ and $\hat{p}_i$ from the joint probability distribution in (3.6).

8. Let

$$L_i = \frac{l_i!}{d_{i1}! \cdots d_{ir}! l_{i+1}!} Q_{i1}^{d_{i1}} \cdots Q_{ir}^{d_{ir}} \left(1 - \sum_{\delta=1}^{r} Q_{i\delta}\right)^{l_{i+1}}.$$

(a) Derive the expectations

$$-E\left(\frac{\partial^2}{\partial Q_{i\delta} \partial Q_{i\varepsilon}} \log L_i\right), \qquad \delta, \varepsilon = 1, \ldots, r.$$

(b) Formulate a matrix **A** with these expectations as its elements, and find the inverse $\mathbf{A}^{-1}$ of the matrix **A**. Show that the diagonal elements of $\mathbf{A}^{-1}$ are given by

$$\frac{1}{E(l_i)} Q_{i\delta}(1 - Q_{i\delta})$$

and the off diagonal elements by

$$-\frac{1}{E(l_i)} Q_{i\delta}Q_{i\varepsilon}, \qquad \delta \neq \varepsilon;\ \delta, \varepsilon = 1, \ldots, r.$$

Matrix **A** is known as the information matrix and $\mathbf{A}^{-1}$ as the variance-covariance matrix of the asymptotic distribution of the maximum likelihood estimators $\hat{Q}_{i\delta}$ in Problem 7. See footnote 1 on page 229 in Chapter 10 regarding the expectations of "mixed" derivatives of $\log L_u$.

9. In writing the first equation in (4.4) we made use of the following elementary result. Let a_i, b_i, $i = 1, \ldots, r$, be positive numbers. If

$$\frac{a_1}{b_1} = \cdots = \frac{a_r}{b_r}$$

then

$$\frac{a_i}{b_i} = \frac{a_1 + \cdots + a_s}{a_1 + \cdots + b_s} \qquad \text{for} \qquad s \leq r.$$

Prove it.

10. Derive formula (4.18) for the variance of $\hat{q}_{i\delta}$ from (4.17).

11. Substitute (4.22), (4.25), and (4.26) in (4.28) and verify the resulting equation.

12. In Tables 5 to 7 are the population sizes, the number of deaths, and deaths from three causes by age for the total United States population, the white population, and nonwhite population in the year 1960. For a particular cause of your choice, compute the crude, net, and partial crude probabilities and construct appropriate life tables to study the effect of the selected cause on the longevity of people in these populations. Prepare short paragraphs to summarize the results and interpret your findings.

Table 5. Population, total deaths, and deaths from three causes by age for total United States population, 1960

Age Interval (in years)	Midyear Population[a]	Deaths from All Causes[b]	Deaths by Cause[b]		
			Major Cardiovascular Renal Diseases[c]	Malignant Neoplasms[d]	Accidents[e]
0–1	4,126,560	110,873	501	298	3,831
1–5	16,195,304	17,682	398	1,762	5,119
5–10	18,659,141	9,163	401	1,391	3,687
10–15	16,815,965	7,374	541	1,024	3,149
15–20	13,287,434	12,185	825	1,023	6,704
20–25	10,803,165	13,348	1,198	977	6,753
25–30	10,870,386	14,214	1,972	1,600	5,025
30–35	11,951,709	19,200	3,911	2,841	4,777
35–40	12,508,316	29,161	8,347	5,373	4,955
40–45	11,567,216	42,942	15,933	9,001	4,903
45–50	10,928,878	64,283	27,556	14,729	5,143
50–55	9,696,502	90,593	43,105	21,535	5,058
55–60	8,595,947	116,753	60,767	27,632	4,686
60–65	7,111,897	153,444	86,206	34,163	4,498
65–70	6,186,763	196,605	117,751	39,711	4,738
70–75	4,661,136	223,707	143,284	38,793	4,951
75–80	2,977,347	219,978	150,321	31,541	5,021
80–85	1,518,206	185,231	133,423	20,698	4,917
85–90	648,581	120,366	90,075	10,108	3,604
90–95	170,653	50,278	38,280	2,838	1,731
95+	44,551	13,882	10,523	528	470
Not stated		720	350	61	86
Total	179,325,657	1,711,982	935,668	267,627	93,806

[a] U.S. Census of Population 1960, U.S. Summary, Detailed Characteristics, Table 155, Bureau of the Census, U.S. Dept. of Commerce.

[b] Vital Statistics of the U.S. 1960, vol. II, Part A, Table 5-11, National Center for Health Statistics, U.S. Dept. of Health, Education, and Welfare.

[c] Category Numbers 330–334, 400–468 and 592–594 of the 1955 Revision of the *International Classification of Diseases, Injuries, and Causes of Death*, WHO, 1957.

[d] Category Numbers 140–205.

[e] Category Numbers E800–E962.

Table 6. ***Population, total deaths, and deaths from three causes by age for United States white population, 1960***

Age Interval (in years)	Midyear Population[a]	Deaths from All Causes[b]	Deaths by Cause[b]: Major Cardiovascular Renal Diseases[c]	Malignant Neoplasms[d]	Accidents[e]
0–1	3,513,798	82,479	360	258	2,672
1–5	13,851,760	13,200	261	1,584	3,809
5–10	16,067,397	7,427	298	1,262	2,868
10–15	14,677,328	6,061	404	924	2,583
15–20	11,666,367	10,209	628	913	5,924
20–25	9,479,847	10,671	895	854	5,830
25–30	9,565,462	10,641	1,370	1,390	4,194
30–35	10,596,064	14,254	2,763	2,401	3,962
35–40	11,170,430	22,199	6,171	4,564	4,082
40–45	10,393,094	33,958	12,377	7,652	4,071
45–50	9,838,979	52,515	22,201	12,660	4,309
50–55	8,777,124	75,412	35,538	18,755	4,283
55–60	7,761,453	98,880	50,820	24,406	4,051
60–65	6,519,629	132,600	73,865	30,582	3,885
65–70	5,668,320	174,560	104,148	36,223	4,221
70–75	4,312,863	204,287	130,675	36,009	4,534
75–80	2,759,955	205,326	140,636	29,688	4,674
80–85	1,420,987	175,656	126,786	19,746	4,709
85–90	603,104	114,511	85,950	9,628	3,460
90–95	157,054	47,660	36,430	2,706	1,649
95+	36,664	12,392	9,492	467	438
Not stated		437	205	40	60
Total	158,837,679	1,505,335	842,273	242,712	80,268

[a] U.S. Census of Population 1960, U.S. Summary, Detailed Characteristics, Table 155, Bureau of the Census, U.S. Dept. of Commerce.

[b] Vital Statistics of the U.S. 1960, vol. II, Part A, Table 5-11, National Center for Health Statistics, U.S. Dept. of Health, Education, and Welfare.

[c] Category Numbers 330–334, 400–468 and 592–594 of the 1955 Revision of the *International Classification of Diseases, Injuries, and Causes of Death*, WHO, 1957.

[d] Category Numbers 140–205.

[e] Category Numbers E800–E962.

Table 7. Population, total deaths, and deaths from three causes by age for United States nonwhite population, 1960

Age Interval (in years)	Midyear Population[a]	Deaths from All Causes[b]	Deaths by Cause[b]: Major Cardiovascular-Renal Diseases[c]	Malignant Neoplasms[d]	Accidents[e]
0–1	612,762	28,394	141	40	1,159
1–5	2,343,544	4,482	137	178	1,310
5–10	2,591,744	1,736	103	129	819
10–15	2,138,637	1,313	137	100	566
15–20	1,621,067	1,976	197	110	780
20–25	1,323,318	2,677	303	123	923
25–30	1,304,924	3,573	602	210	831
30–35	1,355,645	4,946	1,148	440	815
35–40	1,337,886	6,962	2,176	809	873
40–45	1,174,122	8,984	3,556	1,349	832
45–50	1,089,899	11,768	5,355	2,069	834
50–55	919,378	15,181	7,567	2,780	775
55–60	834,494	17,873	9,947	3,226	635
60–65	592,268	20,844	12,341	3,581	613
65–70	518,443	22,045	13,603	3,488	517
70–75	348,273	19,420	12,609	2,784	417
75–80	217,392	14,652	9,685	1,853	347
80–85	97,219	9,575	6,637	952	208
85–90	45,477	5,855	4,125	480	144
90–95	13,599	2,618	1,850	132	82
95+	7,887	1,490	1,031	61	32
Not stated		283	145	21	26
Total	20,487,978	206,647	93,395	24,915	13,538

[a] U.S. Census of Population 1960, U.S. Summary, Detailed Characteristics, Table 155, Bureau of the Census, U.S. Dept. of Commerce.

[b] Vital Statistics of the U.S. 1960, vol. II, Part A, Table 5-11, National Center for Health Statistics, U.S. Dept. of Health, Education, and Welfare.

[c] Category Numbers 330–334, 400–468 and 592–594 of The 1955 Revision of the *International Classification of Diseases, Injuries, and Causes of Death*, WHO, 1957.

[d] Category Numbers 140–205.

[e] Category Numbers E800–E962.

CHAPTER 12

Medical Follow-Up Studies

1. INTRODUCTION

Statistical studies in the general category of medical follow-up and life testing have as their common immediate objective the estimation of life expectancy and survival rates for a defined population at risk. These studies must usually be terminated before all survival information is complete and are therefore said to be truncated. The nature of the problem in an investigation concerned with the medical follow-up of patients is the same as in the life testing of electric bulbs, although difference in sample size may require different approaches. For illustration, we use cancer survival data of a large sample and therefore our terminology is the same as that of the medical follow-up study.

In a typical follow-up study, a group of individuals with some common morbidity experience is followed from a well-defined zero point, such as date of hospital admisssion. The purpose of the study might be to evaluate a certain therapeutic measure by comparing the expectation of life and survival rates of treated patients with those of untreated patients, or by comparing the expectation of life of treated and presumably cured patients with that of the general population. When the period of observation ends, there will usually remain a number of individuals for whom the mortality data is incomplete. First, some patients will still be alive at the close of the study. Second, some patients will have died from causes other than that under study, so that the chance of dying from the specific cause cannot be determined directly. Finally, some patients will be "lost" to the study because of follow-up failure. These three sources of incomplete information have created interesting statistical problems in the estimation of the expectation of life and survival rates. Many contributions have been made to methods of analysis of follow-up data. They include the studies of Greenwood [1925], Frost [1933], Berkson and Gage [1952], Fix and Neyman [1951], Boag [1949], Armitage [1959], Harris, Meier and Tukey [1950],

Dorn [1950], Cutler and Ederer [1958], and Littell [1952]. For the material presented in this chapter, reference may be made to Chiang [1961a].

The purpose of this chapter is to adapt the life table methodology and competing risk theory, presented in Chapters 10 and 11, to the special conditions of follow-up studies. Section 2 is concerned with the general type of study which investigates mortality experience without reference to cause of death. The maximum likelihood estimator of the probability of dying is derived, and a method for computing the observed expectation of life in such truncated studies is suggested. Section 3 extends the discussion to follow-up studies with the consideration of competing risks, and gives formulas for the estimators of the net, crude, and partial crude probabilities. The problem of lost cases is treated in Section 4, where a patient's being lost is considered as a competing risk. In Section 5 application of the theoretical matter is illustrated with empirical data of a follow-up study of breast cancer patients.

2. ESTIMATION OF PROBABILITY OF SURVIVAL AND EXPECTATION OF LIFE

Consider a follow-up program conducted over a period of y years. A total of N_0 patients is admitted to the program at any time during the study period and observed until death or until termination of the study, whichever comes first. The time of admission is taken as the common point of origin for all N_0 patients; thus N_0 is the number of patients with which the study begins, or the number alive at time zero. The time axis refers to the time of follow-up since admission, and x denotes the exact number of years of follow-up. A constant time interval of one year will be used for simplicity of notation, with the typical interval denoted by $(x, x + 1)$, for $x = 0, 1, \ldots, y - 1$. The symbol p_x will be used to denote the probability that a patient alive at time x will survive the interval $(x, x + 1)$, and q_x the probability that he will die during the interval, with $p_x + q_x = 1$.

2.1. Basic Random Variables and Likelihood Function

For each interval $(x, x + 1)$ let N_x be the number of patients alive at the beginning of the interval. Clearly, N_x is also the number of survivors of those who entered the study at least x years before the closing date.[1] The number N_x will decrease as x increases because of deaths and withdrawal

[1] For easy explanation, the common closing date is used here for describing the methods of analysis; however, these methods are equally applicable to data based on the date of last reporting for individual patients.

Table 1. Distribution of N_x patients according to withdrawal status and survival status in the interval $(x, x+1)$

	Withdrawal Status in the Interval		
Survival Status	Total Number of Patients	Number to be Observed for the Entire Interval[a]	Number Due to Withdraw During the Interval[b]
Total	N_x	m_x	n_x
Survivors	$s_x + w_x$	s_x	w_x
Deaths	D_x	d_x	d_x'

[a] Survivors among those admitted to the study more than $(x + 1)$ years before closing date.

[b] Survivors among those admitted to the study less than $(x + 1)$ years but more than x years before closing date.

of patients due to termination of the study. The decrease in N_x is systematically described below with reference to Table 1.

The N_x individuals who begin the interval $(x, x + 1)$ comprise two mutually exclusive groups differentiated according to their date of entrance into the program. A group of m_x patients who entered the program more than $x + 1$ years before the closing date will be observed for the entire interval; a second group of n_x patients who entered the program less than $x + 1$ years before its termination is due to withdraw in the interval because the closing date precedes their $(x + 1)$th anniversary date. Of the m_x patients d_x will die in the interval and s_x will survive to the end of the interval and become N_{x+1}; of the n_x patients d_x' will die before the closing date and w_x will survive to the closing date of the study. The sum $d_x + d_x' = D_x$ is the total number of deaths in the interval. Thus s_x, d_x, w_x, and d_x' are the basic random variables and will be used to estimate the probability p_x that a patient alive at x will survive the interval $(x, x + 1)$, and its complement q_x.

Let $\mu(\tau)$ be the intensity of risk of death (force of mortality) acting on each patient in the study so that

$$p_x = \exp\left\{-\int_x^{x+1} \mu(\tau)\, d\tau\right\}. \tag{2.1}$$

If we assume a constant intensity function within the interval $(x, x + 1)$ with $\mu(\tau) = \mu(x)$ depending only on x for $x < \tau \le x + 1$, then the probability of surviving the interval is given by

$$p_x = e^{-\mu(x)}. \tag{2.2}$$

For the subinterval $(x, x + t)$, $0 < t \leq 1$, we have the corresponding probability

$$p_x(t) = e^{-t\mu(x)} = p_x^{\ t}, \qquad 0 < t \leq 1. \tag{2.3}$$

Consider first the group of m_x individuals each of whom has a constant probability p_x of surviving and $q_x = 1 - p_x$ of dying in the interval $(x, x + 1)$. Thus, the random variable s_x has the probability distribution

$$f_{x1}(s_x;\, p_x) = \binom{m_x}{s_x} p_x^{\ s_x}(1 - p_x)^{d_x}. \tag{2.4}$$

The expected number of survivors and the expected number of deaths are given by

$$E(s_x \mid m_x) = m_x p_x \qquad \text{and} \qquad E(d_x \mid m_x) = m_x(1 - p_x) \tag{2.5}$$

respectively.

The distribution of the random variables in the group of n_x patients depends upon the time of withdrawal. A plausible assumption is that the withdrawals take place at random during the interval $(x, x + 1)$. Under this assumption the probability that a patient will survive to the closing date is

$$\int_x^{x+1} e^{-(\tau - x)\mu(x)}\, d\tau = -\frac{1 - p_x}{\log p_x}, \tag{2.6}$$

which is approximately equal to $p_x^{1/2}$, or

$$-\frac{1 - p_x}{\log p_x} = p_x^{1/2}, \tag{2.7}$$

since the probability p_x of surviving the interval is almost always large. The quantities on both sides of (2.7) have been computed for selected values of p_x, and the results shown in Table 2 justify the approximation.

Table 2. Comparison between $p_x^{1/2}$ and $-(1 - p_x)/\log p_x$

p_x	$p_x^{1/2}$	$-(1 - p_x)/\log p_x$
.70	.837	.841
.75	.866	.869
.80	.894	.896
.85	.922	.923
.90	.949	.949
.95	.975	.975

Consequently, $p_x^{1/2}$ is taken as the probability of surviving to the closing date and $(1 - p_x^{1/2})$ as the probability of dying before the time of withdrawal. Thus the probability distribution of the random variable w_x in the group of n_x patients due to withdraw is

$$f_{x2}(w_x; p_x) = \binom{n_x}{w_x} p_x^{(1/2)w_x}(1 - p_x^{1/2})^{d_x}. \tag{2.8}$$

The expected number of survivors and the expected number of deaths are given by

$$E(w_x \mid n_x) = n_x p_x^{1/2} \quad \text{and} \quad E(d_x' \mid n_x) = n_x(1 - p_x^{1/2}), \tag{2.9}$$

respectively.

Since the N_x individuals comprise two independent groups according to their withdrawal status, the likelihood function of all the random variables is the product of the two probability functions (2.4) and (2.8), or

$$L_x = f_{x1}(s_x; p_x) f_{x2}(w_x; p_x) = c p_x^{s_x+(1/2)w_x}(1 - p_x)^{d_x}(1 - p_x^{1/2})^{d_x'},$$
$$x = 0, 1, \ldots, y - 1, \tag{2.10}$$

where c stands for the product of the combinatorial factors in (2.4) and (2.8).

2.2. Maximum Likelihood Estimators of the Probabilities p_x and q_x

The maximum likelihood estimators of the probability p_x and its complement q_x are obtained by maximizing the likelihood function (2.10). Taking the logarithm of (2.10), we have

$$\log L_x = \log c + (s_x + \tfrac{1}{2}w_x) \log p_x$$
$$+ d_x \log (1 - p_x) + d_x' \log (1 - p_x^{1/2}). \tag{2.11}$$

Differentiating (2.11) with respect to p_x and setting the derivative equal to zero give

$$(s_x + \tfrac{1}{2}w_x)\hat{p}_x^{-1} - d_x(1 - \hat{p}_x)^{-1} - \tfrac{1}{2}d_x'(1 - \hat{p}_x^{1/2})^{-1}\hat{p}_x^{-1/2} = 0, \tag{2.12}$$

which can be rewritten as

$$(N_x - \tfrac{1}{2}n_x)\hat{p}_x + \tfrac{1}{2}d_x'\hat{p}_x^{1/2} - (s_x + \tfrac{1}{2}w_x) = 0, \tag{2.13}$$

a quadratic equation in $\hat{p}_x^{1/2}$. Since $\hat{p}_x^{1/2}$ cannot take on negative values, (2.13) has the unique solution

$$\hat{p}_x = \left[\frac{-\tfrac{1}{2}d_x' + \sqrt{\tfrac{1}{4}d_x'^2 + 4(N_x - \tfrac{1}{2}n_x)(s_x + \tfrac{1}{2}w_x)}}{2(N_x - \tfrac{1}{2}n_x)}\right]^2 \tag{2.14}$$

with the complement[2]

$$\hat{q}_x = 1 - \hat{p}_x, \qquad x = 0, 1, \ldots, y - 1. \tag{2.15}$$

The maximum likelihood estimator (2.14) is not unbiased, but is consistent in the sense of Fisher; when the random variables s_x, w_x, and d_x' are replaced with their respective expectations as given by (2.5) and (2.9), the resulting expression is identical with the probability p_x:

$$p_x = \left[\frac{-\frac{1}{2}n_x(1 - p_x^{1/2}) + \sqrt{\frac{1}{4}n_x^2(1 - p_x^{1/2})^2 + 4(N_x - \frac{1}{2}n_x)(m_x p_x + \frac{1}{2}n_x p_x^{1/2})}}{2(N_x - \frac{1}{2}n_x)}\right]^2. \tag{2.16}$$

To show the identity (2.16), we replace m_x with $N_x - n_x$ and write

$$m_x p_x + \tfrac{1}{2}n_x p_x^{1/2} = (N_x - \tfrac{1}{2}n_x)p_x + \tfrac{1}{2}n_x p_x^{1/2}(1 - p_x^{1/2}). \tag{2.17}$$

Then the quantity under the square root sign may be simplified to

$$[\tfrac{1}{2}n_x(1 - p_x^{1/2}) + 2(N_x - \tfrac{1}{2}n_x)p_x^{1/2}]^2, \tag{2.18}$$

and the right side of (2.16) is reduced to

$$\left[\frac{-\frac{1}{2}n_x(1 - p_x^{1/2}) + \frac{1}{2}n_x(1 - p_x^{1/2}) + 2(N_x - \frac{1}{2}n_x)p_x^{1/2}}{2(N_x - \frac{1}{2}n_x)}\right]^2 = p_x, \tag{2.19}$$

proving (2.16).

The exact formula for the variance of the estimator $\hat{p}_x$ in (2.14) is unknown, but an approximate formula is derived below for practical applications. Using the asymptotic properties of maximum likelihood estimators, we first find the expectation of the second derivative of the

---

[2] In a follow-up study of children with tuberculosis, W. H. Frost [1933] introduced the concept of "person year" and used the quantity (in the present notation)

$$N_x - \tfrac{1}{2}(w_x + D_x)$$

as the "person-year of life experience" and the ratio

$$\frac{D_x}{N_x - \frac{1}{2}(w_x + D_x)} = M_x$$

as death rate. Using this formula and relation (3.7) in Chapter 9 between M_x and q_x with $a_x = \frac{1}{2}$, we have $q_x = D_x/(N_x - \frac{1}{2}w_x)$ (See Berkson and Gage [1950] and Cutler and Ederer [1958]).

logarithm of the likelihood function in (2.11). A straightforward computation gives

$$E\left(\frac{\partial^2}{\partial p_x^{\ 2}} \log L_x\right) = -\left(\frac{M_x}{p_x q_x} + \pi_x\right), \tag{2.20}$$

where

$$M_x = m_x + n_x(1 + p_x^{1/2})^{-1} \tag{2.21}$$

and

$$\pi_x = \frac{n_x}{4(1 + p_x^{1/2})^2 p_x^{3/2}} (1 - p_x^{1/2}). \tag{2.22}$$

When the bias of the estimator is neglected, an approximate formula for the variance of $\hat{p}_x$ is given by

$$\sigma_{\hat{p}_x}^{\ 2} = \frac{1}{-E\left(\dfrac{\partial^2}{\partial p_x^{\ 2}} \log L_x\right)} = \left(\frac{M_x}{p_x q_x} + \pi_x\right)^{-1}. \tag{2.23}$$

The quantity π_x usually is small in comparison with the preceding term and may be neglected, and (2.23) is reduced to

$$\sigma_{\hat{p}_x}^{\ 2} = \frac{p_x q_x}{M_x}, \qquad x = 0, 1, \ldots, y - 1. \tag{2.24}$$

The sample variance of $\hat{p}_x$ is obtained by substituting (2.14) and (2.15) in (2.24).

Formula (2.24) is quite similar to the variance of a binomial proportion except that M_x instead of N_x is in the denominator. However, M_x is the more logical choice, since a patient who is to be observed for a fraction of the period $(x, x + 1)$ should be weighted less than one who is to be observed for the entire period. According to Equation (2.21), the experience of each of the m_x patients is counted as a whole "trial," whereas the experience of each of the n_x patients due to withdraw is counted as a fraction $(1 + p_x^{1/2})^{-1}$ of a "trial." The fraction is dependent upon the probability p_x of survival. The smaller the probability p_x, the larger will be the fraction. When $p_x = 0$, $M_x = m_x + n_x$; when $p_x = 1$, $M_x = m_x + \frac{1}{2}n_x$.

Remark 1. If the time of death of each of the D_x patients and the time of withdrawal of each of the w_x patients are known, the probability p_x will be estimated from a different formula. In such cases there will be N_x individual observations, and obviously it is unnecessary to consider the N_x patients as two distinct groups according to their withdrawal status.

Let $t_i \leq 1$ be the time of death within the interval $(x, x + 1)$ of the ith death, for $i = 1, \ldots, D_x$, with the probability density

$$e^{-t_i\mu(x)}\mu(x) = -p_x^{\,t_i} \log p_x. \tag{2.25}$$

Let $\tau_j \leq 1$ be the length of observation in interval $(x, x + 1)$ for the jth patient in the group of w_x surviving to the closing date of the study, for $j = 1, \ldots, w_x$, with the probability density

$$e^{-\tau_j\mu(x)} = p_x^{\,\tau_j}. \tag{2.26}$$

Therefore, the likelihood function is

$$L_x = p_x^{\,s_x}\left(\prod_{i=1}^{D_x} p_x^{\,t_i}\right)\left(\prod_{j=1}^{W_x} p_x^{\,\tau_j}\right)[-\log p_x]^{D_x} \tag{2.27}$$

Maximizing (2.27) with respect to p_x gives the maximum likelihood estimator

$$\hat{p}_x = \exp\left\{-\frac{D_x}{s_x + \sum_{i=1}^{D_x} t_i + \sum_{j=1}^{w_x} \tau_j}\right\}. \tag{2.28}$$

Here the ratio in the exponent, which is an estimate of the intensity function $\mu(x)$, is equal to the number of deaths occurring in the interval $(x, x + 1)$ divided by the total length of time lived in the interval by the N_x patients.

2.3. Estimation of Survival Probability

A life table for follow-up subjects can be readily constructed once $\hat{p}_x$ and $\hat{q}_x$ have been determined from (2.14) and (2.15) for each interval of the study. The procedure is the same as for the current life table. Because of their practical importance, we shall consider only the x-year survival rate and the expectation of life.

The x-year survival rate is an estimate of the probability that a patient will survive from the time of admission to the xth anniversary; it is computed from

$$\hat{p}_{0x} = \hat{p}_0\hat{p}_1 \cdots \hat{p}_{x-1}, \qquad x = 1, 2, \ldots, y. \tag{2.29}$$

The sample variance of $\hat{p}_{0x}$ has the same form as that given in Equation (5.12) of Chapter 9,

$$S^2_{\hat{p}_{0x}} = \hat{p}^2_{0x} \sum_{u=1}^{x-1} \hat{p}_u^{-2}\, S^2_{\hat{p}_u}\,. \tag{2.30}$$

2.4. Estimation of the Expectation of Life

To avoid confusion in notation, let us denote by α a fixed number and by $\hat{e}_\alpha$ the observed expectation of life at time α computed from the following formula[3]:

$$\hat{e}_\alpha = \tfrac{1}{2} + \hat{p}_\alpha + \hat{p}_\alpha \hat{p}_{\alpha+1} + \cdots + \hat{p}_\alpha \hat{p}_{\alpha+1} \cdots \hat{p}_{y-1} + \hat{p}_\alpha \hat{p}_{\alpha+1} \cdots \hat{p}_y + \cdots. \tag{2.31}$$

In a study covering a period of y years, if no survivors remain from the patients who entered the program in its first year, $\hat{p}_{y-1}$ will be zero, and $\hat{e}_\alpha$ can be computed from (2.31). However, usually there will be w_{y-1} survivors who were admitted in the first year of the program and are still living at the closing date. In such cases (2.14) shows that $\hat{p}_{y-1}$ is greater than zero, and the values of $\hat{p}_y, \hat{p}_{y+1}, \ldots$ are not observed within the time limits of the study. Consequently, $\hat{e}_\alpha$ cannot be obtained from Equation (2.31).

Nevertheless, $\hat{e}_\alpha$ may be computed with a certain degree of accuracy if w_{y-1} is small. Suppose we rewrite Equation (2.31) in the form

$$\hat{e}_\alpha = \tfrac{1}{2} + \hat{p}_\alpha + \hat{p}_\alpha \hat{p}_{\alpha+1} + \cdots + \hat{p}_\alpha \hat{p}_{\alpha+1} \cdots \hat{p}_{y-1} + \hat{p}_{\alpha y}(\hat{p}_y + \hat{p}_y \hat{p}_{y+1} + \cdots), \tag{2.32}$$

where $\hat{p}_{\alpha y}$ is written for $\hat{p}_\alpha \hat{p}_{\alpha+1} \cdots \hat{p}_{y-1}$. The problem is to determine $\hat{p}_y, \hat{p}_{y+1}, \ldots$ in the last term, since the preceding terms can be computed from the data available.

Consider a typical interval $(z, z+1)$ beyond time y with the survival probability of

$$p_z = \exp\left\{-\int_z^{z+1} \mu(\tau)\, d\tau\right\}, \qquad z = y, y+1, \ldots. \tag{2.33}$$

If the intensity function is constant beyond y, so that $\mu_\tau = \mu$, the probability of surviving the interval $(z, z+1)$ becomes

$$p_z = e^{-\mu} = p, \qquad z = y, y+1, \ldots, \tag{2.34}$$

and is independent of z. Under this assumption, we may replace the last term of (2.32) with $\hat{p}_{\alpha y}(\hat{p} + \hat{p}^2 + \cdots)$, which converges to $\hat{p}_{\alpha y}\hat{p}/(1-\hat{p})$, or

$$\hat{p}_{\alpha y}(\hat{p} + \hat{p}^2 + \cdots) = \hat{p}_{\alpha y} \frac{\hat{p}}{1-\hat{p}}. \tag{2.35}$$

[3] For simplicity, we assume $a_x = \tfrac{1}{2}$ and $c_x = 1$ in the formula for $\hat{e}_\alpha$.

As a result, we have

$$\hat{e}_\alpha = \tfrac{1}{2} + \hat{p}_\alpha + \hat{p}_\alpha\hat{p}_{\alpha+1} + \cdots + \hat{p}_\alpha\hat{p}_{\alpha+1}\cdots\hat{p}_{y-1} + \hat{p}_{\alpha y}\left(\frac{\hat{p}}{1-\hat{p}}\right). \quad (2.36)$$

Clearly, $\hat{p}$ may be set equal to $\hat{p}_{y-1}$ if the intensity function is assumed to be constant beginning with time $(y - 1)$ instead of time y. In order to have small sample variation, however, the estimate of $\hat{p}$ should be based on as large a sample as possible. Suppose there exists a time t, for $t < y$, such that $\hat{p}_t, \hat{p}_{t+1}, \ldots$ are approximately equal, thus indicating a constant intensity function after time t. Then, $\hat{p}$ may be set equal to $\hat{p}_t$, and we have the formula for the observed expectation of life,

$$\hat{e}_\alpha = \tfrac{1}{2} + \hat{p}_\alpha + \hat{p}_\alpha\hat{p}_{\alpha+1} + \cdots + \hat{p}_\alpha\hat{p}_{\alpha+1}\cdots\hat{p}_{y-1} + \hat{p}_{\alpha y}\left(\frac{\hat{p}_t}{1-\hat{p}_t}\right), \quad (2.37)$$

for $\alpha = 0, \ldots, y - 1$.

Although formula (2.37) holds true for $\alpha = 0, \ldots, y - 1$, it is apparent that the smaller the value of α, the smaller the value of $\hat{p}_{\alpha y}$. When $\hat{p}_{\alpha y}$ is small, the error in assuming a constant intensity function beyond y and in the choice of $\hat{p}_t$ will have but little effect on the value of $\hat{e}_\alpha$.

2.5. Sample Variance of the Observed Expectation of Life

In Chapter 10 we proved that the estimated probabilities of surviving any two nonoverlapping intervals have a zero covariance; hence, the sample variance of the observed expectation of life may be computed from

$$s_{\hat{e}_\alpha}^{\ 2} = \sum_{x \geq \alpha}\left\{\frac{\partial}{\partial \hat{p}_x}\hat{e}_\alpha\right\}^2 s_{\hat{p}_x}^{\ 2}. \quad (2.38)$$

The derivatives, taken at the observed point $\hat{p}_x$, $x \geq \alpha$, are given by

$$\left\{\frac{\partial}{\partial \hat{p}_x}\hat{e}_\alpha\right\} = \hat{p}_{\alpha x}[\hat{e}_{x+1} + \tfrac{1}{2}], \qquad x \neq t \quad (2.39)$$

and

$$\left\{\frac{\partial}{\partial \hat{p}_t}\hat{e}_\alpha\right\} = \hat{p}_{\alpha t}\left[\hat{e}_{t+1} + \frac{1}{2} + \frac{\hat{p}_{ty}}{(1-\hat{p}_t)^2}\right], \qquad \alpha \leq t$$

$$= \frac{\hat{p}_{\alpha y}}{(1-\hat{p}_t)^2}, \qquad \alpha > t \quad (2.40)$$

where $\hat{p}_{\alpha x} = \hat{p}_\alpha \hat{p}_{\alpha+1} \cdots \hat{p}_{x-1}$. Substituting (2.39) and (2.40) in (2.38) gives the sample variance of $\hat{e}_\alpha$,

$$S_{\hat{e}_\alpha}{}^2 = \sum_{\substack{x=\alpha \\ x\neq t}}^{y-1} \hat{p}_{\alpha x}{}^2[\hat{e}_{x+1} + \tfrac{1}{2}]^2 S_{\hat{p}_x}{}^2 + \hat{p}_{\alpha t}{}^2\left[\hat{e}_{t+1} + \tfrac{1}{2} + \frac{\hat{p}_{ty}}{(1-\hat{p}_t)^2}\right]^2 S_{\hat{p}_t}{}^2, \qquad \alpha \leq t, \quad (2.41)$$

and

$$S_{\hat{e}_\alpha}{}^2 = \sum_{x=\alpha}^{y-1} \hat{p}_{\alpha x}{}^2[\hat{e}_{x+1} + \tfrac{1}{2}]^2 S_{\hat{p}_x}{}^2 + \frac{\hat{p}_{\alpha y}{}^2}{(1-\hat{p}_t)^4} S_{\hat{p}_t}{}^2, \qquad \alpha > t. \quad (2.42)$$

The value of $\hat{p}_x$ and the sample variance of $\hat{p}_x$ are obtained from formulas (2.14) and (2.24), respectively.

3. CONSIDERATION OF COMPETING RISKS

Most follow-up studies are conducted to determine the survival rates of patients affected with a specific disease. These patients are also exposed to other risks of death from which some of them may eventually die. In a study determining the effectiveness of radiation as a treatment for cancer, for example, some patients may die from heart disease. In such cases, the theory of competing risks is indispensable, and the crude, net, and partial crude probabilities all play important roles.

Let us assume, as in Chapter 11, that r risks, denoted by $R_1, \ldots, R_r$, are acting simultaneously on each patient in the study. For risk R_δ there is a corresponding intensity function (force of mortality) $\mu(\tau, \delta)$, $\delta = 1, \ldots, r$, and the sum

$$\mu(\tau; 1) + \cdots + \mu(\tau, r) = \mu(\tau) \quad (3.1)$$

is the total intensity function. Within the time interval $(x, x+1)$ we assume a constant intensity function for each risk, $\mu(\tau; \delta) = \mu(x; \delta)$, which depends only on the interval $(x, x+1)$ and the risk R_δ; for all risks, $\mu(\tau) = \mu(x)$ for $x < \tau \leq x+1$.

Consider a subinterval $(x, x+t)$ and let $Q_{x\delta}(t)$ be the crude probability that an individual alive at time x will die prior to $x+t$, $0 < t \leq 1$, from R_δ in the presence of all other risks in the population. It follows directly from Equation (2.6) in Chapter 11 that

$$Q_{x\delta}(t) = \int_x^{x+t} e^{-(\tau-x)\mu(x)} \mu(x; \delta)\, d\tau. \quad (3.2)$$

Integrating (3.2) gives

$$Q_{x\delta}(t) = \frac{\mu(x;\delta)}{\mu(x)}[1 - e^{-t\mu(x)}] = \frac{\mu(x;\delta)}{\mu(x)}[1 - p_x(t)],$$
$$0 < t \le 1; \quad \delta = 1, \ldots, r. \quad (3.3)$$

From (3.1) we see that the sum of the crude probabilities in (3.3) is equal to the complement of $p_x(t)$, or

$$Q_{x1}(t) + \cdots + Q_{xr}(t) + p_x(t) = 1, \qquad 0 < t \le 1. \quad (3.4)$$

For $t = 1$, we abbreviate $Q_{x\delta}(1)$ to $Q_{x\delta}$, etc. When $t = \frac{1}{2}$, we have the subinterval $(x, x + \frac{1}{2})$ and the corresponding crude probabilities

$$Q_{x\delta}(\tfrac{1}{2}) = \frac{\mu(x;\delta)}{\mu(x)}[1 - e^{-\mu(x)/2}] = Q_{x\delta}[1 + p_x^{1/2}]^{-1}, \qquad \delta = 1, \ldots, r. \quad (3.5)$$

Equation (3.4) implies that

$$Q_{x1}[1 + p_x^{1/2}]^{-1} + \cdots + Q_{xr}[1 + p_x^{1/2}]^{-1} + p_x^{1/2} = 1,$$
$$x = 0, 1, \ldots, y - 1. \quad (3.6)$$

The assumption of a constant relative risk [see Equation (2.3) in Chapter 11] required for the relations between the crude, net, and partial crude probabilities derived in Section 2, Chapter 11, is obviously satisfied in this case. These relations are

$$q_{x\delta} = 1 - p_x^{Q_{x\delta}/q_x}, \quad (3.7)$$

$$q_{x.\delta} = 1 - p_x^{(q_x - Q_{x\delta})/q_x}, \qquad \delta = 1, \ldots, r; \quad (3.8)$$

$$Q_{x\delta.1} = \frac{Q_{x\delta}}{q_x - Q_{x1}}[1 - p_x^{(q_x - Q_{x1})/q_x}], \qquad \delta = 2, \ldots, r; \quad (3.9)$$

and

$$Q_{x\delta.12} = \frac{Q_{x\delta}}{q_x - Q_{x1} - Q_{x2}}[1 - p_x^{(q_x - Q_{x1} - Q_{x2})/q_x}], \qquad \delta = 3, \ldots, r, \; x = 0, 1, \ldots, y - 1. \quad (3.10)$$

Our immediate problem is to estimate $Q_{x\delta}$, p_x, and q_x.

3.1. Basic Random Variables and Likelihood Functions

Identification of the random variables in the present case follows directly from the discussion in Section 2.1, except that deaths are further divided by cause, as shown in Table 3.

Table 3. Distribution of N_x patients according to withdrawal status, survival status, and cause of death in the interval $(x, x + 1)$

	Withdrawal Status in the Interval		
	Total Number of Patients	Number to be Observed for the Entire Interval[a]	Number Due to Withdraw During the Interval[b]
Total	N_x	m_x	n_x
Survivors	$s_x + w_x$	s_x	w_x
Deaths (all causes)	D_x	d_x	d'_x
Deaths due to cause			
R_1	D_{x1}	d_{x1}	d'_{x1}
.	.	.	.
.	.	.	.
.	.	.	.
R_r	D_{xr}	d_{xr}	d'_{xr}

[a] Survivors among those admitted to the study more than $(x + 1)$ years before closing date.

[b] Survivors among those admitted to the study less than $(x + 1)$ years but more than x years before closing date.

The m_x patients to be observed for the entire interval $(x, x + 1)$ will be divided into $r + 1$ mutually exclusive groups, with s_x surviving the interval and $d_{x\delta}$ dying from cause R_δ in the interval, $\delta = 1, \ldots, r$. Since the sum of the corresponding probabilities is equal to unity [Equation (3.4)] the random variables $s_x, d_{x1}, \ldots, d_{xr}$ have the joint probability distribution

$$f_{x1}(s_x, d_{x\delta}; p_x, Q_{x\delta}) = \binom{m_x}{d_{x1}, \ldots, d_{xr}} p_x^{s_x} Q_{x1}^{d_{x1}} \cdots Q_{xr}^{d_{xr}}, \qquad (3.11)$$

where $s_x + d_{x1} + \cdots + d_{xr} = m_x$. Their mathematical expectations are given by

$$E(s_x \mid m_x) = m_x p_x \qquad \text{and} \qquad E(d_{x\delta} \mid m_x) = m_x Q_{x\delta} \qquad (3.12)$$

respectively.

In the group of n_x patients due to withdraw in interval $(x, x + 1)$, w_x will be alive at the closing date of the study and $d'_{x\delta}$ will die from R_δ before the closing date. Each of the n_x individuals has the survival probability $p_x^{1/2}$ [cf. Equation (2.7)] and the probability of dying from risk R_δ before the closing date

$$Q_{x\delta}(\tfrac{1}{2}) = Q_{x\delta}(1 + p_x^{1/2})^{-1}, \qquad \delta = 1, \ldots, r. \tag{3.13}$$

Since $p_x^{1/2}$ and the probabilities in (3.13) add to unity, as shown in (3.6), the random variables $w_x, d'_{x1}, \ldots, d'_{xr}$ have the joint probability distribution

$$f_{x2}(s_x, d'_{x\delta}; p_x, Q_{x\delta}) = \binom{n_x}{d'_{x1}, \ldots, d'_{xr}} p_x^{(1/2)w_x} \prod_{\delta=1}^{r} [Q_{x\delta}(1 + p_x^{1/2})^{-1}]^{d'_{x\delta}}, \tag{3.14}$$

where $w_x + d'_{x1} + \cdots + d'_{xr} = n_x$. The mathematical expectations are

$$E(w_x \mid n_x) = n_x p_x^{1/2} \qquad \text{and} \qquad E(d'_{x\delta} \mid n_x) = n_x Q_{x\delta}(1 + p_x^{1/2})^{-1} \tag{3.15}$$

respectively. Because of the independence of the two groups, the likelihood function of all the random variables in Table 3 is the product of (3.11) and (3.14):

$$\begin{aligned} L_x &= f_{x1}(s_x, d_{x\delta}; p_x, Q_{x\delta}) f_{x2}(w_x, d'_{x\delta}; p_x, Q_{x\delta}) \\ &= c p_x^{s_x + (1/2)w_x} \prod_{\delta=1}^{r} Q_{x\delta}^{d_{x\delta}} [Q_{x\delta}(1 + p_x^{1/2})^{-1}]^{d'_{x\delta}}, \end{aligned} \tag{3.16}$$

where c stands for the product of the combinatorial factors in (3.11) and (3.14). Equation (3.16) may be simplified to give the final form of the likelihood function

$$L_x = c p_x^{s_x + (1/2)w_x} (1 + p_x^{1/2})^{-d'_x} \prod_{\delta=1}^{r} Q_{x\delta}^{D_{x\delta}}. \tag{3.17}$$

3.2. Estimation of Crude, Net, and Partial Crude Probabilities

We again use the maximum likelihood principle to obtain the estimators of the probabilities $p_x, Q_{x1}, \ldots, Q_{xr}$. The logarithm of the likelihood function (3.17) is

$$\log L_x = \log c + (s_x + \tfrac{1}{2}w_x) \log p_x - d'_x \log (1 + p_x^{1/2}) + \sum_{\delta=1}^{r} D_{x\delta} \log Q_{x\delta}, \tag{3.18}$$

which is to be maximized subject to the condition

$$p_x + Q_{x1} + \cdots + Q_{xr} = 1. \tag{3.19}$$

Using the Lagrange method we maximize[4]

$$\Phi_x = \log c + (s_x + \tfrac{1}{2}w_x)\log p_x - d'_x \log(1 + p_x^{1/2})$$
$$+ \sum_{\delta=1}^{r} D_{x\delta} \log Q_{x\delta} - \lambda\left(p_x + \sum_{\delta=1}^{r} Q_{x\delta} - 1\right). \quad (3.20)$$

Differentiating Φ_x with respect to $p_x, Q_{x1}, \ldots, Q_{xr}$ and setting the derivatives equal to zero yield equations

$$\frac{\partial}{\partial p_x}\Phi_x = \frac{s_x + \tfrac{1}{2}w_x}{\hat{p}_x} - \frac{\tfrac{1}{2}d'_x}{(1 + \hat{p}_x^{1/2})\hat{p}_x^{1/2}} - \lambda = 0 \quad (3.21)$$

and

$$\frac{\partial}{\partial Q_{x\delta}}\Phi_x = \frac{D_{x\delta}}{\hat{Q}_{x\delta}} - \lambda = 0, \qquad \delta = 1, \ldots, r. \quad (3.22)$$

Since (3.22) holds true for each δ, we have

$$\lambda = \frac{\sum_{\delta=1}^{r} D_{x\delta}}{\sum_{\delta=1}^{r} \hat{Q}_{x\delta}} \qquad \text{or} \qquad \lambda = \frac{D_x}{\hat{q}_x} \quad (3.23)$$

with $\hat{q}_x = 1 - \hat{p}_x$, and hence,

$$\frac{D_{x\delta}}{\hat{Q}_{x\delta}} = \frac{D_x}{\hat{q}_x}. \quad (3.24)$$

---

[4] Once upon a time a rich man willed eleven urns of gold to his three sons. After his death it was revealed that by the terms of the will one half of the gold was to go to the eldest son, one fourth to the second son, and one sixth to the youngest son; the entire fortune must be divided up among the sons, but the gold in each urn must remain intact. The sons found their intelligence no match for the terms of the will, and they could not find a way to divide the gold according to the instructions. Amid the excitement there came an old man; he was learned and generous. Touched by the sons' anxiety, he offered an urn of gold of his own. The three sons accepted the offer with gratitude because it brought the problem to their level of intelligence: the eldest son received six urns; the second son, three urns; and the youngest son, two urns. The wise old man recovered his urn of gold, which was left over after the three sons had taken their share.

The old man could be Lagrange; his urn played the role of Lagrange coefficient. And it was immaterial whether his urn contained gold or anything else, just as the exact value of the Lagrange coefficient is of no importance in the solution of a practical problem.

Note that if there were no restriction of indivisibility on the gold within the urns, the sons would have to divide the gold an infinite number of times, and in the end they would have received the same amount of gold as they did by the old man's subterfuge.

Using the second equation in (3.23), we find from (3.21) the quadratic equation

$$(N_x - \tfrac{1}{2}n_x)\hat{p}_x + \tfrac{1}{2}d'_x\hat{p}_x^{1/2} - (s_x + \tfrac{1}{2}w_x) = 0, \qquad (3.25)$$

which is identical to Equation (2.13) in Section 2.2 where death is investigated without specification of cause, as might be anticipated. Hence the estimators

$$\hat{p}_x = \left[\frac{-\tfrac{1}{2}d'_x + \sqrt{\tfrac{1}{4}d'^2_x + 4(N_x - \tfrac{1}{2}n_x)(s_x + \tfrac{1}{2}w_x)}}{2(N_x - \tfrac{1}{2}n_x)}\right]^2,$$

$$x = 0, 1, \ldots, y - 1, \quad (3.26)$$

and $\hat{q}_x$ will have the same values as those in Section 2. Substituting (3.26) in (3.24) yields the estimators of the crude probabilities,

$$\hat{Q}_{x\delta} = \frac{D_{x\delta}}{D_x}\hat{q}_x, \qquad \begin{array}{l} \delta = 1, 2, \ldots, r, \\ x = 0, 1, \ldots, y - 1. \end{array} \qquad (3.27)$$

We now use (3.26) and (3.27) in formulas (3.7) to (3.10) to obtain the following estimators of the net and partial crude probabilities:

$$\hat{q}_{x\delta} = 1 - \hat{p}_x^{D_{x\delta}/D_x}, \qquad (3.28)$$

$$\hat{q}_{x\cdot\delta} = 1 - \hat{p}_x^{(D_x - D_{x\delta})/D_x}, \qquad \delta = 1, \ldots, r, \qquad (3.29)$$

$$\hat{Q}_{x\delta\cdot 1} = \frac{D_{x\delta}}{D_x - D_{x1}}[1 - \hat{p}_x^{(D_x - D_{x1})/D_x}], \qquad \delta = 2, \ldots, r, \qquad (3.30)$$

and

$$\hat{Q}_{x\delta\cdot 12} = \frac{D_{x\delta}}{D_x - D_{x1} - D_{x2}}[1 - \hat{p}_x^{(D_x - D_{x1} - D_{x2})/D_x}],$$

$$\delta = 3, \ldots, r; x = 0, 1, \ldots, y - 1. \quad (3.31)$$

These too are maximum likelihood estimators and consistent in Fisher's sense. Consider, for example, the estimator $\hat{Q}_{x\delta\cdot 1}$ in formula (3.30) of the partial crude probability. We have seen in (2.16) that $\hat{p}_x$ is consistent in Fisher's sense. When the other random variables are replaced with the corresponding expectations, the right side of (3.30) may be simplified to

$$\frac{Q_{x\delta}}{q_x - Q_{x1}}[1 - p_x^{(q_x - Q_{x1})/q_x}], \qquad (3.32)$$

which is identical to $Q_{x\delta\cdot 1}$ in formula (3.9), proving the consistency.

3.3. Approximate Formulas for the Variances and Covariances of the Estimators

We may again determine approximate formulas for the variances and covariances of the estimators by using the asymptotic property of maximum likelihood estimators. Because the probabilities p_x and $Q_{x\delta}$, $\delta = 1, \ldots, r$, are not independent, we use (3.19) and make the substitution of

$$Q_{x1} = 1 - p_x - Q_{x2} - \cdots - Q_{xr} \tag{3.33}$$

in (3.18). When the bias is neglected the inverse of the asymptotic covariance matrix of the estimators, $\hat{p}_x, \hat{Q}_{x2}, \ldots, \hat{Q}_{xr}$, is given by

$$\mathbf{A} = \left[\begin{array}{c|c} \left(-E\left\{\dfrac{\partial^2}{\partial p_x^{\,2}} \log L_x\right\}\right) & \left(-E\left\{\dfrac{\partial^2}{\partial p_x\, \partial Q_{x\varepsilon}} \log L_x\right\}\right) \\ 1 \times 1 & 1 \times (r-1) \\ \hline \left(-E\left\{\dfrac{\partial^2}{\partial Q_{x\delta}\, \partial p_x} \log L_x\right\}\right) & \left(-E\left\{\dfrac{\partial^2}{\partial Q_{x\delta}\, \partial Q_{x\varepsilon}} \log L_x\right\}\right) \\ (r-1) \times 1 & (r-1) \times (r-1) \end{array}\right] \tag{3.34}$$

in which the elements are obtained by differentiating formula (3.18) and taking expectations. Direct calculation gives the following mathematical expectations

$$-E\left\{\frac{\partial^2}{\partial p_x^{\,2}} \log L_x\right\} = M_x\left(\frac{1}{p_x} + \frac{1}{Q_{x1}}\right) + \pi_x, \tag{3.35}$$

$$-E\left\{\frac{\partial^2}{\partial p_x \partial Q_{x\delta}} \log L_x\right\} = \frac{M_x}{Q_{x1}}, \tag{3.36}$$

$$-E\left\{\frac{\partial^2}{\partial Q_{x\delta} \partial Q_{x\varepsilon}} \log L_x\right\} = \frac{M_x}{Q_{x1}}, \qquad \delta \neq \varepsilon, \tag{3.37}$$

and

$$-E\left\{\frac{\partial^2}{\partial Q_{x\delta}^{\,2}} \log L_x\right\} = M_x\left(\frac{1}{Q_{x1}} + \frac{1}{Q_{x\delta}}\right), \qquad \delta, \varepsilon = 2, \ldots, r \tag{3.38}$$

where M_x and π_x are defined by Equations (2.21) and (2.22), respectively.

Substituting the respective expectations in (3.34) we have

$$\mathbf{A} = \left[\begin{array}{c|cccc} M_x\left(\dfrac{1}{p_x}+\dfrac{1}{Q_{x1}}\right)+\pi_x & M_x/Q_{x1} & M_x/Q_{x1} & \cdots & M_x/Q_{x1} \\ \hline M_x/Q_{x1} & M_x\left(\dfrac{1}{Q_{x1}}+\dfrac{1}{Q_{x2}}\right) & M_x/Q_{x1} & \cdots & M_x/Q_{x1} \\ M_x/Q_{x1} & M_x/Q_{x1} & M_x\left(\dfrac{1}{Q_{x1}}+\dfrac{1}{Q_{x3}}\right) & \cdots & M_x/Q_{x1} \\ \vdots & \vdots & \vdots & & \vdots \\ M_x/Q_{x1} & M_x/Q_{x1} & M_x/Q_{x1} & \cdots & M_x\left(\dfrac{1}{Q_{x1}}+\dfrac{1}{Q_{xr}}\right) \end{array}\right] \tag{3.39}$$

with its determinant

$$|A| = \frac{M_x^{\ r}}{Q_{x1}\cdots Q_{xr}p_x} + \frac{\pi_x M_x^{r-1}(1-p_x)}{Q_{x1}\cdots Q_{xr}}. \tag{3.40}$$

Denoting the cofactors of $\mathbf{A}$ by $A_{\delta\varepsilon}$, $\delta, \varepsilon = 1, 2, \ldots, r$, the approximate formulas for the asymptotic variance and covariance are given by

$$\sigma_{\hat{p}_x}^{\ 2} = \frac{A_{11}}{|A|} = p_x q_x\left[\frac{1}{M_x + \pi_x p_x q_x}\right] \tag{3.41}$$

$$\operatorname{Var}(\hat{Q}_x) = \frac{A_{\delta\delta}}{|A|} = Q_{x\delta}(1-Q_{x\delta})\left[\frac{1+\pi_x p_x(q_x - Q_{x\delta})/(1-Q_{x\delta})M_x}{M_x+\pi_x p_x q_x}\right] \tag{3.42}$$

$$\operatorname{Cov}(\hat{p}_x, \hat{Q}_{x\delta}) = \frac{A_{1\delta}}{|A|} = -p_x Q_{x\delta}\left[\frac{1}{M_x+\pi_x p_x q_x}\right] \tag{3.43}$$

and

$$\operatorname{Cov}(\hat{Q}_{x\delta}, \hat{Q}_{x\varepsilon}) = \frac{A_{\delta\varepsilon}}{|A|} = -Q_{x\delta}Q_{x\varepsilon}\left[\frac{1+\pi_x p_x/M_x}{M_x+\pi_x p_x q_x}\right],$$

$$\delta \neq \varepsilon;\ \delta, \varepsilon = 2, \ldots, r. \tag{3.44}$$

Since the term Q_{x1} was not explicitly included in (3.34), the formulas for the variance of $\hat{Q}_{x1}$ and the covariances between $\hat{Q}_{x1}$ and other estimators were not presented. By reason of symmetry, however, the above expressions for $\hat{Q}_{x\delta}$ are also true for $\hat{Q}_{x1}$, which is to say that formulas (3.42), (3.43), and (3.44) hold true also for $\delta = 1$.

The quantities inside the square brackets in formulas (3.41) through (3.44) may be approximated with $1/M_x$. These formulas are then reduced to familiar expressions in the multinomial distribution:

$$\sigma_{\hat{p}_x}{}^2 = \frac{1}{M_x} p_x q_x \tag{3.45}$$

$$\operatorname{Var}(\hat{Q}_{x\delta}) = \frac{1}{M_x} Q_{x\delta}(1 - Q_{x\delta}) \tag{3.46}$$

$$\operatorname{Cov}(\hat{p}_x, \hat{Q}_{x\delta}) = -\frac{1}{M_x} p_x Q_{x\delta} \tag{3.47}$$

and

$$\operatorname{Cov}(\hat{Q}_{x\delta}, \hat{Q}_{x\varepsilon}) = -\frac{1}{M_x} Q_{x\delta} Q_{x\varepsilon}, \qquad \begin{array}{l} \delta \neq \varepsilon;\ \delta, \varepsilon = 1, \ldots, r, \\ x = 0; 1, \ldots, y - 1. \end{array} \tag{3.48}$$

We can obtain approximate formulas for the asymptotic variance and covariance of the estimators of the net and partial crude probabilities. Formulas for the variance of $\hat{q}_{x.\delta}$ and $\hat{Q}_{x\delta.1}$ are given below.

$$\sigma^2_{\hat{q}_{x\cdot\delta}} = \frac{(1 - q_{x.\delta})^2}{M_x p_x q_x} [p_x \log(1 - q_{x\delta}) \log(1 - q_{x\cdot\delta}) + (q_x - Q_{x\delta})^2]$$

$$\delta = 1, \ldots, r, \tag{3.49}$$

$$\operatorname{Var}(\hat{Q}_{x\delta.1}) = \frac{(q_x - Q_{x1} - Q_{x\delta})}{M_x(q_x - Q_{x1})Q_{x\delta}} Q^2_{x\delta\cdot 1} + \frac{[Q_{x\delta\cdot 1}(q_x - Q_{x1}) - Q_{x\delta}]^2}{M_x p_x q_x (q_x - Q_{x1})}$$

$$\times \left[(q_x - Q_{x1}) + p_x Q_{x1}\left(\frac{\log p_x}{q_x}\right)^2\right] \qquad \delta = 2, \ldots, r. \tag{3.50}$$

4. LOST CASES

Every patient in a medical follow-up is exposed not only to the risk of dying but also to the risk of being lost to the study because of follow-up failure. Untraceable patients have caused difficulties in determining survival rates, as have patients withdrawing due to the termination of a study. However, lost cases and withdrawals belong to entirely different categories. In a group of N_x patients beginning the interval $(x, x + 1)$, for example, everyone is exposed to the risk of being lost, but only n_x patients are subject to withdrawal in the interval. Therefore, it is incorrect to treat lost cases and withdrawals equally in estimating probabilities of survival or death. For the purpose of determining the probability of dying from a specific cause, patients lost due to follow-up failure are not different from those dying of causes unrelated to the study. Being lost, therefore, should be considered as a competing risk, and the survival experience of lost cases should be evaluated by using the methods discussed in the preceding section. In this approach to the problem all formulas in Section 3 will remain intact, the solution requiring only a different interpretation of the symbols.

Suppose we let R_r denote the risk of being lost; for the time element $(\tau, \tau + \Delta)$ in the interval $(x, x + 1)$ let

$$\mu(x; r)\Delta + o(\Delta) = \Pr\{\text{a patient will be lost to the study in } (\tau, \tau + \Delta) \text{ due to follow-up failure}\}, \quad x < \tau < x + 1. \tag{4.1}$$

The following are a few examples of the new interpretation:

$$p_x = \Pr\{\text{a patient alive at time } x \text{ will remain alive and under observation at time } x + 1\}; \tag{4.2}$$

$$\begin{aligned} q_x &= 1 - p_x \\ &= \Pr\{\text{a patient alive at time } x \text{ will either die or be lost to the study due to follow-up failure in interval } (x, x + 1)\}; \end{aligned} \tag{4.3}$$

$$Q_{xr} = \Pr\{\text{a patient alive at time } x \text{ will be lost to the study in } (x, x + 1)\}; \tag{4.4}$$

$$\begin{aligned} q_{x.r} &= 1 - p_x^{(q_x - Q_{xr})/q_x} \\ &= \Pr\{\text{a patient alive at time } x \text{ will die in interval } (x, x + 1) \text{ if the risk } R_r \text{ of being lost is eliminated}\}; \end{aligned} \tag{4.5}$$

$$\begin{aligned} 1 - q_{x.r} &= p_x^{(q_x - Q_{xr})/q_x} \\ &= \Pr\{\text{a patient alive at } x \text{ will survive to time } x + 1 \text{ if the risk } R_r \text{ of being lost is eliminated}\}; \end{aligned} \tag{4.6}$$

$$\begin{aligned} Q_{x\delta.r} &= \frac{Q_{x\delta}}{q_x - Q_{xr}} [1 - p_x^{(q_x - Q_{xr})/q_x}] \\ &= \Pr\{\text{a patient alive at } x \text{ will die in } (x, x + 1) \text{ from cause } R_\delta \text{ if the risk } R_r \text{ of being lost is eliminated}\}. \end{aligned} \tag{4.7}$$

The probabilities in (4.5) (4.6), and (4.7) are equivalent to q_x, p_x, and $Q_{x\delta}$, respectively, if there is no risk of being lost.

The symbol d_{xr} in Table 3 now stands for the number of lost cases among the m_x patients and d'_{xr} for the number of lost cases among the n_x patients; the sum $D_{xr} = d_{xr} + d'_{xr}$ is the total number of cases lost in the interval. The probabilities in (4.2) through (4.7) can be estimated from formulas (3.26) through (3.30) in Section 3.

5. AN EXAMPLE OF LIFE TABLE CONSTRUCTION FOR THE FOLLOW-UP POPULATION

Application of the methods developed in this chapter is illustrated with data collected by the Tumor Registry of the California State Department of Public Health.[5] The material selected consists of 20,858 white female patients admitted to certain California hospitals and clinics between January 1, 1942, and December 31, 1962, with a diagnosis of breast cancer. For the purpose of this illustration, the closing date is December 31, 1963, and the date of entrance to follow-up for each patient is the date of hospital admission. Each patient was observed until death or until the closing date, whichever came first. Deaths were classified according to one of $r = 4$ causes: R_1, breast cancer; R_2, other cancer; R_3, other causes; and R_4, lost to the study because of follow-up failure.

The first step is to construct a table similar to Table 4, showing the survival experience of the patients grouped according to their withdrawal status and cause of death for each time period of follow-up. The interval length selected (Column 1) will depend upon the nature of the investigation; generally a fixed length of one year is used. The total number of patients admitted to the study is entered as N_0 in the first line of Column 2, which is 20,858. Since in this example the closing date (December 31, 1963) was at least one year after the last admission (December 31, 1962) all the N_0 patients were observed for the entire first year. Therefore 20,858 is also equal to m_0 (Column 3), the number of patients observed for the entire interval (0, 1). Of the m_0 patients, s_0 (17,202, Column 4) survived to their first anniversary and d_0 (3656, Column 5) died during the first year of follow-up. The deaths were divided by cause into d_{01} deaths (2381, Column 6) due to breast cancer (R_1), d_{02} deaths (56, Column 7) due to other cancer (R_2), d_{03} deaths (649, Column 8) due to other causes (R_3), and d_{04} cases (570, Column 9) lost to the study (R_4). In general, there will be n_0 (Column 10) patients due to withdraw in the interval (0, 1), of which w_0 (Column 11) survive to the closing date and d_0' (Column 12) die before the closing date. These deaths are again divided by cause into d_{01}' (Column 13), d_{02}' (Column 14), d_{03}' (Column 15), and d_{04}' (Column 16). The second interval began with the $s_0 = 17{,}202$ survivors from the first interval, which is entered as N_1 in Column 2 of line 2. The N_1 patients were again divided successively by withdrawal status, by survival status,

[5] I express my appreciation to George Linden and the California Tumor Registry, California Department of Public Health, for making their follow-up data available for use in this book.

and by cause of death. Of the N_1 patients, m_1 (16,240, Column 3) were the survivors of those admitted prior to January 1, 1962, and hence were observed for the entire interval (1, 2); n_1 (962, Column 10) were the survivors of those admitted during the year 1962 and hence were due to withdraw during the interval. At the beginning of the final interval (21, 22) there were $N_{21} = 49$ survivors of the patients admitted in 1942; all were due to withdraw during the last interval, or $n_{21} = 49$ (last line, Column 10). Of the 49 patients, $w_{21} = 40$ (Column 11) were alive at the closing date. This means that $\hat{p}_{21}$ is greater than zero, and $\hat{p}_z$ for $z \geq 22$ cannot be observed.

The material has been used to compute estimates of various probabilities and to construct a life table for the breast cancer patients. Table 5 shows the estimates of p_x (Column 2), the probability that a patient alive at time x will remain alive and under observation at time $x + 1$, $q_x = 1 - p_x$ (Column 3), the probability that a patient alive at time x will either die or be lost to the study within interval $(x, x + 1)$, and the estimates of various net probabilities (Columns 4 to 6) and partial crude probabilities (Columns 7 to 8).

The main life table functions and the corresponding sample standard errors are shown in Table 6. The basic quantity in the table is $\hat{q}_{x\cdot 4}$, the estimate of the probability of death if the risk of being lost is eliminated. The figures in Column 4 were computed from formula (2.29) with the substitution of $1 - \hat{q}_{x\cdot 4}$ for $\hat{p}_x$. The observed expectation of life $\hat{e}_{x\cdot 4}$ in Column 6 was determined from formula (2.37) with $t = 16$ and $\hat{p}_x$ replaced by $(1 - \hat{q}_{x\cdot 4})$. The sample variances of $\hat{q}_{x\cdot 4}$ and $\hat{e}_{x\cdot 4}$ were computed from (3.49), (2.41) and (2.42), respectively.

PROBLEMS

1. Verify equation (2.6).

2. Table 2 on page 272 shows the numerical difference between $p_x^{1/2}$ and $-(1 - px_2)/\log p_x$. Justify the approximation (2.7) by showing that

$$-\frac{(1 - p_x)}{\log p_x} - p_x^{1/2} = \tfrac{1}{24}q_x^{\,2} + \tfrac{1}{48}q_x^{\,3} + o(q_x^{\,3}).$$

3. Show that the estimator $\hat{p}_x$ in (2.14) is a maximizing value of the likelihood function in (2.10).

4. Verify Equation (2.20).

5. Derive the maximum-likelihood estimator $\hat{p}_x$ in (2.28) from the likelihood function in (2.27).

6. Verify the formula in (2.41) and (2.42) for the sample variance of $\hat{e}_\alpha$.

7. Integrate (3.2) to obtain $Q_{x\delta}(t)$ in (3.3).

8. Justify Equations (3.5) and (3.6).

9. Check the computations in (3.35) through (3.38) by showing intermediate steps.

10. Verify that the determinant of the matrix **A** in (3.39) is given in (3.40).

11. Verify formulas (3.41) through (3.44).

12. During the period from January 1, 1942, to December 31, 1954, in the State of California there were 5,982 registered patients with a diagnosis of cervix uteri. Table 7 shows their survival experience up to December 31, 1954.

(a) Construct a life table for these cancer patients, including the sample standard errors of $\hat{p}_{0x}$, $\hat{q}_x$, and $\hat{e}_x$.

(b) Estimate the crude probabilities of dying in interval $(x, x + 1)$ from cancer of the cervix and from other causes, for $x = 0, 1, \ldots, 12$.

(c) Estimate the net probability of dying in $(x, x + 1)$ if cervix cancer is eliminated as a cause of death. Use the result to construct a life table.

(d) In light of the two life tables in (a) and (c), discuss the impact of cervix cancer in shortening the life span of a woman.

Table 4. Survival experience following diagnosis of

Interval Since Admission (in Years)	Number Living at Beginning of Interval $(x, x+1)$	Number Not Due for Withdrawal in Interval $(x, x+1)$						
		Total Not Due for With-drawal	Number Surviving the Interval	Number of Deaths and Losses in Interval				
				Total	Breast Cancer	Other Cancer	Other Cause	Lost Cases
x to $x+1$	N_x	m_x	s_x	d_x	d_{x1}	d_{x2}	d_{x3}	d_{x4}
(1)	(2)	(3)	(4)	(5)	(6)	(7)	(8)	(9)
0–1	20,858	20,858	17,202	3,656	2,381	56	649	570
1–2	17,202	16,240	14,052	2,188	1,689	35	352	112
2–3	14,052	13,134	11,563	1,571	1,161	28	282	100
3–4	11,563	10,769	9,521	1,248	863	29	257	99
4–5	9,521	8,741	7,881	860	533	33	187	107
5–6	7,881	7,223	6,486	737	432	33	184	88
6–7	6,486	5,880	5,392	488	270	17	131	70
7–8	5,392	4,798	4,385	413	195	24	128	66
8–9	4,385	3,888	3,527	361	177	13	100	71
9–10	3,527	3,120	2,865	255	106	17	85	47
10–11	2,865	2,470	2,293	177	84	10	50	33
11–12	2,293	1,942	1,798	144	44	11	59	30
12–13	1,798	1,499	1,376	123	47	5	46	25
13–14	1,376	1,168	1,071	97	25	12	39	21
14–15	1,071	894	817	77	21	2	36	18
15–16	817	652	599	53	13	1	25	14
16–17	599	460	419	41	16	—	13	12
17–18	419	312	284	28	9	1	9	9
18–19	284	202	192	10	1	—	6	3
19–20	192	130	117	13	—	3	6	4
20–21	117	55	49	6	2	1	2	1
21–22	49	0	0	—	—	—	—	—

Source. California Tumor Registry, California State Department of Public Health.
[a] December 31, 1963 was the common closing date.

breast cancer, cases initially diagnosed 1942–1962[a]

Number Due for Withdrawal in Interval $(x, x+1)$							Number Living at Beginning of Interval $(x, x+1)$	Interval Since Admission (in Years)
Total Due for With-drawal	Number of Live With-drawals	Number of Deaths and Losses in Interval						
		Total	Breast Cancer	Other Cancer	Other Cause	Lost Cases		
n_x	w_x	d'_x	d'_{x1}	d'_{x2}	d'_{x3}	d'_{x4}	N_x	x to $x+1$
(10)	(11)	(12)	(13)	(14)	(15)	(16)	(2)	(1)
—	—	—	—	—	—	—	20,858	0–1
962	745	217	59	1	5	152	17,202	1–2
918	726	192	44	4	11	133	14,052	2–3
794	610	184	31	—	8	145	11,563	3–4
780	596	184	26	3	11	144	9,521	4–5
658	504	154	17	—	9	128	7,881	5–6
606	461	145	17	1	6	121	6,486	6–7
594	438	156	20	1	5	130	5,392	7–8
497	392	105	8	—	3	94	4,385	8–9
407	321	86	7	—	6	73	3,527	9–10
395	313	82	3	—	7	72	2,865	10–11
351	256	95	15	—	3	77	2,293	11–12
299	228	71	4	—	5	62	1,798	12–13
208	157	51	—	—	1	50	1,376	13–14
177	145	32	1	—	2	29	1,071	14–15
165	118	47	3	—	1	43	817	15–16
139	111	28	2	—	3	23	599	16–17
107	82	25	1	—	1	23	419	17–18
82	69	13	—	—	3	10	284	18–19
62	43	19	1	—	3	15	192	19–20
62	45	17	—	1	1	15	117	20–21
49	40	9	—	—	1	8	49	21–22

Table 5. *Survival experience following diagnosis of breast cancer. Maximum likelihood estimates of net and partial crude probabilities*

Interval Since Diagnosis (in Years)	Estimate of Probability of Surviving and Under Observation in Interval $(x, x+1)$	Estimate of Probability of Dying or Being Lost in Interval $(x, x+1)$	Estimate of Net Probability of Dying in Interval $(x, x+1)$ when			Estimate of Partial Crude Probability of Dying in Interval $(x, x+1)$ from	
			Breast Cancer Acting Alone	Breast Cancer and Other Cancers Acting Alone[a]	Risk of Being Lost Is Eliminated	Breast Cancer When Risk of Being Lost Is Eliminated	Breast Cancer or Other Cancers when Risk of Being Lost Is Eliminated[a]
$(x$ to $x+1)$	$1000\hat{p}_x$	$1000\hat{q}_x$	$1000\hat{q}_{x1}$	$1000\hat{q}_{xc}$	$1000\hat{q}_{x\cdot 4}$	$1000\hat{Q}_{x1\cdot 4}$	$1000\hat{Q}_{xc\cdot 4}$
(1)	(2)	(3)	(4)	(5)	(6)	(7)	(8)
0–1	824.72	175.28	117.95	120.55	150.13	115.83	118.55
1–2	856.65	143.35	106.36	108.43	128.68	105.06	107.22
2–3	870.77	129.23	90.24	92.53	113.16	89.12	91.49
3–4	872.30	127.70	81.76	84.29	107.16	80.64	83.26
4–5	886.25	113.75	62.61	66.51	87.64	61.78	65.76
5–6	882.63	117.37	60.97	65.31	90.25	60.03	64.45
6–7	898.24	101.76	47.49	50.40	72.20	46.88	49.82
7–8	889.19	110.81	43.41	48.33	74.10	42.71	47.68
8–9	888.08	111.92	46.03	49.18	73.80	45.36	48.55
9–10	898.07	101.93	35.00	40.16	67.30	34.41	39.59
10–11	903.66	96.34	33.46	37.23	58.46	33.03	36.82
11–12	888.42	111.58	28.78	34.06	63.25	28.27	33.54
12–13	883.61	116.39	32.01	35.09	65.97	31.44	34.53
13–14	884.84	115.16	20.45	30.12	61.67	20.02	29.63
14–15	889.98	110.02	23.25	25.34	64.15	22.76	24.83
15–16	866.07	133.93	22.74	24.15	59.96	22.31	23.71
16–17	871.45	128.55	35.26	35.26	65.55	34.70	34.70
17–18	857.52	142.48	28.59	31.40	59.09	28.14	30.95
18–19	906.63	93.37	4.25	4.25	41.72	4.17	4.17
19–20	807.23	192.77	6.67	26.41	83.32	6.41	25.64
20–21	746.03	253.97	25.16	49.68	85.31	24.37	48.75
21–22	666.39	333.61	0.00	0.00	44.10	0.00	0.00

Source. California Tumor Registry, California State Department of Public Health.

[a] For simplicity a subscript c is used here to denote both risks R_1 (breast cancer) and R_2 (other cancers).

Table 6. Survival experience following diagnosis of breast cancer, the main life table functions and their standard errors (when risk of being lost is eliminated)

Interval Since Diagnosis (in Years)	Estimate of Net Probability of Dying in Interval		x-Year Survival Rate		Observed Expectation of Life at x[a]	
(x to $x+1$)	$1000\hat{q}_{x\cdot 4}$	$1000S_{\hat{q}_{x4}}$	$1000\hat{p}_{0x\cdot 4}$	$1000S_{\hat{p}_{0x\cdot 4}}$	$\hat{e}_{x\cdot 4}$	$S_{\hat{e}_{x\cdot 4}}$
(1)	(2)	(3)	(4)	(5)	(6)	(7)
0–1	150.13	2.47	1000.00	0.00	11.26	.50
1–2	128.68	2.59	849.87	2.47	12.16	.59
2–3	113.16	2.72	740.51	3.08	12.88	.67
3–4	107.16	2.93	656.71	3.39	13.46	.76
4–5	87.64	2.96	586.34	3.59	14.02	.84
5–6	90.25	3.30	534.95	3.70	14.32	.92
6–7	72.20	3.29	486.67	3.80	14.69	1.01
7–8	74.10	3.67	451.53	3.87	14.79	1.09
8–9	73.80	4.06	418.07	3.95	14.94	1.18
9–10	67.30	4.34	387.22	4.03	15.09	1.27
10–11	58.46	4.54	361.16	4.12	15.14	1.36
11–12	63.25	5.29	340.05	4.21	15.05	1.44
12–13	65.97	6.11	318.54	4.34	15.03	1.54
13–14	61.67	6.74	297.53	4.49	15.06	1.64
14–15	64.15	7.81	279.18	4.67	15.02	1.75
15–16	59.96	8.75	261.27	4.88	15.01	1.86
16–17	65.55	10.74	245.60	5.13	14.94	1.98
17–18	59.09	12.32	229.50	5.47	14.95	1.84
18–19	41.72	12.82	215.94	5.87	14.86	1.85
19–20	83.32	21.74	206.93	6.27	14.48	1.85
20–21	85.31	30.00	189.69	7.30	14.75	1.84
21–22	44.10	41.25	173.51	8.77	15.08	1.80
22–			165.86		14.76	

[a] $\hat{p}_{16\cdot 4}$ is used for $\hat{p}_t$ in formula (2.37).

Table 7. Survival experience following diagnosis of cancer of the cervix uteri cases initially diagnosed 1942 to 1954

Interval Since Diagnosis (in Years)	Number Living at Beginning of Interval (x to $x+1$)	Number Not Due for Withdrawal in Interval $(x, x+1)$[a]					Number Due for Withdrawal in Interval $(x, x+1)$[b]				
		Total Not Due for Withdrawal	Number Surviving the Interval	Number Dying in the Interval			Total Due for Withdrawal	Number Living at Time of Withdrawal	Number Dying before Withdrawal		
				Total	Cancer of the Cervix	Other Causes			Total	Cancer of the Cervix	Other Causes
$x - x+1$	N_x	m_x	S_x	d_x	d_{x1}	d_{x2}	n_x	w_x	d'_x	d'_{x1}	d'_{x2}
(1)	(2)	(3)	(4)	(5)	(6)	(7)	(8)	(9)	(10)	(11)	(12)
0–1	5982	5317	4030	1287	1105	182	665	576	89	70	19
1–2	4030	3489	2845	644	557	87	541	501	40	31	9
2–3	2845	2367	2117	250	206	44	478	459	19	15	4
3–4	2117	1724	1573	151	113	38	393	379	14	8	6
4–5	1573	1263	1176	87	61	26	310	306	4	2	2
5–6	1176	918	861	57	24	33	258	254	4	3	1
6–7	861	692	660	32	16	16	169	167	2	2	0
7–8	660	496	474	22	11	11	164	161	3	2	1
8–9	474	356	344	12	5	7	118	116	2	1	1
9–10	344	256	245	11	7	4	88	85	3	2	1
10–11	245	164	158	6	4	2	81	78	3	1	2
11–12	158	76	72	4	1	3	82	80	2	1	1
12–13	72	0	0	0	0	0	72	72	0	0	0

Source. California Tumor Registry, Department of Public Health, State of California.
[a] Survivors of those admitted more than $x+1$ years prior to closing date.
[b] Survivors of those admitted between x and $x+1$ years prior to closing date.

References

Aitken, A. C. [1962]. *Determinants and Matrices.* Oliver and Boyd, Edinburgh and London.

Armitage, P. [1951]. The statistical theory of bacterial populations subject to mutation, *J. Royal Statist. Soc.*, **B14,** 1–33.

Armitage, P. [1959]. The comparison of survival curves, *J. Royal Statist. Soc.*, **A122,** 279–300.

Bailey, N. T. J. [1964]. *The Elements of Stochastic Processes with Applications to the Natural Sciences.* Wiley, New York.

Barlow R. E. and F. Proschan [1965]. *Mathematical Theory of Reliability.* Wiley, New York.

Bartlett, M. S. [1949]. Some evolutionary stochastic processes, *J. Royal Statist. Soc.*, **B11,** 211–229.

Bartlett, M. S. [1956]. *An Introduction to Stochastic Processes.* Cambridge University Press.

Bates, G. E. and J. Neyman [1952]. Contribution to the theory of accident proneness. I. An optimistic model of the correlation between light and severe accidents, *Univer. Calif. Pub. Statist.*, **1,** 215–254, University of California Press, Berkeley.

Bellman, R. [1960]. *Introduction to Matrix Analysis.* McGraw-Hill, New York.

Berkson, J. and L. Elveback [1960]. Competing exponential risks, with particular reference to smoking and lung cancer, *J. Amer. Statist. Assoc.*, **55,** 415–428.

Berkson, J. and R. P. Gage [1952]. Survival curve for cancer patients following treatment, *J. Amer. Statist. Assoc.*, **47,** 501–515.

Berman, S. M. [1963]. A note on extreme values, competing risks and semi-Markov processes, *Ann. Math. Statist.*, **34,** 1104–1106.

Bharcha-Reid, A. T. [1960]. *Elements of the Theory of Markov Processes and their Applications.* McGraw-Hill, New York.

Birbaum, Z. W., J. D. Esary, and S. C. Saunders [1961]. Multi-component systems and structures and their reliability, *Technometrics*, **3,** 55–77.

Birkhoff, G. and S. MacLane [1953]. *A Survey of Modern Algebra* (rev. ed.). Macmillan, New York.

Blackwell, D. [1947]. Conditional expectation and unbiased sequential estimation, *Ann. Math. Statist.*, **18,** 105–110.

Boag, J. W. [1949]. Maximum likelihood estimates of the proportion of patients cured by cancer therapy, *J. Royal Statist. Soc.*, **B11,** 15–53.

Brockmayer, E., H. L. Halstrøm, and A. Jensen [1948]. *The life and work of A. K. Erlang*, Copenhagen Telephone Co., Copenhagen.

Buck, R. [1965]. *Advanced Calculus* (2nd ed.). McGraw-Hill, New York.

Chiang, C. L. [1960a]. A stochastic study of the life table and its applications: I. Probability distributions of the biometric functions, *Biometrics*, **16,** 618–635.

Chiang, C. L. [1960b]. A stochastic study of the life table and its applications: II. Sample variance of the observed expectation of life and other biometric functions, *Human Biol.*, **32,** 221–238.

Chiang, C. L. [1961a]. A stochastic study of the life table and its applications: III. The follow-up study with the consideration of competing risks, *Biometrics*, **17,** 57–78.

Chiang, C. L. [1961b]. Standard error of the age-adjusted, *Vital Statistics, Special Reports Selected Studies*, **47**(9), 275–285. National Center for Health Statistics.

Chiang, C. L. [1964a]. A stochastic model of competing risks of illness and competing risks of death, *Stochastic Models in Medicine and Biology*, J. Gurland, Ed. University of Wisconsin Press, Madison, pp. 323–354.

Chiang, C. L. [1964b]. A birth-illness-death process, *Ann. Math. Statist.*, **35,** 1390–1391, abstract.

Chiang, C. L. [1965]. An index of health: mathematical models, *Vital and Health Statistics*, Ser. 2, No. 5, 1–19. National Center for Health Statistics.

Chiang, C. L. [1966]. On the expectation of the reciprocal of a random variable, *The Amer. Statist.*, **20,** 28.

Chiang, C. L. [1967]. Variance and covariance of life table functions estimated from a sample of deaths, *Vital and Health Statistics*. Ser. 2, No. 20, 1–8. National Center for Health Statistics.

Chiang, C. L. [1968]. *On an invariant property of the matrix of eigenvectors* (unpublished manuscript).

Chiang, C. L., F. D. Norris, F. Olsen, P. W. Shipley, and H. E. Supplee [1961]. Determination of the fraction of last year of life and its variation by age, race, and sex. (An unpublished manuscript presented at the Annual Meeting of the American Public Health Assoc., Detroit, Michigan, November 1961.)

Chung, K. L. [1960]. *Markov Chains with Stationary Transition Probabilities*. Springer-Verlag, Berlin.

Cornfield, J. [1957]. The estimation of the probability of developing a disease in the presence of competing risks, *J. Amer. Public Health Assoc.*, **47,** 601–607.

Cox, D. R. and H. D. Miller [1965]. *The Theory of Stochastic Processes*. Wiley, New York.

Cramér, H. [1946]. *Mathematical Methods of Statistics*. Princeton University Press.

Cramér, H. and M. R. Leadbetter [1967]. *Stationary and Related Stochastic Processes*. Wiley, N.Y.

Cutler, S. J. and F. Ederer [1958]. Maximum utilization of the life table method in analyzing survival, *J. Chronicle Diseases*, **8,** 699–712.

The Department of National Health and Welfare and the Dominion Bureau of Statistics [1960]. *Illness and Health Care in Canada, Canadian Sickness Survey*, 1950–1951. Ottawa, The Queen's Printer and Controller of Stationery.

Doob, J. L. [1953]. *Stochastic Processes*. Wiley, New York.

Dorn, H. [1950]. Methods of analysis for follow-up studies, *Human Biol.* **22,** 238–248.

Elveback, L. [1958]. Estimation of survivorship in chronic disease: The actuarial method, *J. Amer. Statist. Assoc.*, **53,** 420–440.

Epstein, B. and M. Sobel [1953]. Life testing, *J. Amer. Statist. Assoc.*, **48,** 486–502.

Feller, W. [1936]. Zur Theorie der stochastischen Prozesse, *Math. Ann.*, **113,** 113–160

Feller, W. [1940]. On the integrodifferential equations of purely discontinuous Markoff processes, *Trans. Amer. Math. Soc.*, **48,** 488–515.

Feller, W. [1957]. *An Introduction to Probability Theory and its Applications*, Vol. 1 (2nd ed.). Wiley, New York.

Feller, W. [1966]. *An Introduction to Probability Theory and its Applications*, Vol. II. Wiley, New York.

Fisher, R. A. [1922a]. On the dominance ratio, *Proc. Royal Soc.* (*Edinburgh*), **42,** 321–341.

Fisher, R. A. [1922b]. On the mathematical foundations of theoretical statistics, *Phil. Trans. Royal Soc.* (*London*), **A222,** 309–368.

Fisher, R. A. [1930]. The distribution of gene ratios for rare mutations, *Proc. Royal Soc.* (*Edinburgh*), **50,** 204–219.

Fix, E. and J. Neyman [1951]. A simple stochastic model of recovery, relapse, death and loss of patients, *Human Biol.*, **23,** 205–241.

Ford, L. R. [1933]. *Differential Equations.* McGraw-Hill, New York.

Forster, F. G. [1953]. On the stochastic matrices associated with certain queueing problems, *Ann. Math. Statist.*, **24,** 355–360.

Frost, W. H. [1933]. Risk of persons in familiar contact with pulmonary tuberculosis, *Amer. J. Public Health*, **23,** 426–432.

Gantmacher, F. R. [1960]. *The Theory of Matrices*, Vol. I., Chelsea, New York.

Gompertz, B. [1825]. On the nature of the function expressive of the law of human mortality, *Phil. Trans. Royal Soc.* (*London*), **115,** 513–583.

Graunt, J. [1662]. *Natural and Political Observations Made upon the Bills of Mortality.* Reprinted by the Johns Hopkins Press, Baltimore, 1939.

Greenwood, M. [1925]. A report on the natural duration of cancer, *Rep. Public Health Med. Sub.*, **33,** 1–26.

Grenander, U. [1956]. On the theory of mortality measurement, *Skandinavisk Aktuarie-tidskrift*, **39,** 1–55.

Greville, T. N. E. [1943]. Short methods of constructing abridged life tables, *Record Amer. Inst. Actuaries*, **32,** 29–43.

Greville, T. N. E. [1948]. Mortality tables analyzed by cause of death, *Record Amer. Inst. Actuaries*, **37,** 283–294.

Haldane, J. B. S. [1927]. A mathematical theory of natural and artificial selection, Part V: Selection and Mutation, *Proc. Royal Soc.* (*Edinburgh*), **23,** 838–844.

Halley, E. [1693]. An estimate of the degrees of the mortality of mankind, drawn from curious tables of the births and funerals at the city of Breslau, *Philos. Trans. Royal Soc.* (*London*), **17,** 596–610.

Harris, T. E. [1963]. *The Theory of Branching Processes.* Springer-Verlag, Berlin.

Harris, T. E., P. Meier, and J. W. Tukey [1950]. Timing of the distribution of events between observations, *Human Biol.*, **22,** 249–270.

Hochstadt, H. [1964]. *Differential Equations.* Holt, New York.

Homma, T. [1955]. On a certain queueing process, *Repts. Statist. Appl. Research, Union Japan. Scientists and Engrs.*, **4,** 14–32.

Hogg, R. V. and A. T. Craig [1965]. *Introduction to Mathematical Statistics* (2nd ed.). Macmillan, New York.

Irwin, J. O. [1949]. The standard error of an estimate of expectation of life, *J. Hygiene*, **47,** 188–189.

Kaplan, E. L. and P. Meier [1958]. Nonparametric estimation from incomplete observations, *J. Amer. Statist. Assoc.*, **53,** 457–481.

Karlin, S. [1966]. *A First Course in Stochastic Processes.* Academic, New York.

Kendall, D. G. [1948]. On the generalized birth-and-death process, *Ann. Math. Statist.*, **19,** 1–15.

Kendall, D. G. [1952]. Some problems in the theory of queues, *J. Royal Statist. Soc.*, **B3,** 151–185.

Kendall, D. G. [1954]. Stochastic processes occurring in the theory of queues and their analysis by means of the imbedded Markov chain, *Ann. Math. Statist.*, **24,** 338–354.

Kendall, M. G. and A. Stuart [1961], *The Advanced Theory of Statistics*, Vol. 2., Griffin, London.

Kimball, A. W. [1958]. Disease incidence estimation in populations subject to multiple causes of death, *Bull. Internat. Statist. Inst.*, **36,** 193–204.

King, G. [1914]. On a short method of constructing an abridged mortality table, *J. Inst. Actuaries*, **48,** 294–303.

Kolmogorov, A. M. [1931]. Über die analytischen Methoden in der Wahrscheinlichkeitsrechnung, *Mathematische Annln*, **104,** 415–458.

Lehmann, E. [1959]. *Testing Statistical Hypotheses.* Wiley, New York.

Lehmann, E. and H. Scheffé [1950]. Completeness, similar regions and unbiased estimation, *Sankhyã*, **10,** 305–340.

Littell, A. S. [1952]. Estimation of the T-year survival rate from follow-up studies over a limited period of time, *Human Biol.*, **24,** 87–116.

Lundberg, O. [1940]. *On Random Processes and Their Applications to Sickness and Accident Statistics.* Uppsala.

Makeham, W. M. [1860]. On the law of mortality and the construction of annuity tables, *J. Inst. Actuaries*, **8.**

Makeham, W. M. [1874]. On an application of the theory of the composition of decremental forces, *J. Inst. Actuaries*, **18,** 317–322.

Miller, F. H. [1941]. *Partial Differential Equations* (11th printing, 1965). Wiley, New York.

Miller, R. S. and J. L. Thomas [1958]. The effect of larval crowding and body size on the longevity of adult *Drosophila Melanogaster*, *Ecology*, **39,** 118–125.

Mood, A. M. and F. A. Graybill [1963], *Introduction to the Theory of Statistics* (2nd ed.). McGraw-Hill, New York.

Neyman, J. [1935]. Su un teorema concernente le cosiddetti statistiche sufficienti, *Giornale dell' Istituto degli Attuari*, **6,** 320–334.

Neyman, J. [1949]. Contribution to the theory of the χ^2 test, *Proc. of Berkeley Symp. Math. Statist. Probability*, University of California Press, Berkeley, pp. 239–273.

Neyman, J. [1950]. *First Course in Probability and Statistics.* Holt, New York.

Neyman, J. and E. L. Scott [1964]. A stochastic model of epidemics, *Stochastic Models in Medicine and Biology*, J. Gurland, Ed. University of Wisconsin Press, Madison, pp. 45–83.

Parzen, E. [1962]. *Stochastic Processes.* Holden-Day, San Francisco.

Patil, V. T. [1957]. The consistency and adequacy of the Poisson-Markoff model for density fluctuations, *Biometrika*, **44,** 43–56.

Polya, G. and G. Szegö [1964]. *Aufgaben und Lehrsätze aus der Analysis*, Vol. 2. Springer-Verlag, Berlin.

Quenouille, M. H. [1949]. A relation between the logarithmic, Poisson, and negative binomial series, *Biometrics*, **5,** 162–164.

Rao, C. R. [1945]. Information and accuracy attainable in the estimation of statistical parameters, *Bull. Calcutta Math. Soc.*, **37,** 81–91.

Rao, C. R. [1949]. Sufficient statistics and minimum variance estimates, *Proc. Cambridge Phil. Soc.*, **45,** 213–218.

Reed, L. J. and M. Merrill [1939]. A short method for constructing an abridged life table, *Amer. J. Hygiene*, **30,** 33–62.

Riordan, J. [1958]. *An Introduction to Combinational Analysis.* Wiley, New York.

Ruben, H. [1962], Some aspects of the emigration-immigration process, *Ann. Math. Statist.*, **33,** 119–129.

Rudin, W. [1953]. *Principles of Mathematical Analysis.* McGraw-Hill, New York.

Sirken, M. [1964]. Comparison of two methods of constructing abridged life tables, *Vital and Health Statistics*, Ser. 2, No. 4, 1–11, National Center for Health Statistics.

Takács, L. [1960], *Stochastic Processes.* Methuen, London.

Todhunter, I. [1949]. *A History of the Mathematical Theory of Probability.* Chelsea, New York.

Uspensky, J. V. [1937]. *Introduction to Mathematical Probability.* McGraw-Hill, N.Y.

Weibull, W. [1939]. A statistical theory of the strength of material, *Ing. Vetenskaps Akad. Handl.*, **151.**

Yule, G. [1924]. A mathematical theory of evolution based on the conclusion of Dr. J. C. Willis, F.R.S., *Phil. Trans. Royal Soc.* (*London*), **B213,** 21–87.

Zahl, S. [1955]. A Markov process model for follow-up studies, *Human Biol.*, **27,** 90–120.

Author Index

Subject Index